Luminescence Thermometry

Related Titles

1. *Optofluidics, Sensors and Actuators in Microstructured Optical Fibers,* 978-1-78242-329-4
2. *Nano Optoelectronic Sensors and Devices*, 978-1-4377-3471-3
3. *Radiometric Temperature Measurements, I. Fundamentals, part of the Experimental Methods in the Physical Sciences seri*es, 978-0-12-374021-2

Woodhead Publishing Series in Electronic and Optical Materials

Luminescence Thermometry

Methods, Materials, and Applications

Miroslav Dramićanin

ELSEVIER

WP
WOODHEAD PUBLISHING
An imprint of Elsevier

Woodhead Publishing is an imprint of Elsevier
The Officers' Mess Business Centre, Royston Road, Duxford, CB22 4QH, United Kingdom
50 Hampshire Street, 5th Floor, Cambridge, MA 02139, United States
The Boulevard, Langford Lane, Kidlington, OX5 1GB, United Kingdom

Notices

Knowledge and best practice in this field are constantly changing. As new research and experience broaden our understanding, changes in research methods, professional practices, or medical treatment may become necessary.

Practitioners and researchers must always rely on their own experience and knowledge in evaluating and using any information, methods, compounds, or experiments described herein. In using such information or methods they should be mindful of their own safety and the safety of others, including parties for whom they have a professional responsibility.

To the fullest extent of the law, neither the Publisher nor the authors, contributors, or editors, assume any liability for any injury and/or damage to persons or property as a matter of products liability, negligence or otherwise, or from any use or operation of any methods, products, instructions, or ideas contained in the material herein.

Library of Congress Cataloging-in-Publication Data
A catalog record for this book is available from the Library of Congress

British Library Cataloguing-in-Publication Data
A catalogue record for this book is available from the British Library

ISBN: 978-0-08-102029-6 (print)
ISBN: 978-0-08-102030-2 (online)

Publisher: Matthew Deans
Acquisition Editor: Kayla Dos Santos
Editorial Project Manager: Joshua Bayliss
Production Project Manager: Sojan P. Pazhayattil
Designer: Greg Harris

Typeset by Thomson Digital

Contents

Preface ix

1. Introduction to Measurements of Temperature
1.1 Why We Measure Temperature? 1
1.2 The Market Share of Temperature Sensors and Thermometers 1
1.3 Historical Background 2
1.4 Why are Novel Concepts and Methods for Measuring Temperature Needed? 10
References 12

2. Temperature and Ways of Measuring It
2.1 Introduction 13
2.2 Temperature 13
2.3 Primary Thermometers 16
2.4 The International Temperature Scale and the Provisional Low Temperature Scale 20
2.5 Common Thermometers and Their Classification 23
2.6 Measurement Errors (Uncertainties) and Performance Characteristics of Thermometers 27
References 31

3. Luminescence: The Basics, Methods, and Instrumentation
3.1 Introduction 33
3.2 Electromagnetic Radiation and its Interaction with Mater 33
3.3 Luminescence 35
3.4 Luminescence in Solids 39
3.5 Quantum Efficiency of Emission: Temperature Quenching of Luminescence 40
3.6 Types of Photoluminescence Measurements 44
3.7 Instrumentation Used in Photoluminescence Measurements 51
References 60

4. Schemes for Temperature Read-Out From Luminescence
4.1 Introduction 63
4.2 Temperature Sensing from Excitation and Emission Band Positions and Bandwidths 64
4.3 Sensing Temperature from the Intensity of a Single Emission Band 68
4.4 Ratiometric Temperature Sensing 69
4.5 Temperature Sensing Using Decay Time 73
4.6 Temperature Sensing Based on Rise Time 74
4.7 Temperature Sensing from Luminescence Polarization (Anisotropy) 78
4.8 The generic Luminescence Thermometry System 78
4.9 Static (Time-Integrated) Versus Time-Resolved Methods: Pros and Cons 80
References 82

5. Methods of Analysis for Luminescence Thermometry Measurements
5.1 Introduction 85
5.2 Removal of Baseline Offset 86
5.3 Dealing with Noise 91
5.4 Differentiation and Integration of Spectral Data 95
5.5 Quantification of the Features of Luminescence Spectra 98
5.6 Evaluation of Lifetimes from Time-Resolved Measurements 104
5.7 Calculating the Performance of a Luminescence Thermometer 109
References 110

6. Lanthanide and Transition Metal Ion Doped Materials for Luminescence Temperature Sensing
6.1 Introduction 113
6.2 Characteristics of Lanthanide and Transition Metal Ion Luminescence Centers 114
6.3 Downshifting, Upconverting, Scintillating, and Quantum-Cutting Phosphors 124
6.4 Ways to Prepare Rare Earth and Transition Metal Doped Materials 127
6.4.1 Crystal Growth 127
6.4.2 Preparation of Powders 129
6.4.3 Preparation of Thin Films and Coatings 132
6.4.4 Glasses and Glass-Ceramics 134
6.5 Lanthanide Ion Doped Materials for Temperature Sensing 135
6.5.1 Temperature Sensing Via Downshifting Emission 140
6.5.2 Temperature Sensing Via Upconversion Emission 145
6.6 Transition ion Doped Materials for Temperature Sensing 147
References 151

7. Luminescence Temperature Sensing Using Semiconductor Quantum Dots
7.1 Introduction 159
7.2 Semiconductor Nanostructures and Quantum Confinement 160
7.3 Luminescence of Quantum Dots 164
7.4 Temperature Sensing Using Various Luminescence Features of Semiconductor Quantum Dots 166
References 185

8. Luminescence Temperature Sensing Using Organic Materials
8.1 Introduction 189
8.2 Dyes and Organic Pigments 190
8.3 Dyes Incorporated in Functional Polymers 198
8.4 Exciplex-Type Probes 203
8.5 Discrete Metal–Organic Complexes 204
8.6 Metal–Organic Frameworks 207
References 211

9. Applications of Luminescence Thermometry in Engineering
9.1 Introduction 215
9.2 Thermal Imaging of Electronic Circuits and Electrical Machines via Luminescence 216
9.3 Luminescence Thermometry of Fluid Flows 222
9.4 Thermal and Environmental Barrier Coatings 224
9.5 Surface Temperature Measurements Using Temperature-Sensitive Paints and Phosphor Coatings 227
References 230

10. Biomedical Applications of Luminescence Thermometry
10.1 Introduction 235
10.2 Optical Properties of Tissue; Biological Windows 235
10.3 Tissue Fluorescence 238
10.4 Intracellular and Intercellular Temperature Measurements via Luminescence 240
10.5 Using Luminescence for Measurements of In-Tumor Temperature During Thermal Therapy 245
10.6 Using Luminescence Thermometry to Diagnose Ischemia 246
References 249

11. Temperature Measurements at the Nanoscale
11.1 Introduction 251
11.2 Methods of Nanoscale Temperature Measurement 252

11.3 Temperature Mapping of Microfluidic and Nanofluidic Systems 256
11.4 Luminescence Thermal Imaging in Nanoelectronics 260
References 261

12. Achieving Multifunctionality by Combining Thermometry With Other Luminescence Applications

12.1 Introduction 265
12.2 Multifunctional and Smart Materials 265
12.3 Luminescence Sensing of Pressure 268
12.4 Luminescence Sensing of an Electric Field 270
12.5 Luminescence Sensing of a Magnetic Field 275
12.6 Luminescence Sensing of Structural and Compositional Changes in Materials 277
12.7 Luminescence Sensing of Chemical and Biochemical Analytes 279
12.8 Luminescence Biomedical and Thermal Imaging 280
References 283

Index 287

Preface

In the first chapter of the book the author asks the question "Why we measure temperature?" He provides us the answer by not only telling us that temperature is a basic physical quantity but rather how temperature effects our daily life, from the mundane as to what we wear to the importance of temperature in the global climate. Fortunately, the author does not stop here but continues to tantalize the reader to continue reading with a captivating chapter on the history of thermometry. The following chapters in the book are more technical and present a detailed and comprehensive treatment of thermometry; an excellent and multifaceted overview of interest to numerous groups. The book addresses the most basic aspects of thermometry and the necessary background, thus it can serve, first and foremost, as a textbook for graduate and undergraduate students, as well as for accomplished researchers in the field. Each chapter begins with a straightforward physical description and then moves on to a more detailed account. The book contains 12 chapters, clearly written and insightful, spanning the importance of thermometry in everyday life, the basic methods and instruments for measuring temperature, luminescence thermometry using transition and lanthanide ion doped materials, semiconductor quantum dots, organic materials, applications, and nanothermometry.

Chapter 3 initiates the reader to principles of luminescence, which is employed extensively to measure temperature. Chapter 4 outlines different techniques for the read-out of temperature using luminescence based on the following parameters: intensity, bandwidth, bandshape, spectral shift, polarization, and lifetimes. An in-depth discussion of their advantages and disadvantages is provided. Chapter 5 supplies the reader insight into a plethora of methods, which have been developed and are being used to analyze data obtained from luminescence thermometry measurements. The next three chapters discuss three principle classes of materials used in luminescence thermometry. For each genre of materials, a foundation is provided that allows the reader to understand the essential physical concepts on their own without the need to resort to references. For example, topics, such as quantum confinement and transition metal and lanthanide luminescence centers are dealt with in a simplified fashion and can be understood with limited appeal to quantum mechanics. These chapters also discuss different synthetic methods employed in producing these types of materials. The author provides the reader with two chapters, Chapters 9 and 10, on the applications of thermometry in engineering and biomedicine. These chapters cover varied and disparate applications from fluid flows to using luminescence thermometry in disease diagnosis. Chapters 11 and 12 cover the emerging fields of temperature measurements at the nanoscale and the importance of multifunctionality, which combines thermometry with other luminescence applications, respectively.

All chapters are sufficiently detailed and provide a most useful and interesting one-volume compendium for the scientific community (chemistry, physics, engineering, biology, and medicine). The book affords the reader an excellent and balanced list of references from the recent literature. The book can be proudly recommended to students, teachers, and seasoned researchers without any restriction or hesitation.

John A. Capobianco
Center for NanoScience Research,
Concordia University, Montreal,
Quebec, Canada

Chapter 1

Introduction to Measurements of Temperature

1.1 WHY WE MEASURE TEMPERATURE?

Temperature is important because it is something we feel and because it has an effect on the smallest parts of our daily life, from how to adjust our housing and clothing, to what we eat for supper. We recognize a fever caused by an infection, for example, if someone has a body temperature over 38°C. In almost all fields of science, such as physics, chemistry, and biology, temperature is recognized as one of the basic variables. Aerodynamics, thermodynamics, heat transfer, and atmospheric sciences would be completely lost without temperature measurements.

For global climate patterns temperature is a major determining factor; a tiny change in average temperature may exert formidable effects on the environment worldwide. Plants would have lower yields or wilt when overall temperature changes by just few degrees. An increase in temperature can allow invasive species to take hold, and crop diseases to appear. Water temperature has a crucial influence on the biological activity and growth of aquatic organisms, since most of them are cold-blooded and are not able to internally regulate their body temperature. If water falls below, or rises above, certain temperatures, the number of individual organisms decreases until finally they disappear. Temperature is also important because of its influence on water chemistry, particularly on oxygen content.

1.2 THE MARKET SHARE OF TEMPERATURE SENSORS AND THERMOMETERS

Temperature is by far the most commonly measured physical quantity; sensors of temperature account for 80% of all sensors worldwide comprising a market of USD5.13 billion in 2016. Temperature sensors are used across a broad spectrum of human activities, such as in medicine, home appliances, meteorology, agriculture, industry, and military contexts.

At present, Europe represents the largest regional temperature sensor market worldwide, while Asia-Pacific ranks as the fastest growing market. It is expected that the market will grow at a compound annual growth rate (CAGR)

Luminescence Thermometry. http://dx.doi.org/10.1016/B978-0-08-102029-6.00001-4

of 4.8% between 2016 and 2022, to reach USD6.8 billion by 2022. Among the market's different segments, it is estimated that the oil and gas segment alone will account about 37% of the total market revenue by 2020. The key market drivers are: (1) legislation of stringent environmental and safety standards; (2) the increasing demand for temperature sensors among industrial end-users such as the food and beverage industry, oil and gas industry, paper industry, plastics industry, metal industry, and glass industry, among others; (3) the growing use of multifunctional temperature sensors in smartphones; (4) the increasing demand for miniature and smart temperature sensors in the consumer electronics industry; (5) the growing demand for noncontact temperature sensors; and (6) implementation of automated process monitoring and control in the industrial sector.

Among different types of temperature sensors, the demand for thermistors has been, and will remain to be, higher than for other types. Other key products are thermocouples, resistance temperature detectors, bimetallic temperature sensors, and fiber optic sensors. A significant growth in demand is expected in the near future for contactless temperature sensors, which are not only easy to use, but are less complex and more accurate than contact temperature sensors.

1.3 HISTORICAL BACKGROUND

Philo of Byzantium (born c. 280 BC, died c. 220 BC), also known as Philo Mechanicus, an ancient Greek engineer, was probably the first human performing comprehensive experiments with heating and cooling of objects. He invented a device, a predecessor of a middle age thermoscope [1], Fig. 1.1, which consisted of a tube exiting from a hollow sphere that was extended over a jug filled with water. Upon warming the globe with a candle, or when it was exposed to the sun, the air inside the globe expanded and pushed its way through the tube into the jar containing water, producing bubbles of escaping air. Upon cooling, the air contracted and water was drawn back into the tube.

The first "degrees" characterizing a heated body were mentioned by Hero of Alexandria (born c. AD 10, died c. AD 70, shown on the left of Fig. 1.2), Greek engineer and mathematician, the greatest experimentalist of antiquity, in descriptions of his famous steam-powered device and windwheel. Sometime later, a Greek Galen (Claudius Galenus, born c. AD 129, died c. AD 199, shown on the right of Fig. 1.2), a leading physician and surgeon in the Roman Empire at that time, introduced four degrees of coldness according to which the effects of different medicines should be classified. He also introduced a "neutral temperature," that of the mixture of ice and boiling water, having a value of zero degrees, something which depended on geographical latitude according to him. Medicines were divided into four categories in respect to their potency to cool down or warm up patients. To achieve the prescribed degree, medicines were prepared by mixing, and the term "temperature" is actually derived from the Latin word "temperatura," which means "due" or "proper measure." The

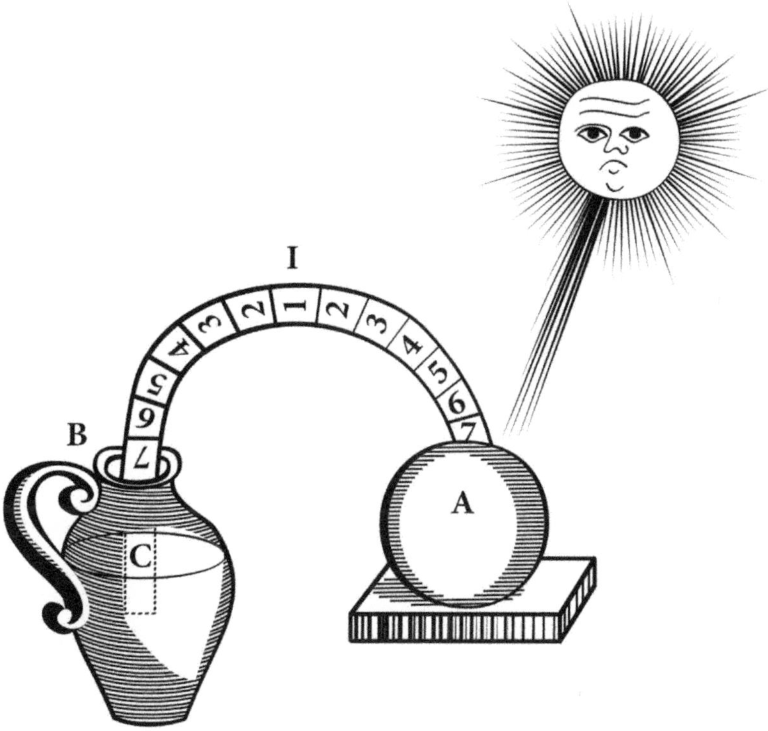

FIGURE 1.1 Philo's thermometers as depicted by Robert Fludd in 1638.

FIGURE 1.2 Hero of Alexandria (left) and Galen (right).

current meaning of the word was presumably established in the beginning 17th century, approximately at the same time the word "thermometer" appeared for the first time in a book *La Récréation Mathématique* (by J. Leurechon in 1624).

There was no progress in the "science of thermometry" until the end of 16th century when the first thermoscope was built. It is unknown for certain who made it. Most of the evidence suggests Sanctorius of Padua, but the invention is also attributed to a London physician Robert Fludd, to Galileo, to Cornelius Drebbel of Alcmar, and to father Paul of Cracow. This device today, known as an "air-thermoscope," is made of a glass tube with a bulb on one end. The other end of a tube is immersed in a colored liquid which is poured into an open vessel, Fig. 1.3A. The initial heating and cooling of the air in the bulb forced some quantity of the liquid to be drawn into the tube. The liquid level changed with temperature fluctuations. However, the level also changed due to ambient pressure changes, and the device was either not graduated or was just arbitrarily marked, so it was of little worth. The absence of "proper" graduation is a reason for referring to this device as a thermoscope rather than a thermometer. In 1632, a physician Jean Rey substituted air with water as the thermometric medium in an open-ended water-in-glass thermometer.

It is believed that the first alcohol thermometer was invented under Galileo's supervision in 1641. By the year of 1654 some these famous spirit-in-glass thermometers, known as the Little Florentine Thermometers had been built, Fig. 1.3B. The Grand Duke Ferdinand II de' Medici (1610–70) founded the first Academy of Sciences in Florence (Italy), named the "Accademia del Cimento" (i.e., Academy of Experiment). Florentine Academicians and Galileo analyzed the density

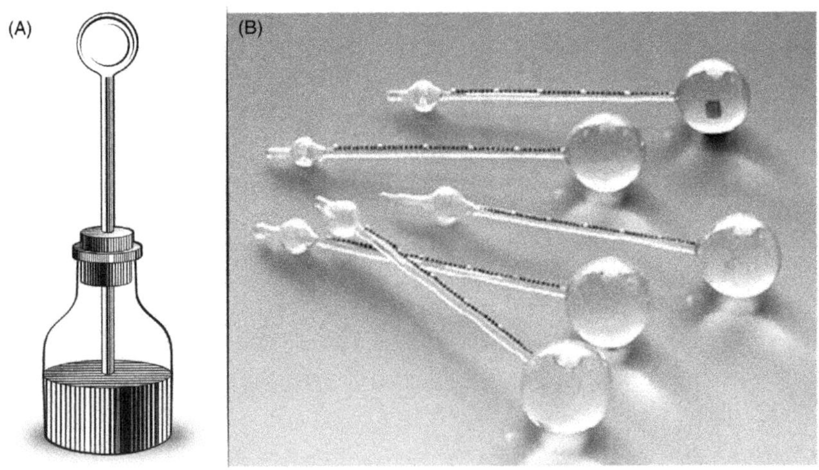

FIGURE 1.3 (A) Schematic of an air-thermoscope. (B) Little Florentine Thermometers. *((Reproduced from Camuffo D, Bertolin C. The world's earliest instrumental temperature records, from 1632 to 1648, claimed by G. Libri, are reality or myth? Clim Change 2013;119(3):647–57, with permission of Springer.))*

of different liquids, and chose to use a pure spirit (ethyl alcohol or ethanol) for their thermometer since they found that it was quite sensitive to temperature changes, did not adhere to the glass tube, and did not freeze in winter (as water does). The spirit was obtained with several repeated distillations of wine grapes or wine to achieve the percentage of alcohol (80% or more), and they named this concentrated alcohol solution "acquarzente." The Florentine Thermometers were built by Antonio Alamanni and Jacopo Giuseppe Mariani [2,3], and consisted of a sealed glass tube with a bulb, filled with the spirit, with 50 "degree" markings on the stem. The air from thermometers was not removed before filling with them with the spirit though. It is interesting to note that Florentine academicians in 1657 also experimented with mercury thermometers, but they concluded that ones containing spirit were more suitable than those containing mercury.

During 1654–70, Little Florentine Thermometers were used by the Medici Network for observation of temperature at 11 geographical locations, and these observations can be regarded as the first temperature measurements in climatology [3]. The Medici Network practiced an identical methodology and used identical thermometers, observed temperature changes after exposure of the thermometers in the shade and sunshine, and collected other climatology data.

Sir Robert Southwell brought the Little Florentine Thermometer to England in 1661, and shortly after, Robert Hook, a curator of the Royal Society, improved the thermometer's design by adding a red dye into the spirit as well as devising a new scale. He demonstrated that comparable thermometers could be made even if thermometer dimensions were not exactly the same. The method he employed was based on the equal rise of a volume of liquid starting from the freezing point of water. He introduced a scale ranging from −7 degrees (extreme cold) to +13 degrees (for the highest summer heat). In 1665, Robert William Boyle, an Anglo-Irish scientist, best known for Boyle's law, emphasized the importance of the comparability of scales of different thermometers. He proposed a scale with just one fixed point, the temperature of congealing oil of aniseed, and graduations based on the fractional increase in a volume of alcohol (thermometric liquid) from the reference point. At that time, it was more than obvious that some scale was necessary, and that fixed points should be defined to make the scales between thermometers comparable. There were many ideas regarding fixed points, the best known are summarized in Table 1.1.

In 1693, the British scientist E. Halley for the first time used mercury as a thermometric liquid. In 1701, Danish astronomer Rømer proposed the first calibrated scale. Initially, Rømer calibrated a zero temperature using a mixture of ice and salt, and a top temperature at the boiling point of water (then defined as 60 degrees). He considered it was much more convenient for calibration to use the freezing point of pure water as its lower point (at that time he defined it as 7.5 degrees). At the same time in Great Britain, Sir Isaac Newton described a temperature scale with 12 degrees. He chose two fixed points and assigned zero degrees at the temperature of water's freezing point and 12 degrees for the temperature of a healthy human being.

TABLE 1.1 Summary of Fixed Points Used by Various Scientists

Scientist	Year	Fixed points ("&" indicates a two-point system)
Sanctorius	c. 1600	Candle flame & snow
Accademia del Cimento	c. 1640	Most severe winter cold & greatest summer heat
Otto von Guericke	c. 1660	First night frost
Robert Hooke	1663	Freezing distilled water
Robert Boyle	1665	Congealing oil of aniseed or freezing distilled water
Christiaan Huygens	1665	Boiling water or freezing water
Honoré Fabri	1669	Snow & highest summer heat
Francesco Eschinardi	1680	Melting ice & boiling water
Joachim Dalencé	1688	Freezing water & melting butter or ice & deep cellars
Edmond Halley	1693	Deep caves & boiling spirit
Carlo Renaldini	1694	Melting ice & boiling water
Isaac Newton	1701	Melting snow & blood heat
Guillame Amontos	1702	Boiling water
Ole Rømer	1702	Ice and salt mixture & boiling water
Philippe de la Hire	1708	Freezing water & Paris Observatory cellars
Daniel Gabriel Farenheit	c. 1720	Ice, water, and salt mixture & ice and water mixture & healthy body temperature
John Fowler	c. 1727	Freezing water & water hottest to be endured by a hand held still
R.A.F. de Réaumur	c. 1730	Freezing water
Joseph-Nicolas De l'Isle	1733	Boiling water
Anders Celsius	by 1741	Melting ice & boiling water
J.B. Micheli du Crest	1741	Paris Observatory cellars & boiling water
Encyclopedia Britanica	1771	Freezing water & congealing wax

Source: Data extracted from Chang H. Inventing temperature: measurement and scientific progress. New York, NY, USA: Oxford University Press Inc.; 2004.

In 1703 in France, a physicist Guillaume Amontons described a new thermometer which measured the change in air pressure upon heating, instead of an increase in air volume. He observed that air pressure increased by approximately one-third between the temperatures of "cold" and the boiling point of water and that temperatures of the "greatest summer heat" and the "greatest winter cold" in Paris have the ratio of about six to five. Then, he went further to conclude that the lowest possible temperature is the one that corresponds to zero air pressure. He attributed this temperature to a value which corresponds to what we now define as −240°C. This brilliant observation, even though it had an enormous quantitative error, was the first on the way to an understanding of the concept of temperature and the derivations of the gas laws, in particular, Gay-Lussac's law. Amontons work showed that temperature could be defined as being proportional to the pressure of a gas, so one fixed point is sufficient to define a scale. Unfortunately, his work was not properly understood, and at that time the science of thermometry took two different paths. One path focused on the design and production of improved thermometers and inventions of practical scales—significant contributions to which were given by Fahrenheit, Celsius, and Réaumur.

The other path, started by Amontons, was continued by French chemist Joseph Louis Gay-Lussac who found that at a constant pressure, the volume of a gas would expand at 1/267 for each degree of temperature rise. A long time passed until, in 1874, a French chemist and physicist Henri Victor Regnault obtained better experimental results, deriving the number to be 1/273. He concluded that pressure would approach zero at 1/273.15°C. This lead to the definition of the ideal gas scale, which, although empirical, precisely equals the thermodynamic scale (absolute scale) since it predicts zero pressure for a gas at −273.15°C.

After visiting Rømer in 1708, Daniel Gabriel Fahrenheit, an instrument maker and inventor born in Gdańsk and working in Holland, began producing thermometers. He substantially improved the existing thermometer design by changing the reservoir to a cylinder, and by introducing mercury as a thermometric fluid instead of spirit. The linear rate of thermal expansion of mercury proved to be of significant benefit for the further development of thermometry. He presented the specification of a mercury-in-glass thermometer to The Royal Society in 1724, along with his new scale, today known as the Fahrenheit scale, which is still used in many countries worldwide, mostly the UK and United States. The scale was based on three fixed-point temperatures: (1) a mixture of ice, water, and ammonium chloride was taken for the zero point, (2) a mixture of ice and water was taken for 32°, and (3) human body temperature was taken for 96°. After adoption of this scale in Great Britain, the boiling point of water was set to a temperature of 212°.

French scientist René Antoine Ferchault de Réaumur introduced a new scale in 1730, and this scale gained considerable recognition. He set 0° and 80° for temperatures of freezing and boiling water. He used diluted ethanol as the thermometric fluid with a concentration such that the fluid expanded in volume by 8% when the temperature reached 80°Ré, the temperature at which ethanol starts to boil. The scale was in use throughout Europe and was abandoned in France in 1790 after the adoption of the Celsius scale and the metric system.

Interestingly, the Réaumur scale is still in use in Italy and Switzerland for monitoring milk temperature during cheese production.

Joseph-Nicolas Delisle, a French cartographer and astronomer, introduced in 1732, while working in St. Petersburgh, the mercury thermometer and his scale that used the temperature of boiling water as its fixed zero point. The scale had a large number of graduations, usually 2400 or 2700. In 1738, a German scientist Josias Weitbrecht, also working in Russia, recalibrated the Delisle thermometer using the temperature of the boiling point of water as 0° and the temperature of the freezing point of water as 150°. He sent this thermometer to many scientists, including Celsius. The thermometer was widely used in Russia for more than 100 years.

The Swedish astronomer and physicist Anders Celsius presented the Celsius scale in 1742. He assigned 0° to the temperature of boiling water and 100° to the temperature of melting ice and divided the range into 100 equal parts (called "centigrades"). Initially, the scale was named the Centigrade scale. The scale was similar to the Delisle scale, with the main difference being the 100° difference between fixed points instead of 150°. In 1745, shortly after Celsius' death, another Swedish scientist Carl Linnaeus reversed the scale for practical reasons.

In 1817, French physicists Pierre Louis Dulong and Alexis Thérèse Petit examined the thermometric properties of various substances. They recognized the arbitrary nature of temperature scales with only gas thermometer scales being comparable.

In 1848, the Belfast born scientist William Thomson, who later assumed the title Lord Kelvin of Largs, Fig. 1.4, introduced "An Absolute Thermometric Scale." Lord Kelvin proposed that the scale's zero point should be an "infinite cold" (absolute zero) which does not depend on the properties of a particular material and at which material can become no colder. He calculated that absolute zero was equivalent to −273°C on the air thermometers of the time. His conclusions were derived from analysis of the Carnot cycle which depends only on hot-body and heat-sink temperatures, not on particular material properties. This absolute scale is known today as the Kelvin thermodynamic temperature scale. In 1954, the Kelvin scale was redefined by designating the triple point of water as its second defining point and assigning its temperature to precisely 273.16 K.

In 1859, the Glasgow physicist William John Macquorn Rankine proposed another thermodynamic scale, today known as the Rankine scale, which also has its zero set to absolute zero temperature, but with its degrees defined as equal to one degree Fahrenheit, rather than the one degree Celsius used by the Kelvin scale. Table 1.2 gives conversion formulas from, and to, the Celsius scale from the other, most frequently used temperature scales.

These final findings may be considered a closing of the epoch of great achievements in thermometry. Research in modern times encompasses the conception and technical perfection of new methodologies, but does not provide novel ideas and concepts to the general science of thermometry [6]. Things

FIGURE 1.4 **Portrait of Lord Kelvin (1824–1907).**

TABLE 1.2 Conversion Between the Celsius Scale and Different Temperature Scales

Scale	From celsius to	To celsius from
Fahrenheit	[°F] = [°C] × 9/5 + 32	[°C] = ([°F] − 32) × 5/9
Kelvin	[K] = [°C] + 273.15	[°C] = [K] − 273.15
Rankine	[R] = ([°C] + 273.15) × 9/5	[°C] = ([R] − 491.67) × 5/9
Delisle	[°De] = (100 − [°C]) × 3/2	[°C] = 100 − [°De] × 2/3
Newton	[°N] = [°C] × 33/100	[°C] = [°N] × 100/33
Réaumur	[°Ré] = [°C] × 4/5	[°C] = [°Ré] × 5/4
Rømer	[°Rø] = [°C] × 21/40 + 7.5	[°C] = ([°Rø] − 7.5) × 40/21

Source: Data from Conversion of units of temperature. Wikipedia; 2017. Available from: https://en.wikipedia.org/wiki/Conversion_of_units_of_temperature.

have, however, started to change with emerging technologies, such as nanotechnology and biotechnology, for which existing methods of thermometry proved insufficient. New thermometry concepts, which are discussed in the next section, are mostly based on noncontact, optical measurements of temperature. Among them, researchers have given special attention to temperature readings from luminescence.

The brief historical outline of the development of luminescence thermometry can be found in the review article of Allison and Gillies [7]. They rightfully concluded that research on luminescence thermometry has been facilitated by extensive research on luminescing materials conducted after the introduction of the fluorescent lamp in 1938. Temperature dependence of luminescence of materials is an important property for many novel applications. By examining this property, scientists have found that temperature can be determined from different luminescence features of heated materials (see Chapter 4: Schemes for Reading Temperature from Luminescence). Many have started to explore and develop this new thermometry concept for an assortment of applications. In 1953, Bradley [8] studied the use of thermographic phosphors mixed with binders and ceramic materials to measure surface temperature in aerodynamics. This study may be considered the first applicative study of the, new at that time, luminescence thermometry method. Later, thermographic phosphors were used for temperature measurements in high-speed wind tunnels and gas turbine engines. Subsequently, reports on the use of luminescence thermometry for remote optical measurements [9] and bio-medical uses [10,11] appeared.

1.4 WHY ARE NOVEL CONCEPTS AND METHODS FOR MEASURING TEMPERATURE NEEDED?

As our civilization advances, scientific and technological developments and awareness of environmental and health problems generate a continual need for the development of novel measurement concepts and instruments. For example, from April 10, 2014, the manufacture or supply of mercury glass tube thermometers became illegal. This fell in line with European regulations aimed to cut the overall spread of mercury-containing products due to the adverse effects of mercury on health and the environment. To meet the market requirement for glass tube thermometers, suppliers had to "reinvent" spirit filled thermometers. Even more importantly, new temperature measurement concepts and temperature probes are required for applications in emerging environments, the most important of which being nanotechnology, biotechnology, and integrated optics. Today, there is an immediate need for noncontact thermometry for moving objects or objects which are sensitive to contact, difficult to access, or in hazardous locations [12]. Temperature measurements obtained from changes in the optical properties of materials is considered a promising route to meet these needs. Optical methods of interest are Raman scattering, near-field scanning optical microscopy, thermoreflectance, optical interferometry and, of course, luminescence spectroscopy/microscopy. To illustrate the growing need for novel temperature measurement concepts, typical problems in several thermometry applications will be briefly outlined emphasizing the importance of finding a way to resolve them.

Surface temperature measurements of moving systems such as pistons in engines or rotors in gas turbines encountered difficulties using traditional invasive or semiinvasive methods (thermocouples, coatings with temperature sensitive paints or thermochromic liquid crystals, etc.) [13]. The use of noninvasive pyrometry needs accurate data for the emissivity of the surface, which may change during operational conditions, and is compromised by interfering light particularly in combustion environments.

Flows of liquid and gas are found in a huge range of devices on which we are increasingly reliant for our energy, safety, comfort, and mobility. These include flows used for the cooling of electronic chips and for air conditioning, flows inside heat exchangers and boilers, and reacting flows in gas turbines and car engines. Such flows are extremely complex, and simultaneous measurements of fluid temperature and velocity—quantities of critical interest—are urgently needed. This may be achieved, for example, by using thermographic phosphor particles [14].

When dealing with very small objects, such as in nanoelectronics, integrated photonics, and biomedicine, traditional thermometry methods are inadequate. In fact, traditional methods cannot provide spatial resolutions greater than one micrometer which represents the resolution needed for these emerging and important applications. Knowledge of the locations of possible "hot-spots" in nanoelectronic devices may ease and increase their development and reliability. Moreover, precise thermal maps of nanoelectronic devices can help us to understand the operation of such devices, and to development theories explaining nanoscale thermal behavior. The refractive index of a material is strongly temperature dependent, and it is the main parameter used for the manipulation of light in integrated photonic devices. Therefore, novel methods are needed for obtaining knowledge of the temperature distribution in microphotonic and nanophotonic devices. It is well recognized that in any biosystem temperature plays a critical role in defining its dynamics and properties [15]. For example, temperature is related to many cellular functions, including gene expression, protein stabilization, enzyme–ligand interactions, and enzyme activity [16]. In medicine, a precise knowledge of tissue temperature is crucial for novel cancer treatments based on induced heating of malignant cells. Finally, we are now witnessing a new phase in material research, where it is expected that materials and structures will perform several functions; it is reasonable to believe that in the near future most, if not all, materials and structures will be "smart," "tailored" multifunctional materials [17]. Polyvalent functions (multifunctionality) may be achieved, for example, with luminescence measurement techniques capable of performing both temperature and chemical sensing, or by sensing temperature along with some other physical property such as pressure, electric field strength, or magnetic field strength. In addition, temperature mapping may be combined with luminescence imaging and/or photothermal therapy, for advanced medical diagnoses and treatments.

REFERENCES

[1] McGee TD. Principles and methods of temperature measurement. New York, NY, USA: Wiley; 1988.

[2] Camuffo D, Bertolin C. The world's earliest instrumental temperature records, from 1632 to 1648, claimed by G. Libri, are reality or myth? Clim Change 2013;119(3):647–57.

[3] Camuffo D, Bertolin C. The earliest temperature observations in the world: the medici network (1654–1670). Clim Change 2012;111(2):335–63.

[4] Chang H. Inventing temperature: measurement and scientific progress. New York, NY, USA: Oxford University Press, Inc.; 2004.

[5] Conversion of units of temperature. Wikipedia. 2017. Available from: https://en.wikipedia.org/wiki/Conversion_of_units_of_temperature.

[6] Barnett MK. A brief history of thermometry. J Chem Educ 1941;18(8):358–64.

[7] Allison SW, Gillies GT. Remote thermometry with thermographic phosphors: Instrumentation and applications. Rev Sci Instrum 1997;68(7):2615–50.

[8] Bradley LCI. A temperature sensitive phosphor used to measure surface temperatures in aerodynamics. Rev Sci Instrum 1953;24(3):219–20.

[9] McCormack JS. Remote optical measurement of temperature using luminescent materials. Electron Lett 1981;17:630–1.

[10] Sholes RR, Small JG. Fluorescent decay thermometer with biological applications. Rev Sci Instrum 1980;51(7):882–4.

[11] Samulski TV, Chopping PT, Haas B. Photoluminescent thermometry based on europium-activated calcium sulphide (hyperthermia application). Phys Med Biol 1982;27(1):107–14.

[12] Antić Ž, Dramićanin MD, Prashanthi K, Jovanović D, Kuzman S, Thundat T. Pulsed laser deposited dysprosium-doped gadolinium–vanadate thin films for noncontact, self-referencing luminescence thermometry. Adv Mater 2016;28(35):7745–52.

[13] Brübach J, Pflitsch C, Dreizler A, Atakan B. On surface temperature measurements with thermographic phosphors: a review. Prog Energy Combust Sci 2013;39(1):37–60.

[14] Abram C, Fond B, Heyes AL, Beyrau F. High-speed planar thermometry and velocimetry using thermographic phosphor particles. Appl Phys B 2013;111(2):155–60.

[15] Jaque D, Vetrone F. Luminescence nanothermometry. Nanoscale 2012;4(15):4301–26.

[16] Brites CDS, Millán A, Carlos LD. Chapter 281 – Lanthanides in luminescent thermometry. In: Bünzli J-C, Pecharsky VK, editors. Handbook on the physics and chemistry of rare earths. Elsevier; 2016. p. 339–427.

[17] Dramićanin MD. Sensing temperature via downshifting emissions of lanthanide-doped metal oxides and salts. A review. Methods Appl Fluoresc 2016;4(4):42001. (23pp).

Chapter 2

Temperature and Ways of Measuring It

2.1 INTRODUCTION

This chapter presents the basic elements of thermometry—the field of science devoted to temperature measurements. It gives answers to the question "what is temperature?" Answers which are derived from our everyday experience of hot and cold, and from the fundamental theories of thermodynamics and statistical mechanics. In both theoretical approaches, the temperature appears as a parameter of an equilibrium situation and is inseparably linked to Boltzmann's constant. The current definition of kelvin, the SI unit of temperature, and expected definition of kelvin in terms of the fixed numerical value of Boltzmann's constant are presented.

The concept and definition of primary thermometry methods are discussed in the Section 2.3, and principles of operation of the most common primary methods are presented. These methods include acoustic gas thermometry (AGT), spectral-band radiometric thermometry, constant-volume gas thermometry, dielectric constant gas thermometry, refractive index thermometry, Johnson noise thermometry, Doppler broadening, and magnetic thermometry.

Section 2.4 is devoted to the current International Temperature Scale (ITS-90) and the Provisional Low Temperature Scale (PLTS-2000). This chapter also presents brief descriptions of thermometers which are the most commonly used: liquid-in-glass thermometers, bimetallic thermometers, thermocouples, resistance thermometers, thermistors, manometric thermometers, and infrared thermometers. This chapter concludes with an explanation of measurement errors (uncertainties) and parameters that describe the performance of a thermometer.

2.2 TEMPERATURE

What is temperature? The qualitative definition is that it is *the measure (or degree) of hotness or coldness*, the very first definition being derived from human sensations of hot and cold. But an exact comprehending of temperature as a fundamental physical quantity must involve principles of either classical thermodynamics or statistical mechanics, both producing exactly equivalent formalisms and results. When defining temperature, classical thermodynamics

Luminescence Thermometry. http://dx.doi.org/10.1016/B978-0-08-102029-6.00002-6

explores how the heating process works, while statistical physics links temperature with properties of matter in equilibrium. In both approaches, temperature appears as a parameter of an equilibrium situation.

While developing classical thermodynamics, many scientists (Kelvin, Joule, Clausius, and others) realized that temperature could be considered as that parameter of the state of a fluid which has the property of always taking the same value in those systems which are in equilibrium with one another. This later becomes the famous zeroth law of thermodynamics, stating that "if two systems are in thermal equilibrium with a third system then they are in thermal equilibrium with each other." Thus, this law defines *temperature* as *the parameter of state that has the same value for any systems which are in thermal equilibrium*, and also underlies any concept of practical temperature measurement. The idea of defining temperature from thermodynamic principles originates from Carnot's study of heat engines and further elaboration of Carnot's theorem by Kelvin and Clausius, which led to the second law of thermodynamics. The Carnot cycle of an ideal heat engine is composed of isothermal heat transfer at a high temperature (T_2) coupled with adiabatic expansion, and isothermal heat transfer at a low temperature (T_1) coupled with adiabatic compression back to the high temperature. The efficiency (η) of such an ideal engine does not depend on the properties of the working substance, but only on the ratio of temperatures between which engine operates:

$$\eta = 1 - \frac{T_1}{T_2}. \tag{2.1}$$

Bearing in mind that efficiency must be less or equal to one, it is easy to conclude that the thermodynamic temperature (T_1) cannot take negative values, and that a theoretical maximal efficiency of 1 is reached when $T_1 = 0$ (absolute zero). Starting from this zero temperature, a thermodynamic temperature can be derived just by choosing the size of the increment. As mentioned in Chapter 1, Introduction to Measurements of Temperature, for this increment Kelvin chose the size of the degree of Celsius, while Rankin chose the Fahrenheit's degree. Thus, if it were possible to construct an "ideal" Carnot engine, then one could define temperature in terms of equal increments of work output needed to complete the isothermal transfer of heat between reservoirs at two temperatures. However, construction of such ideal engine is not possible.

From the axiomatic foundation of thermodynamics, developed by the mathematician Carathéodory [1] starting from the differential form of the first law of thermodynamics [2], the temperature is defined as being the inverse value of the rate of change of entropy (S) with internal energy (U) at constant volume (V):

$$T = \frac{1}{(\partial S / \partial U)_V}. \tag{2.2}$$

Just after the introduction of thermodynamic temperature by Kelvin and Clausius, the first definition of temperature based on the molecular or kinetic theory of gasses appeared based on the work of Maxwell and Boltzmann:

temperature can be assumed as a quantity which characterizes the energy distribution of the molecules in a system. This definition, however, can be applied only to gasses since the underlying theory considers only those systems composed of independent, noninteracting particles. The kinetic approach for defining temperature was generalized by statistical mechanics on the basis of Boltzmann and Gibbs' work. Gibbs provided a fundamental breakthrough by considering the distribution of a whole system of molecules (solid, liquid, or gas) over various states of the system. Carathéodory showed that the probability $P(E)dE$ of finding a closed-volume system which is in thermal equilibrium with a heat container and has an energy between E and $E + dE$ is given by:

$$P(E)dE \propto \Omega(E)\exp(-E/\theta)dE, \tag{2.3}$$

where $\Omega(E)dE$ is the number of available states between E and $E + dE$, and θ is the phenomenological parameter which determines the rate of decrease of the $\exp(-E/\theta)$ function. Eq. (2.3) has a maximum at an energy E_{max} given by:

$$\frac{d(\log\Omega(E))}{dE} = \frac{1}{\theta}. \tag{2.4}$$

One should note that the last equation resembles Eq. (2.2). In fact, the complete correspondence to the definition of temperature from thermodynamics by Eq. (2.2) is obtained by considering $U = E_{max}$, $k \cdot \log\Omega(U) = S(U)$ as the entropy of the system, and $T = \theta / k$ as the thermodynamic temperature, where k is Boltzmann's constant.

The SI unit of thermodynamic temperature is the kelvin (K). In 1954, the thermodynamic temperature of the triple point of water (the triple point of a substance is the condition where vapor, liquid, and solid coexist) was defined to be precisely 273.16 K at 611.657 Pa (0.00603659 atm) pressure. In other words, kelvin has a numerical value of 1/273.16 of the temperature of the triple point of water. Today, the Celsius scale is most frequently used, and its unit is the degree Celsius (°C), which has the same magnitude as kelvin. This scale is defined with two fixed points: 0° for the freezing point of water and 100° for the boiling point of water at 1 atm pressure. Since the freezing point of water is located approximately 0.00993 K (≈0.01 K) below the triple point of water, the zero point of the Celsius temperature scale is thus redefined at 273.15 K.

When a numerical value of the thermodynamic temperature of the triple point of water is defined, then the thermodynamic temperature of other systems can be measured using primary thermometers. For these thermometers, the fundamental relation between the measurand and temperature is explicitly known, so that temperature can be calculated without the introduction of unknown, temperature-dependent constants. Primary thermometers will be described in Section 2.3; at this point, it is important to emphasize fundamental difficulties in thermometry which arise from the current definition of kelvin. Since kelvin is defined from a temperature of the triple point of water, it is linked to a material property. Also, with such a definition, any primary thermometer used to measure a thermodynamic temperature must operate well at both the triple point of water and the unknown temperature [3].

At present, the main activities in the field of metrology are related to defining units of measurement by experimental procedures that link units to fundamental constants. On the 24th meeting in 2011, The General Conference on Weights and Measures (CGPM) agreed that the kilogram, ampere, kelvin, and mole should be defined by fixed numerical values [4] of the Planck constant, elementary charge, Boltzmann constant, and Avogadro constant, respectively [5,6]. While writing this book, researchers from several European metrology centers are making final adjustments to the Boltzmann constant's value. To be used for the new definition of kelvin at the next meeting of CGPM in 2018, the Boltzmann constant must be determined with a relative standard uncertainty of less than 1×10^{-6}, and by at least two fundamentally different methods. So, it is expected that the *kelvin* will be defined *as the value of change of thermodynamic temperature that results in a change of thermal energy kT by* $1.3806488 \times 10^{-23}$ J (the Boltzmann constant k will be set to $1.3806488 \times 10^{-23}$ J/K = kgm^2/s^{2}K). By defining kelvin on the basis of the fixed numerical value of Boltzmann's constant, it will be possible to construct thermometers working in different temperature ranges, without a need to operate at the triple point of water, and also to exploit various equations of state that include temperature to make traceable temperature measurements [3]. As a first consequence, the temperature of the triple point of water will no longer be fixed value.

2.3 PRIMARY THERMOMETERS

Primary thermometry entails primary temperature measurement methods, instrumentation, and standards needed for the determination of the thermodynamic temperature and the founding and realization of an internationally common temperature scale. *The primary thermometer* is established on a well-known physical system for which an explicit relation (equation of state) exists between the measured quantity and temperature and can be written without an inclusion of any unknown or significantly temperature-dependent constant. The second version of the *mise en pratique* ("practical realization"; MeP-K) of the definition of the kelvin, approved by the Consultative Committee for Thermometry in 2013, which will come into force after the redefinition of the kelvin, defines two categories of primary thermometers, absolute and relative [6]. Thermodynamic temperature is measured directly from the defined numerical value of the Boltzmann constant (which determines the thermal energy kT) by *absolute primary thermometers*. These measurements do not require referencing to any temperature fixed point, and all equation of state parameters must be measured or otherwise determined. On the other hand, *relative primary thermometers* indirectly measure thermodynamic temperature, using temperature fixed points to determine values of parameters in the equation of state. Thermodynamic temperature of fixed points and their uncertainties are known from prior absolute or relative primary temperature measurements.

MeP-K recommends two primary thermometry methods: AGT and spectral-band radiometric thermometry, but many others are also important and relevant. These others include: constant-volume gas thermometry, dielectric constant gas thermometry, refractive index thermometry, Johnson noise thermometry, Doppler broadening, and magnetic thermometry.

AGT is based on the relation between the speed of sound of an ideal gas, μ_0, and the thermodynamic temperature:

$$\mu_0 = \sqrt{\frac{\gamma RT}{M}}, \tag{2.5}$$

where $\gamma = c_p / c_v$ is the ratio of specific heat capacities at constant pressure and constant volume, respectively, M is the molar mass, and R is the gas constant ($R = k \cdot N_A$, where N_A is the Avogadro constant and k is the Boltzmann constant). The speed of sound in real gasses shows slight pressure dependence, so Eq. (2.5) should be modified as a virial expansion in pressure using acoustic virial coefficients. In practice, the speed of sound in real gasses, usually Ar and He, is measured at several pressures (p) and then extrapolated to zero pressure to obtain μ_0. The speed of sound is determined [5] from the resonance frequency f and the dimensions of the spherical resonator, D, Fig. 2.1.

In recent times, quasispherical-shaped cavities have been used, since they act as both acoustic and electromagnetic resonators, and the microwave resonances are exploited to determine the diameter of the cavity d. When performed in such way that the acoustic resonance frequencies, pressures, cavity dimensions, and molecular mass of the gas are traceable to the meter, the kilogram, and the second, AGT is an absolute primary thermometer. The method can also be employed as a relative primary thermometer by measuring the speeds of sound at unknown temperatures T_x, as well as at the temperature of a fixed point for which the thermodynamic temperature is known T_{fp}, and by taking the ratio of the square of measured speeds: $T_x = T_{fp} \cdot \frac{\mu_{0x}^2}{\mu_{0fp}^2}$. The range of temperature

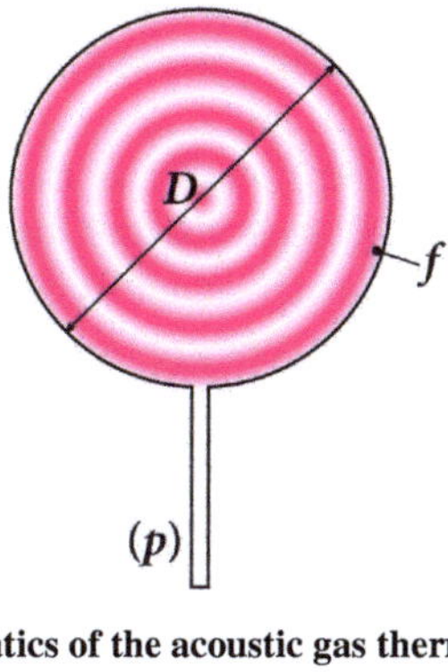

FIGURE 2.1 Simplified schematics of the acoustic gas thermometer.

measurements by AGT spans 7–552 K, but it is expected that this range be extended to cover temperatures from the lambda point of helium $T_\lambda = 2.172$ K to the freezing point of copper $T_{Cu} = 1358$ K [7].

Spectral-band radiometric thermometry is based on the measurements of the spectral radiance of a blackbody. The spectral radiance $L_{b,\lambda}$ is, in fact, the power of radiation emitted by a blackbody at a certain temperature per unit area per solid angle per unit wavelength interval and is given by Planck's law:

$$L_{b,\lambda}(\lambda,T)d\lambda = \varepsilon_b(\lambda,T)\cdot\left(\frac{2hc^2}{n^2\lambda^5}\right)\cdot\frac{d\lambda}{\exp(hc/n\lambda kT)-1}. \tag{2.6}$$

In this equation, $\varepsilon_b(\lambda,T)$ represents the spectral emissivity of the blackbody (which should be well-known), h is the Planck constant, c is the speed of light *in vacuum*, n is the refractive index of the gas along the optical path of detection, and λ is the wavelength of radiation. Hence, a thermodynamic temperature can be directly obtained from the measured spectral radiance, the unit of which is $Wsm^{-2}r^{-1}m^{-1}$. If the measurements are traceable to the definition of the watt and meter the method represents *the absolute primary thermometer*. When used as *the relative primary thermometer*, spectral-band radiometric measurements do not require prior knowledge of the absolute spectral responsivity of the radiometer or the geometric factors of measurements. The spectral radiance is measured relative to those performed at temperatures of one or more fixed-point blackbodies, each with known thermodynamic temperatures with subsequent interpolation and extrapolation of data. This method is mainly used at high temperatures, particularly above silver's melting point.

Constant-volume gas thermometry is one of the most important primary thermometry methods both from a historical and practical point of view. The equation of state for an ideal gas:

$$pV = nRT, \tag{2.7}$$

where V is the constant volume and n is the number of moles of gas in V, tells us that thermodynamic temperature can be determined directly by measuring the pressure p. Even though many real gasses show near-ideal behavior, Eq. (2.7) is in practice modified by virial expansion in the gas density (n/V) to account for the particle interaction contribution (second and third-order interactions, and is some cases interactions of higher order). Then, pressure is measured at different gas densities (n/V) and extrapolated to zero density to obtain the value of RT. The absolute primary thermometer is realized with pressure measurements traceable to its SI unit Pa, and volume measurements to the unit of the meter. Relative primary thermometry is achieved by measuring pressure at the unknown temperature and the temperature of the fixed-point, usually the triple point of water, and by taking the ratio of the measured pressures. In such cases, the gas density effect is compensated to the first order of accuracy. Over the years, this method was the dominant primary thermometric method for temperatures up to 900 K.

In *dielectric constant gas thermometry*, the density of a gas in the equation of state is replaced by the dielectric constant:

$$p = kT \cdot \frac{\varepsilon - \varepsilon_0}{\alpha_0}, \tag{2.8}$$

where ε is the dielectric constant (permittivity), ε_0 is the dielectric constant *in vacuum*, and α_0 is the static electric dipole polarizability [5]. Dielectric constant is determined by measurements of the capacitance changes of a capacitor filled with gas at different pressures and constant temperature (measurement of isotherms) [8]. During measurements, it is necessary to compensate changes of the dimensions of the capacitor with pressure. The method requires knowledge of the polarizability α_0 with sufficient accuracy. The *refractive index gas thermometry* uses similar approach as the dielectric constant gas thermometry in order to eliminate the density of gas in the respective equation of state:

$$p = kT \cdot \frac{(n_r^2 - 1)\varepsilon_0}{\alpha_0}, \tag{2.9}$$

where n_r stands for the refractive index of the gas. Here, the precise optical measurements are needed to derive the refraction index and subsequently temperature (temperature is indirectly measured as in dielectric constant gas thermometry). The method also requires knowledge of the polarizability α_0. Both methods are particularly useful for low-temperature measurements.

Johnson noise (thermal noise) arises from the spontaneous thermal fluctuations in electrical circuits. *Johnson noise thermometry* exploits the temperature dependence of the mean-square noise voltage $\langle V^2 \rangle$ of a sensor having resistance R_e which is measured over the Δf bandwidth. The corresponding equation is a well-known Nyquist equation, more precisely, the best-known approximation of the Nyquist theorem, which is valid for frequencies below 1 MHz and for temperatures above 25 K [9]:

$$\langle V^2 \rangle = 4\,kTR_e \cdot \Delta f. \tag{2.10}$$

Contrary to previously mentioned methods, Johnson noise thermometry is a completely electrical method, and is an absolute primary thermometer since thermodynamic temperature is obtained directly from the measured mean-square voltage, with all measurements traceable to SI units: volt, ohm, and second. Relative primary thermometry can be easily set-up since the equation does not involve any temperature-dependent term aside from the mean-square voltage, therefore, temperature can be measured by reference to one fixed-point. The method, however, is not suitable for use at very low or very high temperatures nor is it suitable in volatile environments.

Doppler-broadening thermometry is a relatively new primary thermometric method which is based on the precise spectroscopic measurement of a Doppler-broadened absorption line profile and determination of its width in a gas phase absorbing medium. At low gas pressures (Doppler regime), the motion of atoms causes a frequency shift from the expected transition frequency via the Doppler effect. This leads to the Doppler width of a spectral line which links the thermal energy to an optical frequency, according to the following equation [10]:

$$\Delta \nu_D = \frac{\nu_0}{c_0}\sqrt{2\log 2\frac{kT}{M_a}}, \tag{2.11}$$

where $\Delta \nu_D$ is the Doppler half width at half maximum (HWHM), ν_0 is the line center frequency, c_0 is the speed of light, and M_a is the absorber mass. In principle, the measurement can be accomplished by conventional laser spectroscopic techniques. The advantage of the method is also that it uses gas pressures which are much smaller (from 0.1 to 500 Pa) than those used in acoustic gas and dielectric gas thermometry [5].

Magnetic thermometry is based on the relation between the magnetic susceptibility χ of paramagnetic substance and thermodynamic temperature, which is given by Curie's law: $\chi = C/T$, where C represents the Curie constant. However, in practice, interactions and other effects must be considered, so the relation becomes:

$$\chi = \frac{C}{T + \Delta + \gamma / T}, \tag{2.12}$$

where Δ includes first-order dipole–dipole and exchange couplings and also a shape factor, while γ is due primarily to crystal field splitting of the ground state and second-order interaction effects. Thermodynamic temperature is determined indirectly, by measuring the magnetization M in a constant polarizing field H. When Δ or γ are derived from theory, the method acts as an absolute thermometer; otherwise, the referencing to several fixed-points is needed to determine the required constants. The method is used for measurements of low temperatures.

At the end of this section, one should note that primary thermometers have been significantly improved over the years. Their uncertainties have been reduced by about one order of magnitude every 20 years. However, only in recent times has our ability to measure thermodynamic temperature exceeded the reproducibility of practical temperature scales.

2.4 THE INTERNATIONAL TEMPERATURE SCALE AND THE PROVISIONAL LOW TEMPERATURE SCALE

A common temperature scale was needed to grant meaningful comparisons of temperature measurements made by different people. *The International Temperature Scale* (*ITS-90*), adopted in 1989 by the International Committee for Weights and Measures (CIPM), covers the temperature range from 0.65 K to the highest temperature measurable by practical implementation of the Planck radiation law using monochromatic radiation. *The Provisional Low Temperature Scale* (*PLTS-2000*), adopted by CIPM in 2000, covers the temperature range from 0.9 mK to 1 K. Since ITS-90 and PLTS-2000 differ in their overlapping range (from 0.65 to 2 K), a new *^{3}He vapor pressure scale* (*PTB-2006*) was adopted. This scale passes into the ITS-90 above 2 K, and into the PLTS-2000 below 1 K. These internationally agreed temperature scales are approximations of the thermodynamic temperature scale and are used for defining international Kelvin temperatures (symbols T_{90} or T_{2000}) and international Celsius temperatures (symbol t_{90} or t_{2000}). The relation between T_{90} (T_{2000}) and t_{90} (t_{2000}) is the same as that of their thermodynamic counterparts: t_{90} (or t_{2000}) [°C] = T_{90} (or T_{2000}) [K] − 273.15.

The ITS-90 assigns temperature values, which have been determined by primary thermometry, to a series of highly reproducible states of matter (temperature fixed points or defining points), postulates interpolating or extrapolating instruments for a specific subrange of temperature, and defines any required interpolating or extrapolating equations [6]. ITS-90 specifies 16 temperature fixed points (defining points) which cover a temperature range from 13.8033 to 1357.77 K. Temperature fixed points are well-defined and reproducible states of matter to which precise values of temperature can be assigned. These are listed in Table 2.1 with their designated temperatures. For practical reasons, one should be aware of a set of secondary reference points which cover a

TABLE 2.1 Temperature Fixed Points (Defining Points) of the ITS-90

Equilibrium state	T_{90} (K)	t_{90} (°C)
Triple point of hydrogen	13.8033	−259.3467
Boiling point of hydrogen at a pressure of 33,321.3 Pa	17.035	−256.115
Boiling point of hydrogen at a pressure of 101,292 Pa	20.27	−252.88
Triple point of neon	24.5561	−248.5939
Triple point of oxygen	54.3584	−218.7916
Triple point of argon	83.8058	−189.3442
Triple point of mercury	234.3156	−38.8344
Triple point of water	273.16	0.01
Melting point of gallium	302.9146	29.7646
Freezing point of indium	429.7485	156.5985
Freezing point of tin	505.078	231.928
Freezing point of zinc	692.677	419.527
Freezing point of aluminum	933.473	660.323
Freezing point of silver	1234.93	961.78
Freezing point of gold	1337.33	1064.18
Freezing point of copper	1357.77	1084.62

temperature range from 0.85 (the superconductive transition of zinc) to 3687 K (the melting point of tungsten). The list of selected secondary reference points can be found in [11].

To define temperature in five ranges, the ITS-90 requires helium vapor pressure thermometers, helium gas thermometers, standard platinum resistance thermometers, and monochromatic radiation thermometers, as follows (see Fig. 2.2):

1. 0.65–5 K defined in terms of the vapour pressures of ^{3}He and ^{4}He,
2. 3–24.5561 K using a constant volume helium gas thermometer,
3. 13.8033–273.16 K using a platinum resistance thermometer,
4. 273.15–1234.93 K using a platinum resistance thermometer, and
5. 1234.94 K and above using a monochromatic radiation thermometer.

The PLTS-2000 uses the melting pressure of ^{3}He to provide the basis for temperature measurement. The guidance on the practical methods by which the melting pressures can be measured is beyond the scope of this book. However, a nice presentation of the topic can be found in [12].

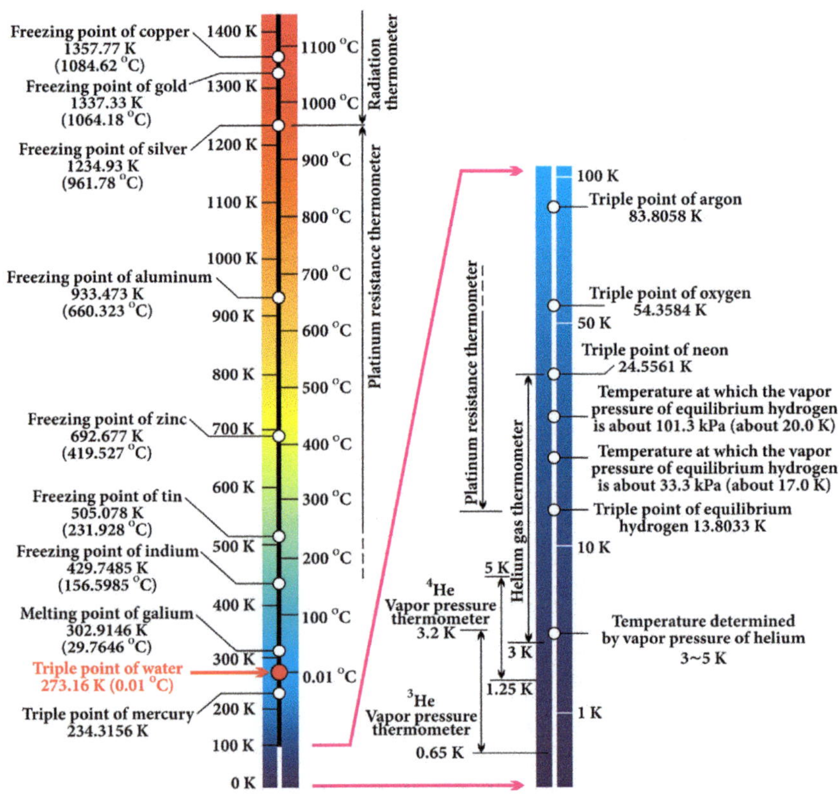

FIGURE 2.2 **Temperature fixed points and thermometers of the International Temperature Scale (ITS-90).**

Both ITS-90 and PLTS-2000 are not only important as an internationally accepted approximation to thermodynamic temperature, but also because they define procedures, fixed points, and thermometer specifications required in order to establish temperature scales. Traceability of all practical temperature measurements ought to be established, developed, and maintained towards ITS-90 or PLTS-2000.

2.5 COMMON THERMOMETERS AND THEIR CLASSIFICATION

The determination of thermodynamic temperature using primary thermometers is a complicated procedure which involves precision instruments and well-controlled measurement conditions. Such measurements are unnecessary for most practical applications for which many different types of thermometers can be used. They can be based on assorted temperature-dependent phenomena, and one should note that the majority of phenomena in nature are temperature-dependent. The assessment of the usability of a particular phenomenon for the establishment of a thermometer is a somewhat arbitrary undertaking. Generally, it is required that the relationship between the physical quantity (the measurand) and the thermodynamic temperature can be precisely specified and that the physical quantity behaves monotonically with temperature.

Thermometers can be classified in a number of ways, but mainly they are grouped according to the nature of physical contact between the sensor and the measured object; as contact and noncontact thermometers. When measuring a variable which characterizes thermal energy flow, temperature sensing can be regarded as *direct*. But when an external energy is needed, as an interrogating medium to capture information about the abilities of the body under measurement to store, dissipate, transmit, or transform energy, temperature measurement is considered to be accomplished by an *inferential method* [13]. The measurement systems may be manually operated or automatic. They can be grouped as electric or nonelectric according to the energy form of the output signal. Frequent classification is one done according to a type of physical variable (the measurand) which is used to sense temperature: radiation thermometers, resistance thermometers, thermometers based on the expansion of solid, liquid, or gas, etc. In practice, the choice of thermometer is usually performed on the basis of measurement temperature range and required precision. Classification of thermometers by measuring range is given in Fig. 2.3.

Today, many types of thermometers are used. These include liquid-in-glass thermometers, bimetallic thermometers, thermocouples, resistance thermometers, thermistors, manometric thermometers, infrared thermometers, etc. The detailed description of their operation and use is beyond the scope of this book. Here, only the physical principles on which their operation is based will be mentioned. More information can be found in many excellent books covering this topic, for example, Ref. [14].

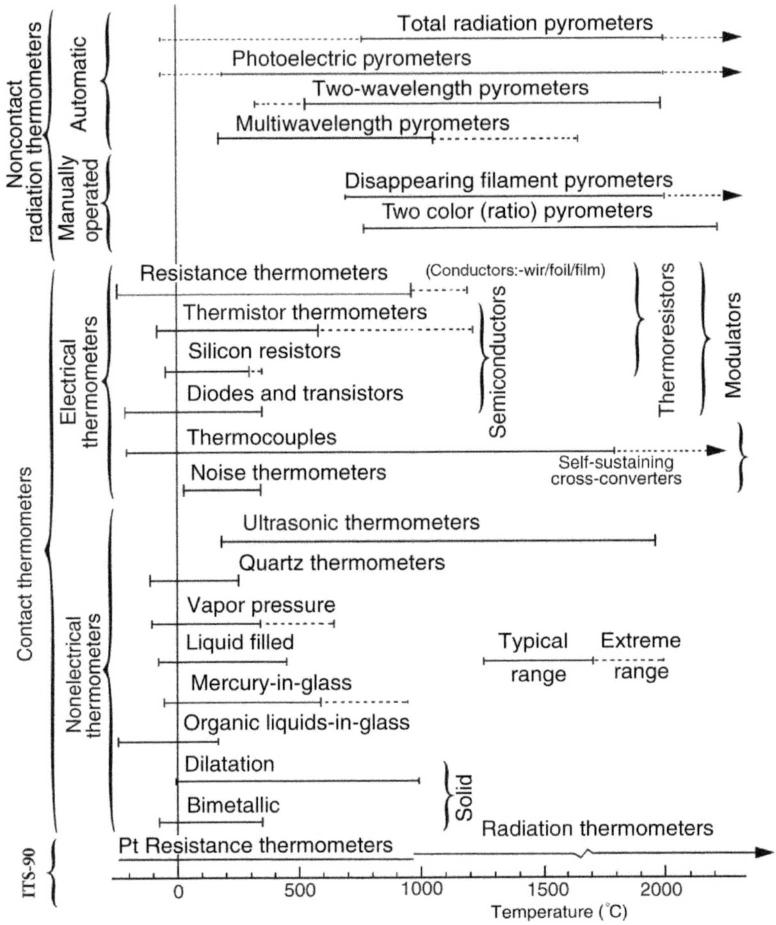

FIGURE 2.3 Classification of thermometers by measuring temperature range. *(Source: Reproduced from Michalski L, Eckersdorf K, Kucharski J, McGhee J. Temperature measurement. 2nd ed. Chichester: John Wiley & Sons; 2001. 501 p. with permission of Wiley).*

The liquid-in-glass thermometer is a contact-type thermometer which measures temperature from changes in the volume of a liquid. Mercury is most commonly used as a thermometric liquid, but also several organic liquids are utilized to cover different temperature ranges (toluene, ethanol, xylene, pentane, etc.). Typically, the solid-stem form of this thermometer consists of the stem, bulb that extends to a narrow tube, thermometric liquid, and scale, see Fig. 2.4A. *Bimetallic thermometers* sense temperature through a mechanical displacement which occurs due to different coefficients of thermal expansion of metal strips bonded together to form a sensing element, see Fig. 2.4B.

To maximize mechanical displacement (bending), materials with considerably different thermal expansion coefficients are used: for high expansion metal

(A)

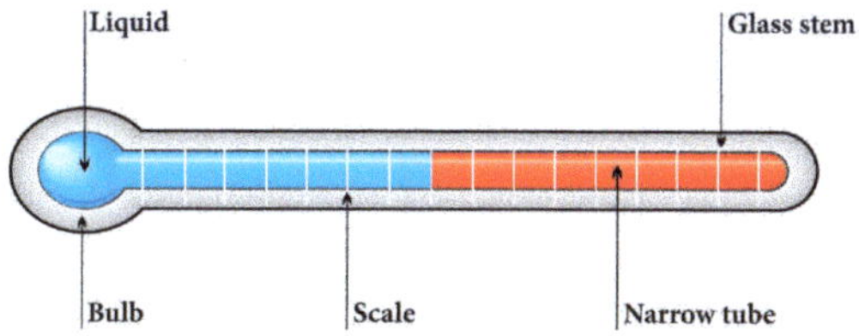

(B)

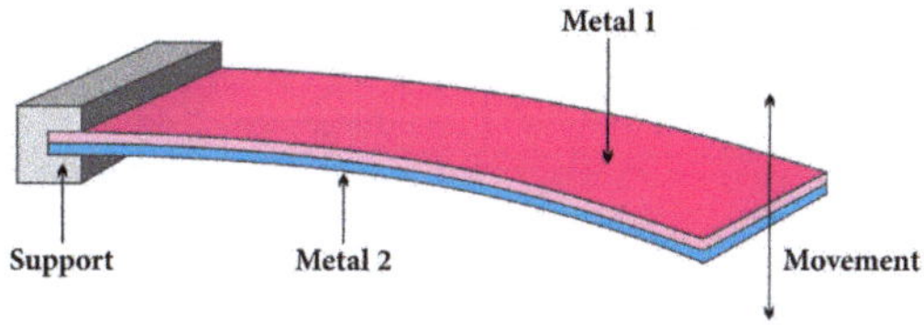

FIGURE 2.4 (A) Liquid-in-glass thermometer. (B) Bimetallic thermometer.

usually brass, copper, iron, or nonmagnetic steel, and for the low thermal expansion material usually invar (an alloy, 64% Fe, 36% Ni). Depending on the geometrical form, bimetal thermometers may be classified as the helix, spiral, cantilever, or flat type. They are widely used in thermostats in household appliances.

The thermocouple is a widely used, relatively inexpensive, electrical, and contact-type temperature sensor that produces a temperature-dependent voltage at the ends of two different conductors whose other ends are welded together. Their principle of operation is based on the physical phenomenon of appearance of an electromotive force due to movements of electrons generated by a heat flow in a conductor (the Seebeck effect). One should note that the temperature of a body (hot point) is always measured against a reference temperature (cold point), Fig. 2.5A. Using different types of thermocouples (different combinations of conductors) it is possible to cover a wide temperature range from –272 °C to 2000 °C. The most frequent combinations of conductors (types) in thermocouples are: nickel–chromium/nickel–alumel (so called type-K), iron/constantan (type-J), copper/constantan (type-T), nickel–chromium/constantan (type-E), nicrosil/nisil (type-N), platinum–rhodium/platinum (type-S 10% Rh, type-R 13% Rh), and platinum–rhodium 30%/platinum–rhodium 6% (type-B).

Resistance thermometers exploit the change in electrical resistance of conductors and semiconductors with temperature. The resistance is usually measured by potentiometric or bridge methods. Copper, nickel, and platinum are mainly used in metal-based devices. Platinum resistance thermometers, Fig. 2.5B, have the largest temperature measuring range and a near linear resistance–temperature relationship. They are a common choice for precise temperature measurements and are used to define the ITS-90. For temperatures lower

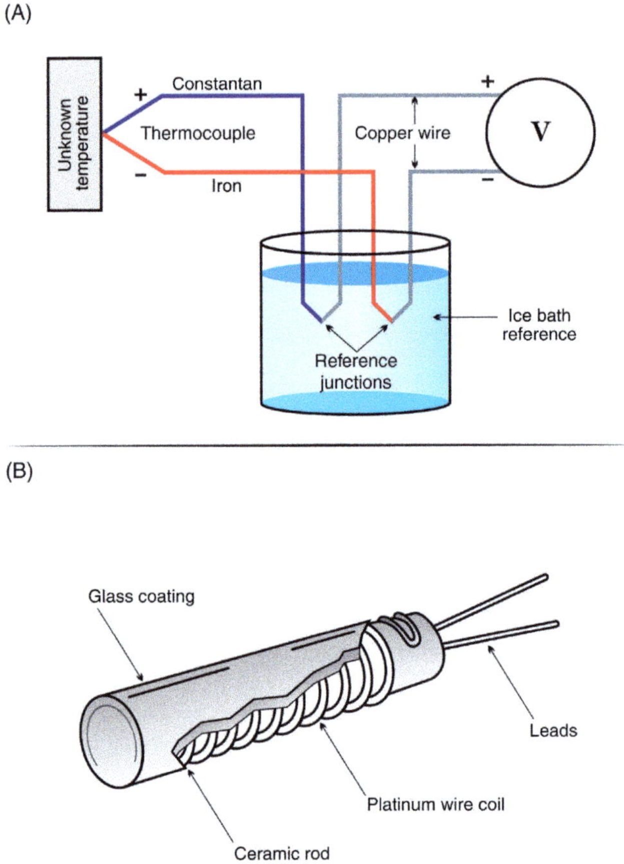

FIGURE 2.5 Schematics of (A) an iron/constantan thermocouple operation and (B) a platinum resistance thermometer.

than 20 K, rhodium–iron, doped germanium, and carbon resistors are frequently utilized. *Thermistors*, that is, thermally sensitive resistors, are made of a ceramic semiconductor whose resistance is sensitive to temperature (with both a negative temperature coefficient or a positive temperature coefficient). They are cheaper, smaller, have a faster response, and present better sensitivity than platinum resistance thermometers, but have much higher levels of uncertainty in measurements and smaller measurement ranges. Current thermistors are ordinarily made of mixtures of oxides, such as nickel, manganese, iron, copper, cobalt, and titanium oxide, along with doped ceramics.

Manometric thermometers, constant volume gas and vapor pressure thermometers, relay on a measurement of pressure as temperature indicators. *Constant volume gas thermometry* can be used for temperature measurements from parts of one kelvin to approximately 1000 K, but mainly for cryogenic temperatures.

The method is used in the ITS-90 and is capable of very low uncertainty. These thermometers can take many forms and most commonly use nitrogen or helium gas as a thermometric medium. *Vapor pressure thermometers* offer even lower uncertainty measurements than gas thermometry. These thermometers are based on the temperature dependence of the saturated vapor pressure of the thermometric liquid, such as helium-3 and helium-4, hydrogen, neon, nitrogen, oxygen, argon, methane, and carbon dioxide. They have good sensitivity, require relatively simple measuring equipment, and are inherently reproducible [14].

Infrared thermometers, also known as *pyrometers* or *radiation thermometers*, are noncontact thermometers, which measure the temperature of a body based on its emitted thermal radiation [13]. In principle, infrared thermometers can be used to measure the temperature of any material, since, regardless of its temperature, all substances will emit electromagnetic radiation. For this reason, applications of infrared thermometry are extensive with a temperature measuring range spanning from 50 to 6000 K. However, primary applications are related to high-temperature measurements, including the definition of ITS-90 above the freezing point of silver (961.78 °C). According to their spectral response and operating method, infrared thermometers can be classified into spectral-band thermometers, total radiation thermometers, ratio thermometers, fibre-optic thermometers, and thermal imagers. Among them, the spectral-band thermometer is the most used. This instrument measures radiant energy over a narrow range of wavelengths between 500 and 25,000 nm. When equipped with silicon or germanium detectors wavelength range is between 500 and 2000 nm. The change of electromagnetic radiation energy with temperature is high in this spectral range, and devices operating in this range have relatively high temperature sensitivity (up to 2%–3% per K).

2.6 MEASUREMENT ERRORS (UNCERTAINTIES) AND PERFORMANCE CHARACTERISTICS OF THERMOMETERS

No physical quantity can be measured with complete certainty since all measurements, however careful, are inevitably accompanied by measurement errors (uncertainties). In science and the majority of applications, it is essential to evaluate uncertainties and to minimize them (since it is not possible to eliminate them entirely). The uncertainty of a measurement is essential for the calibration of a temperature sensor and assessment of its tolerance, and for the evaluation of the performance characteristics of thermometers.

The true value of quantity being measured X_t lies in the interval $X \pm \delta X$, where X represents a value of the measurement and δX its corresponding measurement error. In the simplest consideration, measurement error is a sum of random and systematic (bias) errors: $\delta X = \delta_r X + \delta_b X$ (where r stands for "random" and b for "bias"), see Fig. 2.6. Random errors are always present in a measurement and are caused by changes in experimental conditions, by noise, and by changes of the quantity being measured. They can be treated by statistical

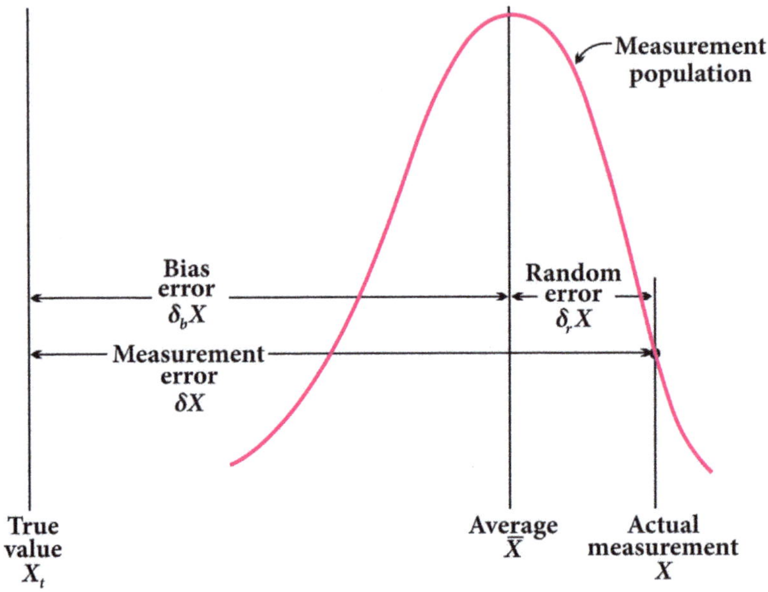

FIGURE 2.6 **Measurement errors: relation between actual measurements and true value.** *(Adapted from Benedict R. Fundamentals of temperature. 3rd ed. New York: John Wiley & Sons; 1984. 532 p. with permission of Willey).*

methods since they show up as random scatter (see full line in Fig. 2.6) around the average of repeated measurements, $\bar{X}$:

$$\bar{X} = \frac{1}{N}\sum_{i=1}^{N} X_i, \tag{2.13}$$

where N stands for the number of measurements. The estimate of the standard deviation of the whole measurement population S_σ is frequently termed the precision index [15]:

$$S_\sigma = \left[\frac{1}{N-1}\sum_{i=1}^{N}(X_i - \bar{X})\right]^{1/2} \tag{2.14}$$

It is an indicator of how close measurements of the same quantity agree with each other, that is, it is an indicator of a measurement's *precision*.

Systematic errors (bias) in measurements are usually caused by measuring instruments, for example, because of a problem with the measuring instrument or data acquisition system, or due to improper use of the measuring system by the operator. In most cases, they appear as a constant difference between measurements and the true value of a quantity, but in some instances, they can be nonconstant (bias due drift). Contrary to random errors, bias can be eliminated, though it is difficult to detect and prevent. Bias can be minimized by calibration to a standard instrument.

When the result of measurement (Q) is derived by calculation from two or more measured quantities (X_1, X_2, …, X_m; m being the number of measured quantities), measurement errors "propagate" through calculations. Then, uncertainty δQ can be calculated from the uncertainties of each measured quantity ($\delta X_1, \delta X_2, \ldots, \delta X_m$) as:

$$\delta Q = \left[\left(\frac{\partial Q}{\partial X_1} \right)^2 + \left(\frac{\partial Q}{\partial X_2} \right)^2 + \ldots + \left(\frac{\partial Q}{\partial X_m} \right)^2 \right]^{1/2}. \tag{2.15}$$

It is generally accepted that for estimating and reporting performance characteristics of thermometers one should use standardized metrological procedures and terminology. Such an approach eases result dissemination and comparison between different devices and theoretical models. It also increases the public's confidence in results. This approach can be accomplished by following universal guidelines for the evaluation of measurement data [16] and by reporting results using internationally accepted terms of metrology. Because of the importance of terminology, some of the terms frequently used for the description of performance characteristics of thermometers will be defined here. The definitions are mostly derived from the International Vocabulary of Metrology (VIM) [17] and assume temperature as the *measurand*—the quantity intended to be measured.

In thermometry measuring systems often do not provide the value of temperature directly, instead they provide the *indication* (Q), which can be, for example, a voltage of a thermocouple or a pressure of a constant volume gas thermometer. Temperature is therefore measured according to a particular *measurement principle*—the physical phenomenon on which measurements using a specific measuring instrument are based. A *measurement method* is a procedure or set of procedures (consistent with the measurement principle implemented) for comparison of the quantity being measured against the unit or scale for the said quantity. The measurement result, that is, the temperature value, is obtained from a *measurement model* which is based on a measurement principle, that is, from the mathematical relation between the temperature, indication, and other quantities involved in a measurement. The measurement model is also used to calculate an *uncertainty* in a value of temperature. *Accuracy of measurement* is the closeness of agreement between the measured value of temperature and the true value. Accuracy is not a quantity and does not have a numerical value; it is a descriptive parameter. For instance, one can say that measurement is more accurate if it presents a smaller measurement error—a deviation of the result from the true value. On the other hand, *precision* is a quantity that is expressed numerically by measures of imprecision, as discussed earlier in this section. *Precision* is always estimated for a specified *condition of measurement*, such as reproducibility or repeatability conditions, as the closeness of agreement between indications or measured temperature values acquired by replicate measurements on the same objects. Accordingly, *repeatability* is the precision of measurement achieved by replicate measurements over a short period of time under a *repeatability measurement condition* that comprises the same measurement procedure, measuring system, operator, operating conditions, and location. Similarly, *reproducibility* is the precision of measurement under a *reproducibility measurement condition* that includes repeated measurements at distinct locations, using different measuring systems and operators.

The difference between the smallest and the largest temperature that can reliably be measured by a thermometer determines the *dynamic range* (measurement range) of the thermometer. The sensitivity and temperature resolution are probably the most important performance characteristics of a thermometer. The *sensitivity* (S) of a thermometer is defined as a quotient of the change in an indication and the change in the temperature value, and it is expressed in terms of the indication unit per kelvin:

$$S = \left|\frac{\partial Q}{\partial T}\right|. \tag{2.16}$$

Since sensitivity depends on the value of the quantity being measured, it is not particularly suitable for the comparison of results obtained with different measuring systems. Sensitivity is frequently termed the *absolute sensitivity* to distinguish it from the term *relative sensitivity* that is defined as the ratio of sensitivity and indication value:

$$S_R = \left|\frac{1}{Q}\frac{\partial Q}{\partial T}\right|. \tag{2.17}$$

Relative sensitivity is in fact the normalized value of sensitivity with respect to the indication value and is expressed in K^{-1}. In some cases relative sensitivity is multiplied by 100 and presented as a percentage change in the indication value per unit change in the temperature, $\%K^{-1}$. *Temperature resolution* (ΔT_{min}) is the smallest change in a temperature that causes a perceptible change in the indication; it is presented in kelvins and is calculated as the ratio of uncertainty (expressed as precision index, see Eq. (2.14)) and sensitivity [18]:

$$\Delta T_{min} = \frac{S_\sigma}{S}. \tag{2.18}$$

Temperature resolution strongly depends on the characteristics of a measuring system (for example, on the noise), but it also depends on the indication value. The *spatial resolution* of a measuring system is defined as the minimum distance between points of a measurement (Δx_{min}) that can be resolved under a temperature resolution of the system [19], and can be calculated as [20]:

$$\Delta x_{min} = \frac{\Delta T_{min}}{|dT/dx|}. \tag{2.19}$$

However, the spatial resolution is restricted by the Rayleigh diffraction limit which can be calculated from the maximal wavelength of the emission used in the measuring system and its numerical aperture. The ability of a luminescence temperature measuring system to follow temporal changes in temperature is presented by its *temporal resolution* ($\Delta t_{\min}$) which is defined as the minimum period of time between measurements capable of resolving a temperature higher than the temperature resolution [19]:

$$\Delta t_{\min} = \frac{\Delta T_{\min}}{|dT/dt|}. \tag{2.20}$$

Selectivity of a measurement method is the extent to which other changes of other physical quantities affect the determination of a measurand.

REFERENCES

[1] Pogliani L, Berberan-Santos MN. Constantin Carathéodory and the axiomatic thermodynamics. J Math Chem 2000;28(1):313–24.

[2] Boer Jde. Temperature as a basic physical quantity. Metrologia 1965;1(4):158–69.

[3] Fischer J, White DR. The Boltzmann constant and the new kelvin. Metrologia 2015;52(5): S213–6.

[4] Mohr PJ, Taylor BN, Newell DB. CODATA recommended values of the fundamental physical constants: 2010. Rev Mod Phys 2012;84(4):1527–605.

[5] Fischer J. Progress towards a new definition of the kelvin. Metrologia 2015;52(5):S364–75.

[6] Fellmuth B, Fischer J, Machin G, Picard S, Steur PPM, Tamura O, et al. The kelvin redefinition and its mise en pratique. Philos Trans R Soc A Math Phys Eng Sci 2016;374(2064):20150037.

[7] Moldover MR, Gavioso RM, Mehl JB, Pitre L, de Podesta M, Zhang JT. Acoustic gas thermometry. Metrologia 2014;51(1):R1–19.

[8] Gaiser C, Fellmuth B, Zandt T. Dielectric-constant gas thermometry and the relation to the virial coefficients. Int J Thermophys 2014;35(3):395–404.

[9] White DR, Galleano R, Actis A, Brixy H, de Groot M, Dubbeldam J, et al. The status of Johnson noise thermometry. Metrologia 1996;33(4):325–35.

[10] Gianfrani L. Linking the thermodynamic temperature to an optical frequency: recent advances in Doppler broadening thermometry. Philos Trans R Soc A Math Phys Eng Sci 2016;374(2064):20150047.

[11] Bedford RE, Bonnier G, Maas H, Pavese F. Recommended values of temperature on the International Temperature Scale of 1990 for a selected set of secondary reference points. Metrologia 1996;33(2):133–54.

[12] Rusby RL, Fellmuth B, Engert J, Fogle WE, Adams ED, Pitre L, et al. Realization of the ^{3}He Melting Pressure Scale, PLTS-2000. J Low Temp Phys 2007;149(3):156–75.

[13] Michalski L, Eckersdorf K, Kucharski J, McGhee J. Temperature measurement. 2nd ed. Chichester: John Wiley & Sons; 2001. p. 501.

[14] Childs PRN. Practical temperature measurement [Internet]. Oxford: Butterworth-Heinemann; 2001. p. 372.

[15] Benedict R. Fundamentals of temperature. 3rd ed. New York: John Wiley & Sons; 1984. p. 532.

[16] Evaluation of measurement data—guide to the expression of uncertainty in measurement. 2008.

[17] JCGM. JCGM 200: 2008 International vocabulary of metrology—basic and general concepts and associated terms (VIM) Vocabulaire international de métrologie—Concepts

fondamentaux et généraux et termes associés (VIM), Vol. 3. Geneva: International Organization for Standardization; 2008. Available from: http://www.bipm.org/utils/common/documents/jcgm/JCGM_200_2008.pdf.

[18] Dramićanin MD. Sensing temperature via downshifting emissions of lanthanide-doped metal oxides and salts. A review. Methods Appl Fluoresc 2016;4(4):42001. pp. 23.

[19] Brites C, Pereira P, João N, Millán A, Amaral V, Palacio F, et al. Organic–inorganic Eu^{3+}/Tb^{3+} codoped hybrid films for temperature mapping in integrated circuits. Front Chem 2013;1. article 9.

[20] Kim K, Jeong W, Lee W, Reddy P. Ultra-high vacuum scanning thermal microscopy for nanometer resolution quantitative thermometry. ACS Nano 2012;6(5):4248–57.

Chapter 3

Luminescence: The Basics, Methods, and Instrumentation

3.1 INTRODUCTION

Before going into the detail of luminescence thermometry methods, materials, and applications, the phenomenon of luminescence and the underlying physics ought to be described. Because the end result of the luminescence process is emission of light, this chapter starts with a brief explanation of the electromagnetic spectrum and the interaction of electromagnetic radiation with matter. It will be shown how different types of luminescence are categorized according to the nature of excitation of electronic states. An emphasis is given to photoluminescence. An explanation of the main physical processes involved in photoluminescence is given using a Jablonski diagram, along with the description of fluorescence, phosphorescence, and delayed luminescence (*DL*). The majority of luminescence temperature probes are made of solid materials, thus, luminescence in solids, both of intrinsic and extrinsic type, is elaborated.

Knowing that luminescence thermometry is based on observations of changes of luminescence features with temperature, mechanisms responsible for the temperature quenching of luminescence are presented in a separate section. These changes may be observed with different types of steady-state and time-resolved luminescence measurements. Thus, basic elements of emission, excitation and synchronous spectroscopy, excitation-emission matrices, total synchronous spectroscopy, emission decay, and luminescence polarization/anisotropy spectroscopy and decay are given. The chapter concludes with an overview of instrumentation commonly used for luminescence measurements, including a description of light sources, monochromators, radiation detectors, filters, and polarizers.

3.2 ELECTROMAGNETIC RADIATION AND ITS INTERACTION WITH MATER

We create models of the world around us by seeing the light bouncing off, bending, or diffracting around objects. However, light is just a small part of the electromagnetic spectrum (see Fig. 3.1), the part we can sense using our eyes and the part which covers an only small range of electromagnetic waves with

Luminescence Thermometry. http://dx.doi.org/10.1016/B978-0-08-102029-6.00003-8

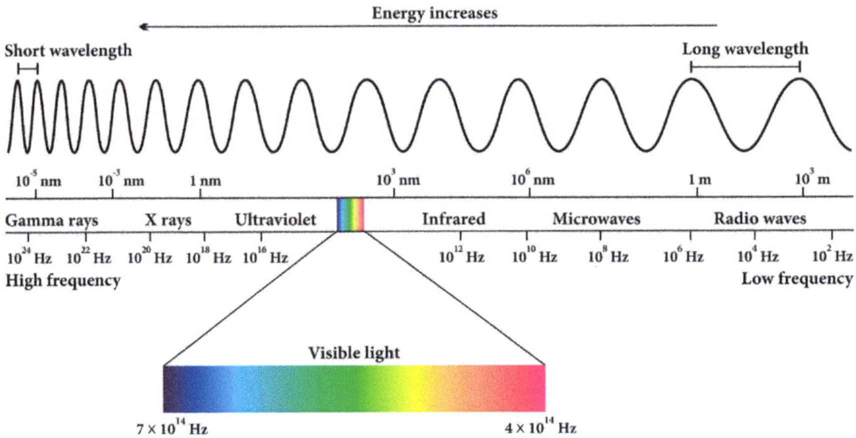

FIGURE 3.1 Electromagnetic spectrum.

different frequencies. Traditionally, the electromagnetic spectrum is divided into seven regions, depending on frequencies or wavelengths of its waves, even though this division is in a way artificial since no other differences apart from their frequency (wavelength, energy) exist between waves in different regions. Accordingly, we recognize gamma-rays, X-rays, ultraviolet, visible, infrared, microwaves, and radio waves, listed here in order of increasing wavelength (decreasing frequency and energy), see Fig. 3.1.

Interaction of electromagnetic waves with matter is the subject of spectroscopy. Many types of spectroscopy are known today and can be categorized depending on the frequency of electromagnetic waves which are utilized: visible, infrared, Raman, nuclear magnetic spectroscopy, etc. Diverse frequencies (energies) used to probe matter and observations of different phenomena provide spectroscopy insight into different properties of matter, because there is always an interaction when electromagnetic waves meet matter. Features of interaction depend on the structure of matter and the frequency of the electromagnetic wave.

For the luminescence phenomena, the interest is mainly in electromagnetic waves that fall into the ultraviolet, visible, and near-infrared (part of the infrared spectral range next to visible) spectral range. When impinging on an object, electromagnetic waves may undergo several fundamental interactions: reflection, refraction, absorption, scattering, and luminescence (see Fig. 3.2). Here we will mention only their basic definitions, except for luminescence, which is the phenomenon underlining the subject this book.

When electromagnetic radiation encounters an object (see Fig. 3.2), part of waves are *reflected* at the interface between the two media and return into the medium from which they originated. Another part is *refracted*, i.e., the waves change their direction of propagation in the new media (the change depends

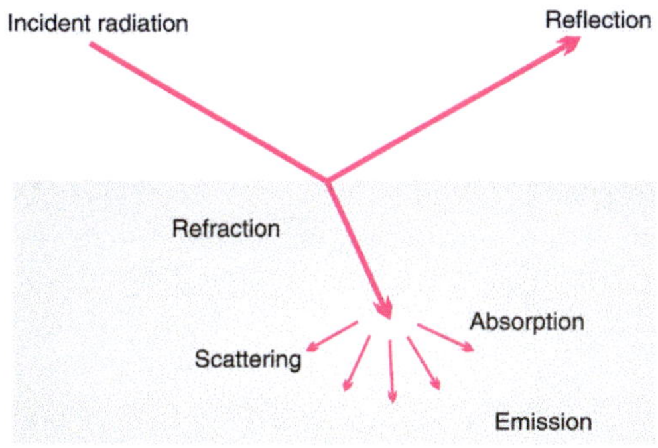

FIGURE 3.2 **Types of interactions of electromagnetic waves with matter.**

on the ratio of the indices of refraction of media and the angle at with which the radiation meets object). While propagating through new media, waves can be *scattered* to different directions, both elastically (without losing energy) or inelastically. Also, the energy of the photon, the quantum of electromagnetic radiation, may be taken up by matter, typically by the electrons of an atom or molecule, in the *absorption* process. In this way, part of the energy of the incident radiation is converted into internal energy of the object. The net effect is manifested by attenuation of the intensity of radiation that propagates through the object. Part of the absorbed energy can be emitted by matter in the form of electromagnetic waves having different frequencies, usually lower frequencies, than the absorbed radiation, i.e., by *photoluminescence*.

3.3 LUMINESCENCE

A German physicist Eilhardt Wiedemann was first to use the term "*luminescenz*" (luminescence) in 1888, but the fascination by this phenomenon dates way back in history to the first human perceptions of "a light without a fire" in nature, such as glowing worms, seashells, the Aurora Borealis, Bologna stone, and the like. Since these first observations, interest in understanding and using luminescence has been growing continuously. Luminescence is frequently defined according to Wiedemann as the light emission from a substance that occurs in addition to thermal radiation or incandescence [1]. This definition is not entirely correct since we know now that some other kinds of radiation can be emitted from a substance beside luminescence, incandescence, and thermal radiation. These are reflected and refracted light, Rayleigh and Raman scattering, and Cherenkov and laser radiation. Luminescence is caused by movements (transitions) of electrons between different energy states which are accompanied by changes in energy. It occurs when excess energy is realized through electron radiative transitions, i.e., by emissions of ultraviolet, visible, and near-infrared photons. Depending on the type of energy used to excite electronic transitions, luminescence is classified to several forms: photo-, chemi-, bio-, lyo-, thermo-, electro-, cathodo-, iono-, crystallo-, mechano-, tribo-, piezo-, sono-, fracto-, etc. These forms are briefly described below.

Photoluminescence is a type of luminescence that is initiated by photoexcitation, i.e., by the absorption of photons (electromagnetic radiation). Different physical processes, radiative (involving photons and indicated by straight arrows) and nonradiative (not involving photons and indicated by squiggly arrows), which comprise photoluminescence are commonly explained using a Jablonski diagram (named after a Polish physicist Aleksander Jabłoński), see Fig. 3.3. Photoluminescence commences by *absorption* of photons (A), a fast radiative process of the order of 10^{-15} s, which provides energy to electrons in the ground state (S_0) to move to higher energy states (S_1, S_2, ...) and a vibrational excited state (v). There, they interact with phonons and undergo *vibrational relaxation* (VR) to the lowest vibrational state of the excited electronic state, nonradiatively loosing part of their energy. They also

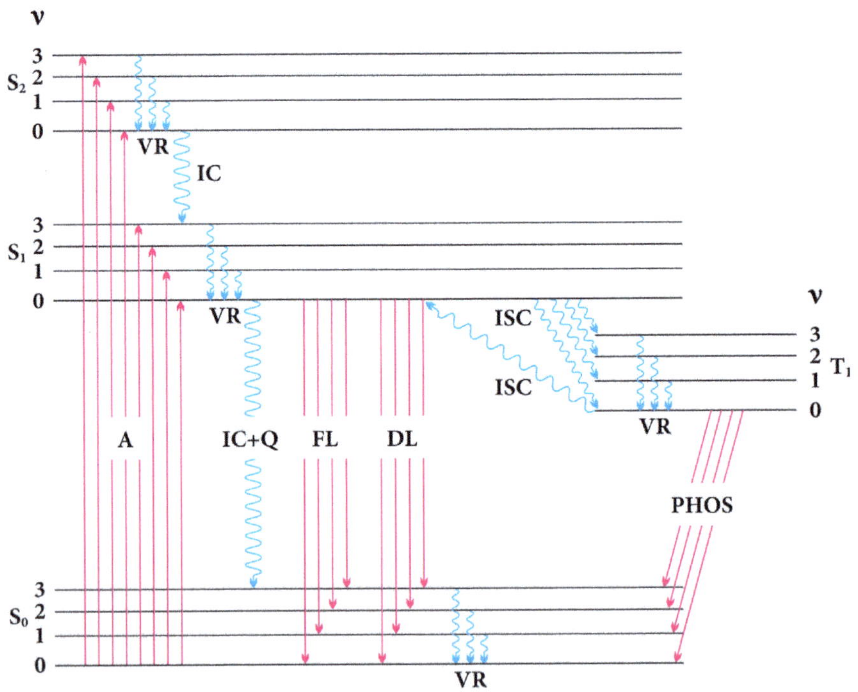

FIGURE 3.3 Illustration of processes involved in photoluminescence (fluorescence, phosphorescence, and delayed luminescence), the so-called Jablonski diagram. *A*, Absorption; *DL*, delayed luminescence; *FL*, fluorescence; *IC*, internal conversion; *ISC*, intersystem crossing; *PHOS*, phosphorescence; *Q*, quenching; *S*, singlet states; *T*, triplet state; *VR*, vibrational relaxation; ν, vibrational states.

nonradiatively lose energy by *internal conversion* (*IC*), which occurs when a vibrational state of an electronically excited state couples to a vibrational state of a lower electronic state, including the ground state. The time period of both processes is of the order of 10^{-12} s. Then, in fluorescence, part of the electrons radiatively deexcite to their ground state, emitting photons of lower energy compared to the photons which provided the excitation, with the other part of the electrons losing energy by some nonradiative process (processes that are sometimes called *luminescence quenching*). Fluorescence (*FL*) is a slow process of the order of 10^{-9}–10^{-7} s, and it is allowed electronic transition since the electron stays in the same multiplicity manifold. In some materials, other paths are possible. *Phosphorescence* (*PHOS*) and *delayed luminescence* (*DL*) are facilitated when electrons in an excited single state undergo *intersystem crossing* (*ISC*), i.e., a nonradiative transition to the state with a different spin multiplicity (triplet state, *T*). This is a slow transition of order of 10^{-8}–10^{-3} s, and is forbidden, i.e., this transition should not occur according to strict electronic selection rules (it is not spin-allowed). However, it becomes weakly allowed due electron–phonon coupling. In this case, electrons may release energy by phosphorescence emission, a very slow forbidden radiative transition of the order of 10^{-4}–10^{-1} s, or nonradiatively by quenching.

Electrons may also undergo *ISC* back to the first excited singlet state, and then to radiatively deexcite to the ground electronic state. This process leads to so-called *delayed luminescence*. One should note that emissions of photons (fluorescence, phosphorescence, or delayed luminescence) occurs only from the lowest energy excited electronic state (either singlet or triplet). This phenomenon is known as Kasha's rule, named in honor of the American physicist Michael Kasha.

In *chemiluminescence* light emission is excited by some chemical or electrochemical reaction, i.e., from the energy of an exothermic reaction. The reaction produces some new molecules in which the electrons are in an excited state, which then deexcite by emission of photons. The most well-known examples of the use of chemiluminescence are the luminol test for the indication of blood (through a reaction with the iron in hemoglobin) and light sticks. *Bioluminescence* is a form of chemiluminescence generated by living organisms. It can be observed, for example, in fireflies, some fungi and bacteria, and in certain marine vertebrates and invertebrates. The main components of the biochemical reaction are luciferin, a light-emitting pigment, and luciferase, the enzyme which catalyzes the oxidation of luciferin. *Lyoluminescence* may also be considered as a form of chemiluminescence, since it describes the light emission which occurs while dissolving a solid into a liquid solvent, usually water. Many materials, both organic and inorganic, exhibit lyoluminescence, and these have been investigated mainly for applications in high-energy radiation dosimetry. Luminescence is caused by free radicals in solution, and its intensity is frequently amplified by the addition of luminol solution. *Crystalloluminescence* is another form of chemiluminescence which proceeds during crystallization.

Cathodoluminescence is a kind of luminescence that is excited by the impact of high-energy electrons (cathode rays) on different materials, usually on semiconductors. The earliest record of the use of cathodoluminescence was by an English chemist and physicist Sir William Crookes in the 1880s, who used it to analyze properties of lead glasses. Today, it is mostly utilized for the analysis of impurities in different materials, particularly semiconductors, and is a standard attachment to imaging instruments, such as the scanning probe microscope. *Ionoluminescence* is similar to cathodoluminescence, but, in this case, emission of light from a material is excited by an ion beam.

Thermoluminescence is the emission of light upon heating of a material following the material's earlier absorption of energy from radiation. It differs from the incandescence since it is reemission of the previously absorbed energy. Electronic excited states, trap states, localized defects, or imperfections may be created in the crystal lattice of material by high-energy radiation. These are not energetically stable, and when a material is heated, they interact with phonons and decay into lower energy states by emission of photons.

Electroluminescence is the direct, nonthermal generation of light resulting from the application of an electric field to a material [2], and has two basic forms: the injection electroluminescence and the high-field electroluminescence. Emission of light occurs during the passage of an electric current through a material in the injection luminescence due to the recombination of the electrons with recombination centers or positive holes. In a high-field electroluminescence, an electric field promotes electrons from donor levels into the conduction band and accelerates them until they collide with luminescent centers and ionize them. Light emission occurs due to a recombination of electrons with an ionized atom of a center.

Mechanoluminescence is fostered by some mechanical action on materials, for example, by grinding, rubbing, cutting, cleaving, shaking, scratching, compressing, or crushing [2]. It can also be induced by rapid heating or cooling of materials (thermal shock). A great number of materials exhibit mechanoluminescence, both inorganic and organic, no matter whether they are crystalline or noncrystalline. Mechanoluminescence may be considered as a general term for a range of similar luminescence types. *Triboluminescence*, for example, is a term commonly used to describe the emission of light due to the breaking of chemical bonds upon mechanical action. Such processes are caused by a separation and reunification of electrical charges. Quite similar to this is *fractoluminescence* which is the luminescence that occurs when crystals are fractured. *Piezoluminescence* (elasticoluminescence) occurs by nondestructive dynamic pressure upon certain solids [3], and can repeatedly be used for mechano-optical conversion. Irradiated salts (NaCl, KCl, KBr), polycrystalline chips of LiF, and rare earth doped piezoelectric solids have been found to exhibit piezoluminescent properties.

In *sonoluminescence*, light flashes, usually shorter than few hundred picoseconds, are generated by intense sound fields. The phenomenon is in fact conversion of mechanical to electromagnetic energy accompanied with immense amplification of the energy density. The flashes are created from hot spots (of extremely high temperatures which are difficult to measure) that form inside bubbles that nucleate, expand, and crash in response to the traveling sound wave. Hot spots range from 10 to 100 nm in size. Sonoluminescence can be classified as single-bubble and multi-bubble, depending on the number of bubbles in the cavitation fields.

3.4 LUMINESCENCE IN SOLIDS

The majority of sensors, including those used in luminescence thermometry, are made of solid materials. In this section, the nature of the electronic transitions responsible for generating photoluminescence in solids is outlined. More details on the luminescence of different types of materials are given in subsequent sections dealing with the applications of various material types for luminescence temperature sensing.

Luminescence in solids can be categorized as *intrinsic luminescence* and *extrinsic luminescence*, depending on the spatial location of the electronic transitions responsible for luminescence in the host material. Thus, intrinsic luminescence comes from electronic transitions of the intrinsic host material while extrinsic luminescence originates from electronic transitions of impurities introduced (usually intentionally) into the host material. Recombination of an electron in the conduction band with a hole in the valence band gives rise to an intrinsic *band-to-band luminescence*. This type of luminescence is characteristic of semiconductor lasers and some light-emitting diodes (LEDs). *Cross-luminescence* is also intrinsic luminescence that comes from the recombination of an electron in the valence band with a hole in the outermost core band [4]. Intrinsic exciton luminescence is produced by exciton recombination (Wannier–Mott or Frenkel). In organic solids, intrinsic luminescence is caused by electronic transitions between lowest unoccupied molecular orbital and highest occupied molecular orbital (LUMO and HOMO) and from intermolecular complexes that are formed in their excited states (excimers and exciplexes).

Electronic transitions within *localized* or *unlocalized luminescent centers* in host materials give rise to extrinsic luminescence. When localized centers are intentionally created by doping materials with impurities, these impurities are frequently termed activators. Impurities that absorb excitation radiation and transfer energy to activators are called sensitizers. Luminescence activators in semiconductors are donors and acceptors.

There are two types of luminescence transitions which give rise to unlocalized extrinsic luminescence: the transition of a free carrier to a bound carrier, and the transition of a bound electron at a donor to a bound hole at an acceptor, which is also called donor–acceptor pair transition [4]. Unlocalized extrinsic

TABLE 3.1 Types of Localized Luminescence Centers Summarized According to Their Electronic Transitions [5]

Electronic transition	Luminescence centers
$1s \rightleftharpoons 2p$	*F* center (an electron trapped at an anion v acancy), Tl^0
$ns^2 \rightleftharpoons nsnp$	Tl^+ type ions: Ga^+, In^+, Tl^+, Ge^{2+}, Sn^{2+}, Pb^{2+}, Sb^{3+}, Bi^{3+}, Cu^-, Ag^-, Au^-, etc.
$3d^{10} \rightleftharpoons 3d^9 4s$	Ag^+, Au^+, Cu^+, etc.
$3d^n \rightleftharpoons 3d^n$	The first row of transition metal ions
$4d^n \rightleftharpoons 4d^n$	The second row of transition metal ions
$4f^n \rightleftharpoons 4f^n$	Lanthanide ions
$4f^n \rightleftharpoons 4f^{n-1}5d$	Divalent lanthanide ions and some trivalent lanthanide ions such as Ce^{3+} and Pr^{3+}
$5f^n \rightleftharpoons 5f^n$	Actinide ions
Charge transfer	VO_4^{3-}, WO_4^{2-}, MoO_4^{2-}
$\pi \rightleftharpoons \pi^*$ $n \rightleftharpoons \pi^*$	Some organic molecules, here *n* denotes a nonbonding electron, and has different meaning than in previous rows

luminescence also can appear from Wannier–Mott excitons bound to impurities and isoelectronic traps. Compared to unlocalized centers, localized luminescence centers present a larger variety of transitions in many different ions. They can be classified according to their electronic transitions, as summarized in Table 3.1 [5].

3.5 QUANTUM EFFICIENCY OF EMISSION: TEMPERATURE QUENCHING OF LUMINESCENCE

The energy of an electronically excited state is lost by a combination of radiative and nonradiative processes. The intensity of the emission I is proportional to the population density of electrons in excited states $I \propto N$, the temporal evolution of which is described by:

$$\frac{dN}{dt} = -(k_R + k_{NR}) \cdot N(t), \tag{3.1}$$

where k_R and k_{NR} are the rates of radiative and nonradiative transitions, respectively. Photon emission is created through radiative deexcitation pathways while nonradiative ones are accompanied by the release of vibrational energy. As a consequence, the electron population in the excited state and the luminescence intensity exponentially decay with a time constant τ that is commonly termed *the lifetime of the excited state*:

$$\tau = \frac{1}{k_R + k_{NR}}. \tag{3.2}$$

The inverse value of the radiative transition rate is named *the radiative lifetime or natural lifetime*:

$$\tau_R = \frac{1}{k_R}, \tag{3.3}$$

and can be calculated, in many instances rather well, from the absorption and emission spectra [6,7], and from the ratio of the measured lifetime and *the internal quantum efficiency of emission* (also termed as *quantum yield*) η:

$$\tau_R = \frac{\tau}{\eta}, \tag{3.4}$$

the latter defined as the fraction of electrons that underwent radiative decay over the total number of deexcited electrons (the number of absorbed photons):

$$\eta = \frac{k_R}{k_R + k_{NR}}. \tag{3.5}$$

One should distinguish the internal quantum efficiency from *the external quantum efficiency* (also known as the *brightness*), which is defined as the ratio of the number of photons emitted over the total number of photons that impinge material (which also includes photons that are reflected, transmitted, or scattered). Note that material may be very bright due to a high probability of absorption of photons while possessing a modest quantum yield. Radiative lifetime values calculated from Eq. (3.4), and those obtained from spectroscopy measurements, differ in many cases [6]. However, with respect to the applications of luminescence in thermometry, the more important thing is that radiative lifetime can be considered as temperature independent for most elaborations.

The intensity of luminescence can be decreased by different processes, and such decreases are termed *quenching*. These include excited-state reactions, molecular rearrangements, energy transfer, ground-state complex formation, collisional quenching, etc. [8]. Quenching may be considered as an undesirable process since it reduces luminescence intensity. However, numerous useful applications have been derived so far based on this phenomenon, mostly in the fields of chemical and biological sensing. Also, quenching is a very useful tool for the investigation of various interactions on atomic and molecular levels.

When quenching occurs due to an increase in the temperature of the material, it is referred as *the temperature quenching of luminescence*, and it is the underlying phenomena of all luminescence thermometry measurement principles. The temperature dependence of the luminescence emission intensity results from the temperature dependence of the nonradiative transition rate:

$$I(T) = C \cdot k_R \cdot \tau(T) = C \cdot \eta(T) = C \cdot \frac{k_R}{k_R + k_{NR}(T)}, \tag{3.6}$$

where C is a constant comprised of a number of physical parameters that are characteristic of the measuring system and investigated material (such as the

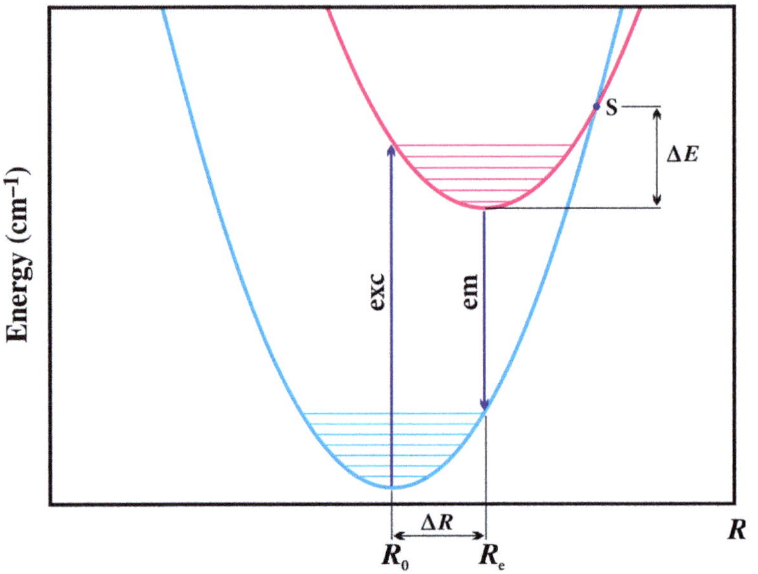

FIGURE 3.4 Configuration coordinate diagram of a luminescence center. According to the Mott–Seitz theory, nonradiative transition from the excited state to the ground state is possible via the crossover S. *(Reproduced from Dramićanin MD. Sensing temperature via downshifting emissions of lanthanide-doped metal oxides and salts. A review. Methods Appl Fluoresc 2016;4(4):42001 (23pp), with permission from the Institute of Physics Publishing).*

intensity of excitation, the detection geometry, concentration of luminescent species in a material and their absorption coefficient, and so on) and, in principle, does not change with temperature.

Theoretical descriptions of temperature-quenching effects are, in fact, linked to the explanation of the temperature induced changes in the nonradiative transition rate. This description is usually given by a Mott–Seitz theory, which is built on a configuration coordinate model that consists of two parabolas shifted with respect to each other [9], Fig. 3.4. These parabolas represent potential curves, each of which represents the total energy of the system in its ground or excited state as a function of the configurational coordinate R.

Based on this model, an electron in the excited state, assisted by thermal energy, nonradiatively deexcites to the ground state through the parabola intersection S with the temperature dependent rate [11]:

$$k_{NR}(T) = A \cdot \exp(-\Delta E / k_B T), \tag{3.7}$$

where ΔE is the energy difference between the lowest level of the excited state and the intersection point of parabolas S, k_B is a Boltzmann constant, and A is a frequency constant (in fact a quantity only weakly dependent on temperature) of the order of 10^{13} s^{-1}. The nonradiative transition rate is lower for larger ΔE, and increases with temperature increase. Expressions for the temperature

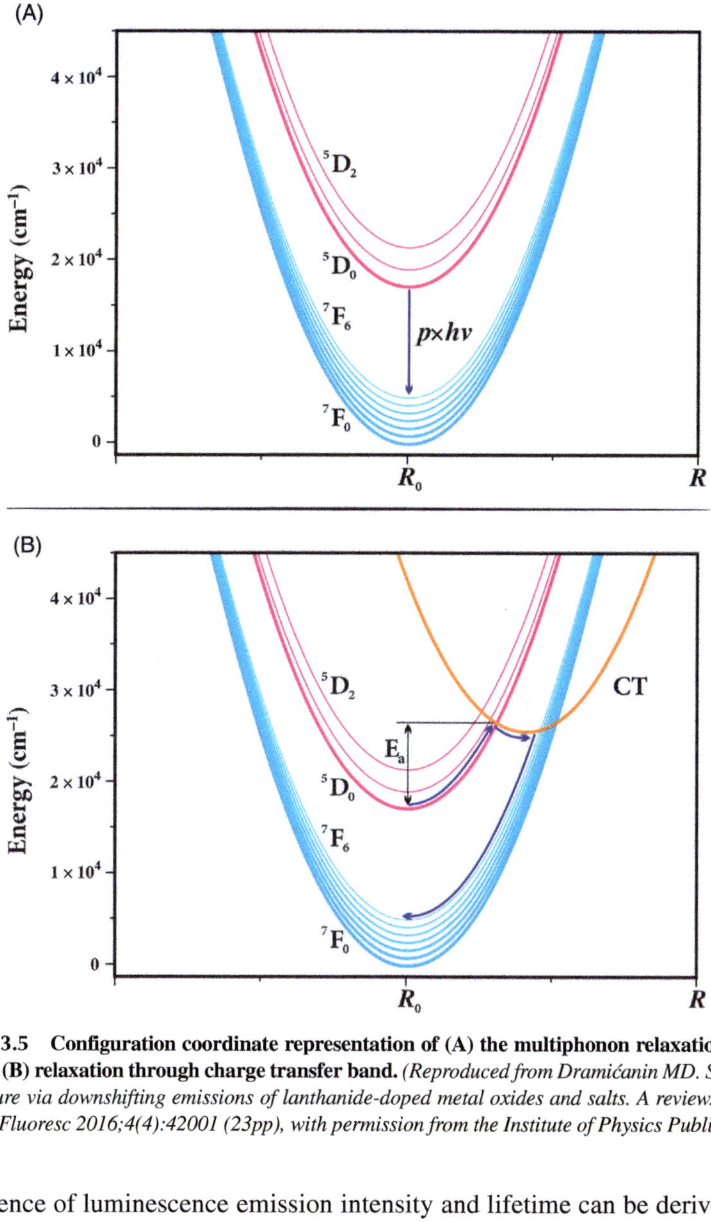

FIGURE 3.5 Configuration coordinate representation of (A) the multiphonon relaxation process, and (B) relaxation through charge transfer band. *(Reproduced from Dramićanin MD. Sensing temperature via downshifting emissions of lanthanide-doped metal oxides and salts. A review. Methods Appl Fluoresc 2016;4(4):42001 (23pp), with permission from the Institute of Physics Publishing).*

dependence of luminescence emission intensity and lifetime can be derived by combining Eqs. (3.6) and (3.7):

$$I(T) = C \cdot \frac{1}{1 + \frac{A}{k_R} \cdot \exp(-\Delta E / k_B T)},$$
$$\tau(T) = \frac{1}{k_R + A \cdot \exp(-\Delta E / k_B T)}. \tag{3.8}$$

The Mott–Seitz theory is suitable for the explanation of broadband emission quenching, where ΔR is large, for example, in molecular fluorescence, see Fig. 3.4. For quenching of narrow-line f–f emissions, which are characteristics of trivalent lanthanide ions (with $\Delta R \approx 0$), some other processes need to be taken into account. Such cases have been explained in early literature reports [12–14] by phonon emission to the lattice mode, so-called multiphonon relaxation, schematically depicted in Fig. 3.5A, and described by the following formula:

$$k_{NR}(T) = k_{NR}(0) \cdot \left[\frac{\exp(h\nu / k_B T)}{\exp(h\nu / k_B T) - 1} \right]^{\frac{\Delta E}{h\nu}}, \tag{3.9}$$

where ΔE is the energy difference from the excited level to the next lowest level, $h\nu$ is the dominant phonon energy of the lattice, and $k_{NR}(0)$ is the value of the nonradiative transition rate at zero temperature that depends on the electron-lattice coupling strength. The $\Delta E/h\nu$ value represents the number of effective phonons that need to be emitted to bridge the energy gap difference. However, the higher the number of phonons the lower the probability of the process is; it is believed that processes involving more than seven effective phonons are quite improbable. In the case of Eu^{3+} doped $GdVO_4$ [15], the multiphonon relaxation process, schematically illustrated in Fig. 3.5A, would require at least 14 phonons of maximal energy (of 885 cm^{-1} [16]) to bridge the gap between $^5D_0$ and $^7F_6$ levels (~12,500 cm^{-1}).

Blasse reported the first observation on the role of the charge transfer state in the temperature quenching of Eu^{3+} luminescence [17]. This nonradiative de-excitation pathway is illustrated in Fig. 3.5B. The minimum of the CT state potential curve has an offset in the configurational coordinate to both excited-state and ground-state potential curves of Eu^{3+}, therefore crossing with them and thus creating a path for electrons in the excited state to nonradiatively reach the ground state [18,19]. The rate equation for this process is given by [20]:

$$k_{NR}(T) = \frac{k_{NR}(0)}{T^{*1/2}} \cdot \exp(-E_a / k_B T^*), \tag{3.10}$$

$$T^* = \frac{h\nu}{2k_B} \cdot \coth(h\nu / 2k_B T),$$

where E_a is the activation energy of the process.

3.6 TYPES OF PHOTOLUMINESCENCE MEASUREMENTS

Luminescence measurements can be classified as steady state (implemented with constant illumination and observation) and time-resolved (performed with pulsed excitation) subject to the temporal nature of excitation and detection. Steady-state measurements are simpler than time-resolved since they require less complex instrumentation, and they are more commonly utilized. However, as they are simply an average of time-resolved phenomena, steady-state measurements do not account for all information imparted to luminescence. Typical examples are information on the distribution of emission decays or the interactions responsible for emission quenching.

Luminescence measurements can also be classified according to the type of observed data (so-called luminescence observables). These include emission intensity, various types of luminescence spectra (emission, excitation, synchronous, excitation-emission matrix), anisotropy, emission intensity decay, anisotropy decay, etc. The simplest form of observable is emission intensity, which is a measurement of the intensity of an emission for fixed energies of excitation and emission. The *emission spectrum* is obtained by recording emission intensity over the spectral range of interest for the fixed energy of excitation, and provides data on the spectral distribution of the emitted radiation by a sample. An *excitation spectrum* is created by measuring the variation of emission intensity at a fixed energy (wavelength) while changing excitation energy in the desired spectral region. In molecular spectroscopy, these spectra are symmetric (a mirror image) because the same transitions are involved in both absorption and emission, and similar vibrational levels are present in both ground and excited states. This phenomenon is known as the *mirror image rule* (see Fig. 3.6).

Even though such behaviour can be observed for many types of materials, some exceptions occur, for example in the case of emissions from transitions

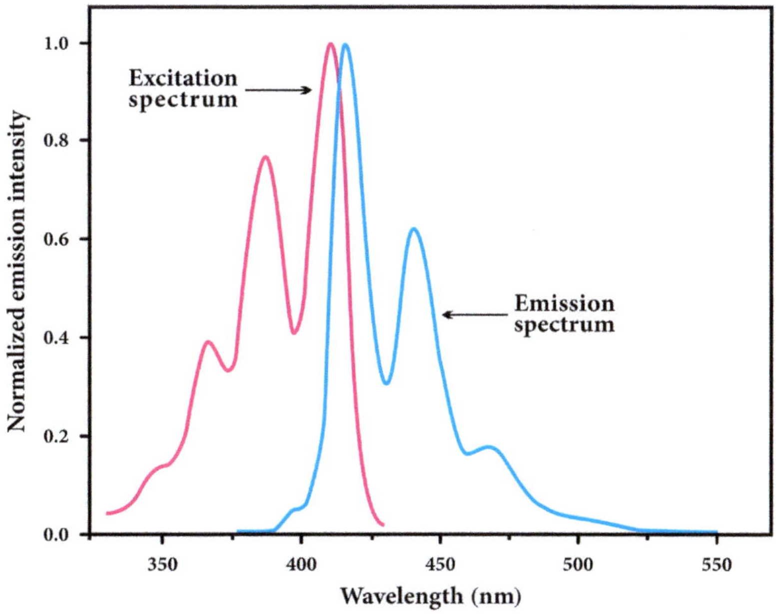

FIGURE 3.6 Excitation and emission spectra of pyrene ($C_{16}H_{10}$; polycyclic aromatic hydrocarbon).

of lanthanide doped materials. Excitation spectroscopy is mainly used for the determination of the positions of excited states, but only those states whose absorption results in the detectable emission of light. One should note this difference in respect to absorption spectroscopy since dissimilarities between the absorption and excitation spectra can provide valuable data.

A *synchronous luminescence spectrum* (SLS) is obtained by simultaneously scanning both emission and excitation wavelengths while keeping the interval constant between emission and excitation (constant-wavelength mode) or keeping the frequency gap constant (constant-energy mode). When this interval equals the difference between the excitation and emission band maxima, the emission intensity is maximally intensified in the measured synchronous spectrum, see Fig. 3.7.

Consequently, synchronous luminescence spectra have sharper spectral characteristics than conventional luminescence spectra, characteristics that are particularly valuable for analyses of multicomponent samples. Also, variable-angle mode can be employed, which in some cases provides better selectivity for multicomponent analysis. This measurement is achieved by scanning the excitation and emission monochromators at different speeds (the wavelength or frequency difference between them is not constant).

By measuring emission spectra at different excitation wavelengths, a two-dimensional *excitation-emission matrix* (EEM) is obtained. This measurement is frequently employed with samples that contain many luminescent species, i.e., for multicomponent samples, when measurement of the single emission spectrum is not sufficient to collect the necessary data for the analysis. In such cases, EEM may be considered as a "fingerprint" of the measured sample and can be effectively used for characterization of fluorophores in complex systems. Multidimensional luminescence measurements can also be obtained with the so-called *three-dimensional total synchronous luminescence spectroscopy* (3D-TSLS) which is acquired as a sequence of synchronous luminescence spectra measured for a range of synchronous intervals. The term "three-dimensional" refers to the space defined by the excitation or emission wavelength, synchronous interval, and the luminescence emission intensity. To ease visualization of data, EEMs and 3D-TSLS are frequently displayed in the form of contour graphs, where intensities of emission are presented by contour lines and different colors (see Fig. 3.8).

Multidimensional luminescence measurements can also be produced using time-resolved emission measurements, i.e., by measuring emission spectra at different time intervals after pulsed excitation [23].

Measurements of the *excited state lifetime* are performed either by the frequency domain method (see Fig. 3.9A) or by the time domain method (see Fig. 3.9B).

In the frequency domain measurements, excitation radiation is harmonically modulated (usually sinusoidally). Emitted radiation follows excitation frequency with a phase lag of $\Delta\Phi$. The value of the excited state lifetime can be

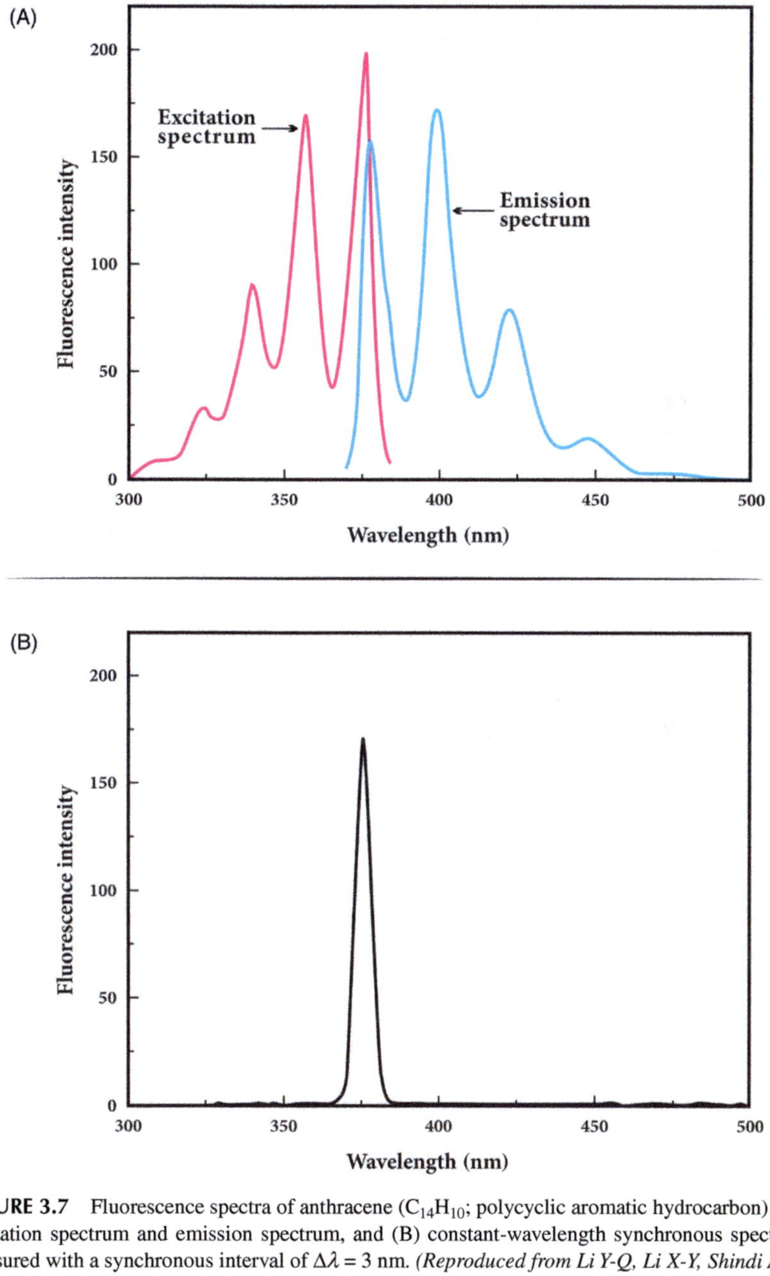

FIGURE 3.7 Fluorescence spectra of anthracene ($C_{14}H_{10}$; polycyclic aromatic hydrocarbon): (A) excitation spectrum and emission spectrum, and (B) constant-wavelength synchronous spectrum measured with a synchronous interval of $\Delta\lambda = 3$ nm. *(Reproduced from Li Y-Q, Li X-Y, Shindi AAF, Zou Z-X, Liu Q, Lin L-R, Li N. Synchronous fluorescence spectroscopy and its applications in clinical analysis and food safety evaluation. In: Geddes CD, editor. Reviews in fluorescence 2010. New York, NY: Springer New York; 2012. p. 95–117, with permission from Springer).*

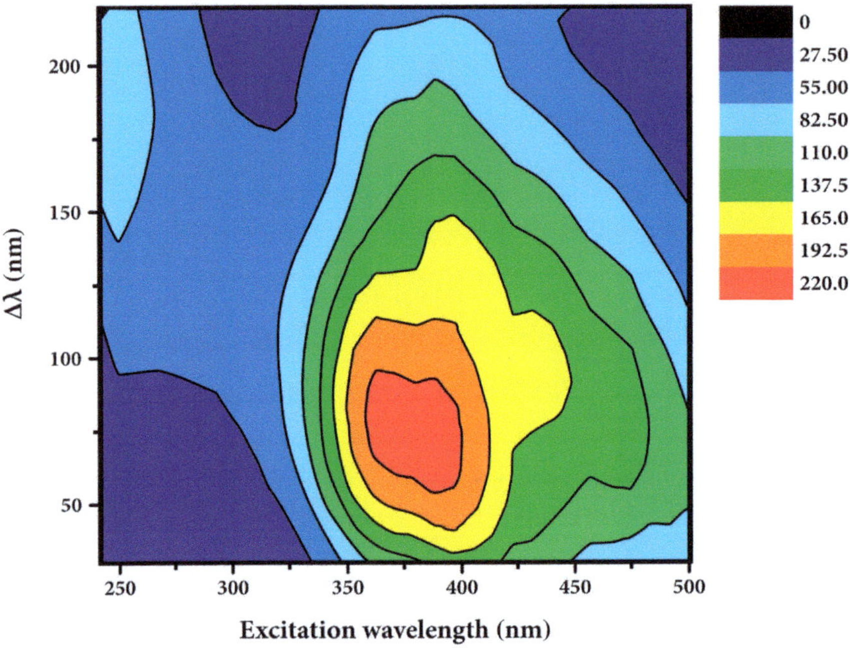

FIGURE 3.8 3D-TSLS of linden honey presented in a form of contour graph. *(Reproduced from Lenhardt L, Zeković I, Dramićanin T, Dramićanin MD, Bro R. Determination of the botanical origin of honey by front-face synchronous fluorescence spectroscopy. Appl Spectrosc 2014;68(5):557–63).*

determined from the phase lag and/or from the relative changes of emission and excitation amplitudes. In time domain measurements a material is excited by near-square shaped pulsed excitation. The lifetime of the excited state is derived from the decay of emission intensity which occurs after the excitation pulse.

Measurement of *quantum yield* gives the probability of the excited state being deactivated by emission of radiation rather than by a nonradiative mechanism, and, as such, provides much useful information concerning radiationless processes in materials. The determination of quantum yield can be accomplished by absolute or relative methods. The later method, also known as the comparative method [25], is common in practice and involves the use of standard samples with known quantum yield values. It assumes that the same excitation wavelength, gain, and slit bandwidths are utilized for measurements of the sample with unknown quantum yield η and for the reference sample with known quantum yield η_R:

$$\eta = \eta_R \cdot \frac{n^2}{n_R^2} \cdot \frac{A_R}{A} \cdot \frac{I}{I_R}. \quad (3.11)$$

Here, n is the refractive index of a sample, A is the absorbance of the material at the excitation wavelength, and I is the emission intensity (index R denotes

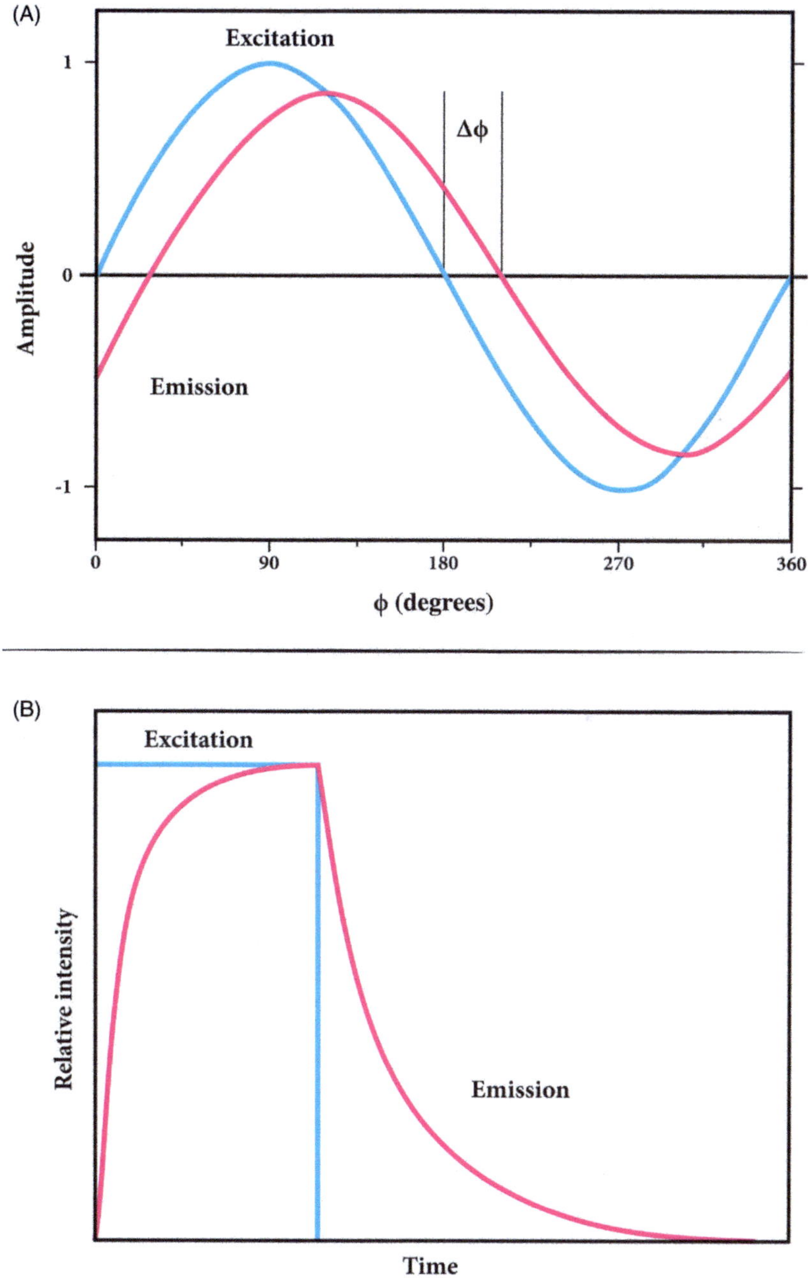

FIGURE 3.9 Excited state lifetime measurements using (A) frequency domain and (B) time domain methods *(Reproduced from Wang X, Wolfbeis OS, Meier RJ. Luminescent probes and sensors for temperature. Chem Soc Rev 2013;42(19):7834–69, with permission from The Royal Society of Chemistry.)*

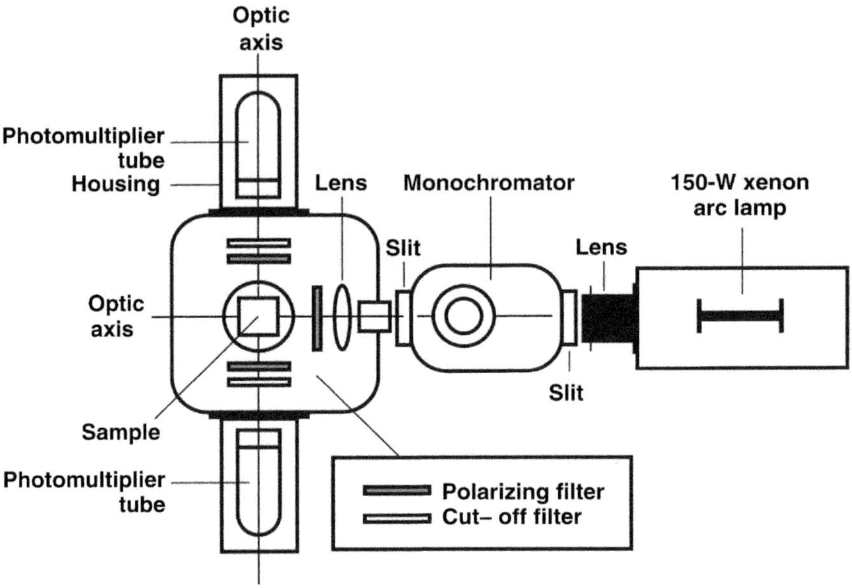

FIGURE 3.10 **Schematic diagram of a fluorescence polarimeter.** *(Reproduced from Bigger SW, Craig RA, Ghiggino KP, Scheirs J. Fluorescence anisotropy measurements in undergraduate teaching. J Chem Educ 1993;70(9):A234–9, copyright (1993) American Chemical Society).*

values for the reference sample). For measurements in solution, it is optimal to dilute samples so that absorbance values are within the range 0.02–0.05 to avoid inner filter effects and ensure a linear response to the intensity.

Fluorescence anisotropy/polarization measurements are both steady-state or time-resolved measurements of a sample's intensity of emission along different axes of polarization of light. Measurements require the use of a polarimeter, such as the one schematically presented in Fig. 3.10 [26]. The method is mainly exploited for analyses of the rotational dynamics of molecules in solution.

To determine an anisotropy r (which is a dimensionless quantity), one needs to measure the intensity of emission I when the emission polarizer is oriented parallel (||) to the direction of the polarized excitation, and when the polarizer is perpendicular ($\perp$) to the excitation:

$$r = \frac{I_{\parallel} - I_{\perp}}{I_{\parallel} + 2I_{\perp}}. \tag{3.12}$$

Steady-state anisotropy measurements give an average anisotropy value. The anisotropy decay, i.e., the time-resolved anisotropy measurement, is more informative, and, also provides residual anisotropy and rotational correlation values. Fluorescence anisotropy, among its many applications, can be used to measure the binding constants and kinetics of reactions that cause a change in the rotational time of the molecules [27].

One should be aware that there is no ideal luminescence measuring instrument. Those available do not yield true excitation or emission spectra. This is because of the nonuniform spectral output of the light sources and the wavelength-dependent efficiency of the monochromators, optics, and detectors. Therefore, it is important to keep in mind the distinction between the technical (uncorrected for instrument characteristics) and the true (corrected) spectrum. Today, modern instruments can produce both types of spectra, implementing built-in or experimentally derived instrument corrections. In many instances, technical emission spectra are sufficient for analysis and presentation. However, to facilitate comparison of emission spectra obtained on different instruments, corrected spectra must be used. Excitation spectra, SFS, EEMs, and 3D-TSLS, are more instrument-dependent, being mainly influenced by the spectral characteristic of excitation sources. These spectra ought to be analyzed and presented after corrections. Also, spectra may contain features that originate from physical processes apart from fluorescence, like Rayleigh and Raman scattering, which should be excluded from subsequent analyses.

3.7 INSTRUMENTATION USED IN PHOTOLUMINESCENCE MEASUREMENTS

A generic set-up for the measurement of photoluminescence (see Fig. 3.11) consists of five components: (1) light source, (2) wavelength selector on the excitation and emission optical paths, (3) sample holder/positioner, (4) polarizers and filters, and (5) detector. Depending on their intended use, practical instruments may have many types of additional elements.

Different types of light sources are currently used to excite luminescence in materials. They can provide polychromatic light over the range of continuous

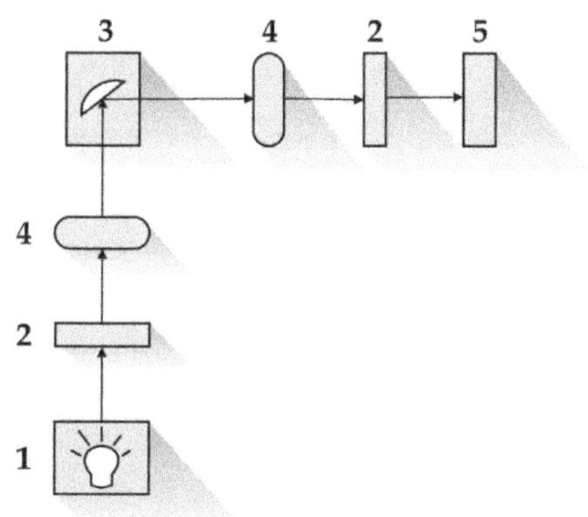

FIGURE 3.11 **The main components of a generic photoluminescence measurement set-up.**

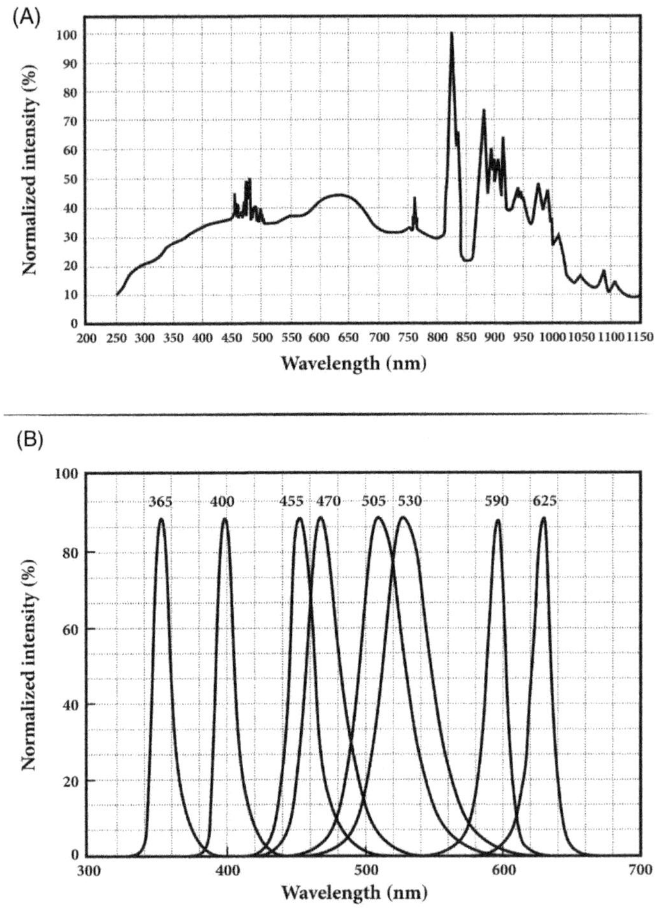

FIGURE 3.12 Spectral distribution of radiation from (A) a Xe arc lamp and (B) light-emitting diodes with different emission maximums.

wavelengths (lamps, tunable wavelength lasers, LED assemblies) or monochromatic light (lasers, laser diodes). Temporal behavior of radiation from sources also differs. The intensity of radiation may be continuous in time for use in steady-state measurements, or pulsed or modulated, which is required for time-resolved measurements.

The most versatile light source, and thus one which is commonly used, is a high-pressure xenon (Xe) arc lamp. It offers a relatively continuous light output over the 250–700 nm wavelength range (see Fig. 3.12A), disturbed with sharp lines occurring near 450 nm and above 800 nm due to ionized Xe atoms in excited states. An output of Xe lamps has strong wavelength dependence, which presents the major source of distortion in excitation. Other types of lamps, such as deuterium, tungsten–halogen, mercury, or metal halide lamps, are less commonly utilized. A light source with a continuous range of wavelengths can be fabricated from white LEDs or several colored LEDs which have slightly overlapped bands peaking at different wavelengths (see Fig. 3.12B).

Similar continuous wavelength excitation radiation can be achieved with tunable [28], supercontinuum [29], and free-electron lasers [30]. However, these are rather expensive devices. Among tunable lasers, some devices can be continuously tuned over a certain range, whereas others can access only discrete wavelengths or at least exhibit mode hops when being tuned over a larger range. The best known tunable lasers are titanium–sapphire and dye lasers, however many other types can be found in spectroscopy, such as Cr:ZnSe and Cr:ZnS lasers, ytterbium-based rare earth doped fiber lasers, etc. Although they are not lasers strictly speaking, optical parametric oscillators (OPOs) are nowadays quite common devices for wideband tuning excitation in various spectral regions.

Intense monochromatic excitations at selected wavelengths (energies) are performed with various lasers and laser diodes, and, in such cases, there is no need for the wavelength selecting element on the excitation path. The drawback is that excitation spectra cannot be measured. Lasers can be classified according to their active medium: gas lasers, liquid lasers, and solid-state lasers [31]. They can also be classified according to the spectral region in which they emit radiation: ultraviolet, visible, near-infrared, and infrared. Selected examples may include: the KrF excimer (248 nm), the 4th harmonic of Nd:YAG (335 nm), nitrogen (337.1 nm), HeCd (441.6 nm), Ar^+ (488 nm), Ar^+ (514.5 nm), the 2nd harmonic Nd:YAG (532 nm), HeNe (632.8 nm), and ryby (694.3 nm). Different lasers along with their radiation wavelengths are illustrated in Fig. 3.13. Laser diodes that are readily available on the market cover a wavelength range of 405–1500 nm . They can operate in all types of temporal regimes, continuous, pulsed, or amplitude modulated (up to several GHz), and are suitable for both steady state and time-resolved photoluminescence measurements.

To excite material with a radiation of a specific wavelength, a polychromatic or white light should be dispersed and then directed to the material. A similar procedure should be applied to radiation that is emitted from the material before it reaches a detector. Absorption (band-pass) optical filters can be used for applications and measurements that do not demand narrow wavelength bands for excitation or emission detection. In other cases, the process is accomplished using monochromators, which can be considered an essential component of any spectroscopic instrument. Monochromators can be based on optical prisms or diffraction gratings. The later ones are mostly used in modern spectrometers.

The simple prism monochromator, also known as the Bunsen prism monochromator, is illustrated in Fig. 3.14A. It consists of the entrance and exit slits, collimating and focusing lenses, and prism. After passing through an entrance slit, radiation illuminates the collimating lens which spreads radiation on the first surface of the prism. Then, the radiation is refracted in the way that its shorter wavelengths (higher energies) are bent more than its longer wavelengths (lower energies). A focusing lens collects dispersed radiation and focuses it on

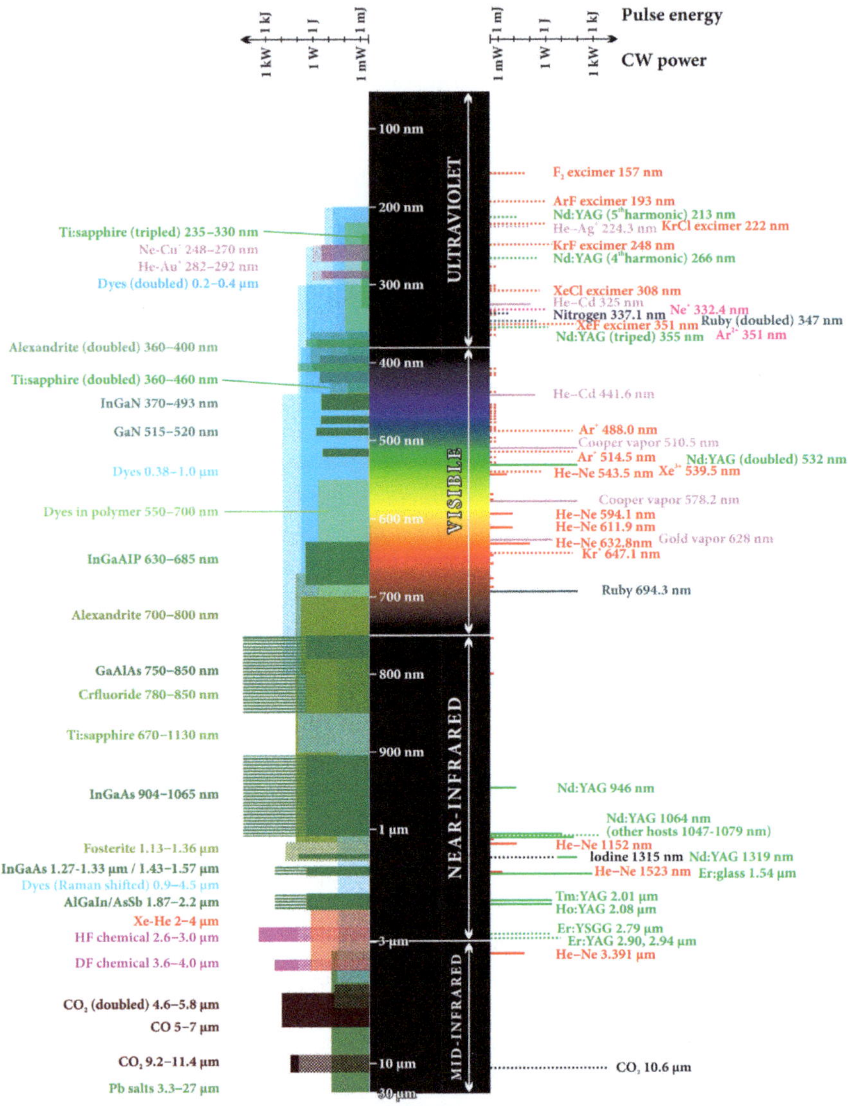

FIGURE 3.13 **Wavelengths of commercially available lasers.** Laser types with distinct laser lines are shown above the wavelength bar, while below are shown lasers that can emit in a wavelength range. The height of the lines and bars gives an indication of the maximal power/pulse energy commercially available, while the color codifies the type of laser material: for semiconductor lasers a solid fill shows the power achievable with a single emitter, while horizontal lines indicate bars or stacks of bars. *Gas lasers* ▮, Excimer laser ▮, Noble gas laser ▮, Metal vapor laser ▮, Chemical laser ▮, Molecular gas laser ▮, Far infrared laser *Solid-state lasers* ▮ Lanthanides ▮ Semiconductor laser ▮, Transition metals ▮, Organic dye in polymer host, *Liquid lasers* ▮, Dyes CW emission pulsed emission ≡ laser diode bars or stacks (see the figure description for details) *(Adapted from Wikipedia. Commercial laser lines. Available from: https://en.wikipedia.org/wiki/Laser#/media/File:Commercial_laser_lines.svg).*

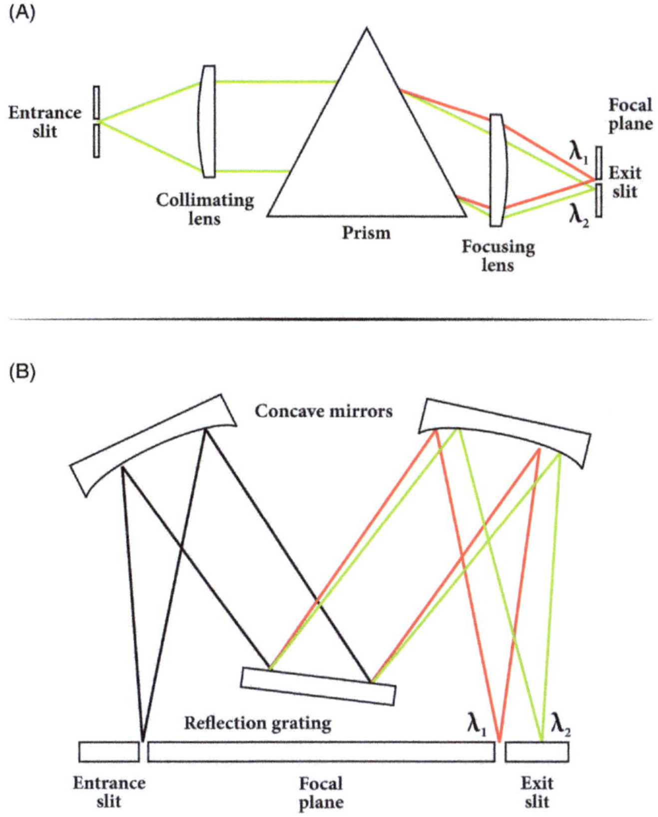

FIGURE 3.14 Schematics of (A) the Bunsen prism and (B) the Czerny–Turner grating monochromator.

the exit slit and further to the sample or detector. Desired wavelengths are selected by rotating the prism. The purpose of the two monochromator slits are to control the size and "position" of the beam of light passing through the slit. The entrance slit provides such a geometrical condition that only a small area of the input beam passes inside the monochromator, and, by varying the slit size, the spectral resolution of the measurement is adjusted. The larger the slit size, the higher the amount of radiation passing through it, but the spectral resolution of the measurement becomes lower. Grating monochromators have similar components as prism monochromators, but use the grating instead of a prism. The construction of a Czerny–Turner grating monochromator is schematically presented in Fig. 3.14B. Incident radiation passes through an entrance slit and falls onto a collimating concave mirror which reflects a parallel polychromatic beam onto a grating which reflects and disperses the beam components. Part of the dispersed radiation of the narrow band of wavelengths is collected by a focusing concave mirror and is directed to an exit slit and further to a detector or sample. As in the case of prism monochromators, different wavelengths are obtained from the monochromator by rotating the grating.

Gratings separate polychromatic radiation into its components because each wavelength of radiation is reflected by the grating at a different angle. The major advantages of gratings over prisms are that they uniformly and linearly disperse radiation along the focal plane. Gratings are manufactured from a master grating which is a hard, optically flat, surface that has a large number of parallel and closely spaced grooves (typically they have 300, 600, 1200, or 1800 grooves/mm when used for UV or visible light). Gratings with larger numbers of grooves present better spectral resolution but have a smaller spectral width. Monochromators can have planar or concave gratings; the concave grating can serve as both the diffraction and focusing element, resulting in a reduction of monochromator elements. Also, the shapes of the grooves in gratings may be different. Even though they are less efficient than ruled gratings, holographic gratings, which are concave gratings with sinusoidal grooves, are preferred in monochromators used in photoluminescence measurements because they lead to much less stray light (the light that passes through the monochromator besides the desired wavelength). Reduction of stray light and improved cut-off is commonly achieved using two monochromators connected in series, with their mechanical systems operating in tandem so that they both select the same wavelength.

The radiation detector can either be single-channel or multichannel. A single-channel detector, usually a photomultiplier tube (PMT) or semiconductor device, detects the intensity of one wavelength at a time. Multichannel detectors, like charge-coupled device (CCD) cameras, record the intensity of emission over a range of wavelengths simultaneously.

Even though more than 80 years have passed since their development, PMTs are the detectors of choice in luminescence instruments. A PMT is a vacuum tube in which a photosensitive cathode, several electron-multiplying dynodes, and an anode are placed. The anode is held at a high positive potential (usually at +950 V; the voltage depends on the PMT type) relative to the photocathode. A great variety of PMTs are available today, and they differ in sensitivity, quantum efficiency, response time, and spectral range they cover. In principle, the spectral range of the PMT depends on the material from which the photocathode is made (see Fig. 3.15A) and the type of transparent material used for the PMT window. Most photocathode materials show good sensitivity in the shorter wavelength regions of the visible and ultraviolet spectrum. PMTs with bialkali photocathodes have a higher quantum efficiency and lower noise level in the green spectral region, while PMTs whose photocathodes are made from multialkali materials respond better in the red and near-infrared. Bialkali photocathodes are composed of an alloy containing antimony and several alkali metals, such as rubidium, potassium, and cesium. The sensitivity of PMTs is governed by the level of dark current, i.e., by the PMT output when no light is falling on PMT; thus, it depends on the construction of the PMT and materials from which it is composed. The dark current can be reduced by cooling

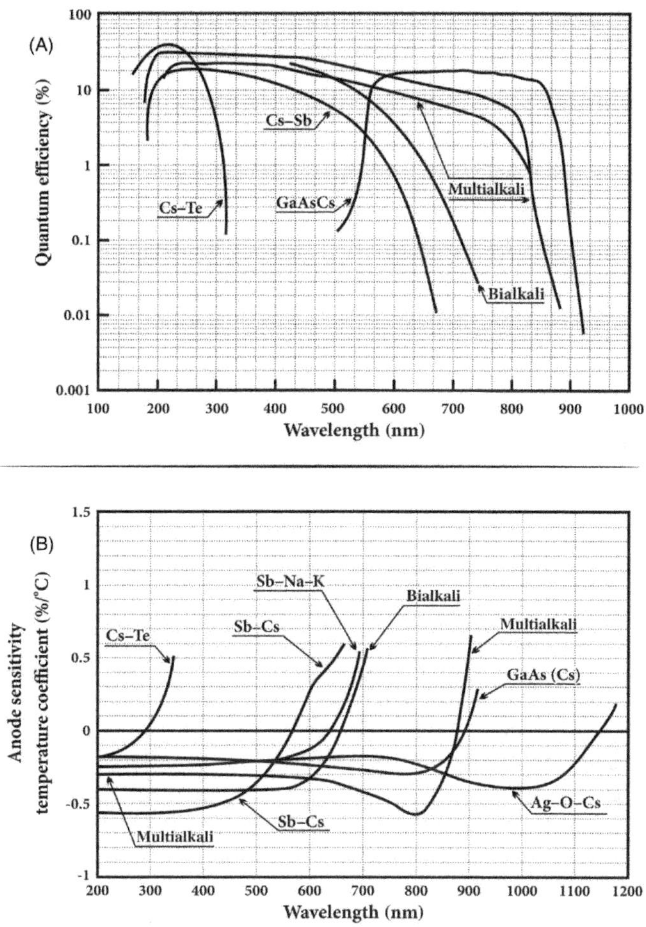

FIGURE 3.15 (A) Spectral response curves and (B) temperature coefficients of typical photomultiplier tube photocathodes.

the PMT. The response time of PMTs is quite fast and is principally set by the time a photoelectron emitted from photocathode needs to reach the anode. It mainly depends on dynode type and supplied voltage. PMTs can operate in two regimes. In the first, so-called analog operating regime, the radiation pulses arrive at the anode and produce a DC current that is proportional to the number of photons reaching the photocathode. In the second, more sensitive approach, single pulses generated from photons interacting with the photomultiplier (PM) cathode are counted (the so-called photon counting regime).

The PMT is sensitive to changes in ambient temperature. The temperature sensitivity of several photocathode materials is given in Fig. 3.15B. Therefore, a PMT should be kept at a fixed temperature when performing precision measurements. Since the interior of the tube is a vacuum, the heat transfer inside the tube

TABLE 3.2 Semiconductor Materials in Photodiodes and Their Spectral Ranges

Semiconductor	Wavelength range (nm)
Silicon	190–1100
Germanium	400–1700
Indium gallium arsenide	800–2600
Lead(II) sulfide	<1000–3500
Mercury cadmium telluride	400–14000

is slow, something which should be considered when performing temperature control. As a rule of thumb, PMTs should be allowed at least an hour to reach ambient temperature, so that their characteristics become stable.

Radiation detection using solid state semiconductor detectors is based on a generation of pairs of charge carriers in a semiconducting material due to absorption of photons. Under voltage, these carriers travel through the material and produce a pulse of current.

There are several solid-state devices available for optical detection such as ordinary photodiodes and phototransistors (photodiodes with some internal amplification), PIN diodes, avalanche photodiodes (APDs), CCDs, etc. Photodiodes have quite good quantum efficiencies, and their current is directly proportional to the intensity of the incident light. Well-manufactured silicon photodiodes can reach 80% quantum efficiency. Depending on the semiconducting material used for their production, photodiodes can be used as detectors over different spectral ranges, see Table 3.2. Spectral response and dark current of photodiodes are temperature sensitive. For example, with temperature increase, the silicon photodiode shifts its spectral response curve toward longer wavelengths and the intensity of dark current doubles for each 8–10°C.

The PIN type of diodes, where there is an intrinsic region between heavily doped p-regions and n-regions, has a faster response than conventional PN type diodes. Although similar in construction, an avalanche photodiode differs from a PIN photodiode by providing internal photo-electronic signal gain due to impact ionization (avalanche effect) under strong reverse bias. Large arrays of APDs may present similar gain to traditional PMTs offering as an advantage the low operating voltage, robustness, and tolerance to excess/ambient light.

The CCD consists of a large number of pixels (typically more than 10^6 pixels) where each pixel acts as a miniature detector accumulating charge proportional to its exposure to radiation. Since the charge at each pixel point can be read out when desired, it is possible to obtain a two-dimensional fluorescence image [33]. The main problem with semiconductor devices is a relatively large

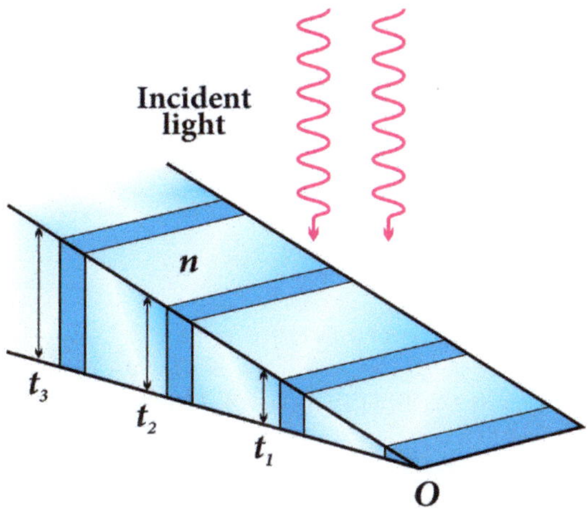

FIGURE 3.16 Schematic of an interference wedge. Wavelength range of transmitted radiation depends on the thickness of the wedge (t_1, t_2, t_3) at which the radiation beam is positioned.

dark current, so their cooling is necessary, particularly if they operate a photon counting regime.

Optical filters are readily used along with monochromators while selecting a specific wavelength of radiation either for excitation or for detection. Filters may block transmission of radiation on both sides of the transmission band (*band pass filters*), or only on one side (*cut off filters*: long-wavelength pass and short-wavelength pass filters). *Neutral-density filters* are used to attenuate light equally at all wavelengths. The main characteristics of filters are their spectral dependences of transmittance, reflectance, and absorbance. Depending on the material and the construction, filters can be classified as color glass filters and interference filters. *Absorption filters* are not expensive and are usually fabricated from colored glass or plastic. One should be careful though not to excite fluorescence from the filter itself since many of these filters are luminescent under ultraviolet excitation. *Interference filters* (also known as Fabry–Perot filters) are composed of a transparent dielectric sandwiched between two semitransparent metallic films and two protective glass plates. They are based on the interference of waves, thus reflect some wavelengths of radiation while transmitting a very narrow bandwidth of others. An *interference wedge* is a dielectric of variable thickness (see Fig. 3.16) that can transmit a broad range of wavelengths. By choosing the appropriate position on the wedge, variable bandwidths of ~20 nm can be isolated.

A polarizer is kind of optical filter that can transmit light of a particular polarization and block light of other polarizations. The common types are linear polarizers and circular polarizers. The best known polarizer used in fluorescence is a Glan–Thompson polarizer which is made from a birefringent material (usually calcite) in the form of a prism. To ensure that one polarized component of light undergoes total internal reflection at the interface, and the other continues along its optical path, the crystal is suitably cut. Types of polarizers include absorptive polarizers (dichroic polarizers) such as polarizing sheets and glasses, thin-film plate polarizers which consist of a dielectric coating on a glass substrate, and polarizing beam splitters based on birefringence such as the Glan–Thompson, the Glan–Taylor, and the Glan–Foucault prisms.

REFERENCES

[1] Williams F. Overview and trends of luminescent research. In: Di Bartolo B, Godberg V, Pacheco D, editors. Luminescence of Inorganic Solids. Boston, MA: Springer US; 1978. p. 1–13.

[2] Electroluminescence. In: Gumlich H-E, Zeinert A, Mauch R, Vij DR, editors. Lumin Solids. Boston, MA: Springer US; 1998. p. 221–69.

[3] Atari NA. Piezoluminescence phenomenon. Phys Lett A 1982;90(1):93–6.

[4] Shionoya S. Photoluminescence. In: Vij DR, editor. Luminescence of Solids. Boston, MA: Springer US; 1998. p. 95–133.

[5] Nakazawa E. Fundamentals of luminescence. In: Yen WM, Shionoya S, Yamamoto H, editors. Phosphor Handbook. Boca Raton: CRC Press; 2006.

[6] Introduction to fluorescence. In: Lakowicz JR, editor. Principles of fluorescence spectroscopy. Boston, MA: Springer US; 2006. p. 1–26.

[7] Strickler SJ, Berg RA. Relationship between absorption intensity and fluorescence lifetime of molecules. J Chem Phys 1962;37(4):814–22.

[8] Quenching of fluorescence. In: Lakowicz JR, editor. Principles of fluorescence spectroscopy. Boston, MA: Springer US; 2006. p. 277–330.

[9] Blasse G. The influence of charge-transfer and rydberg states on the luminescence properties of lanthanides and actinides. In: Braterman PS, Blasse G, Muller A, Baran EJ, Carter RO, editors. Spectra and chemical interactions. Berlin Heidelberg: Springer Berlin Heidelberg; 1976. p. 43–79.

[10] Dramićanin MD. Sensing temperature via downshifting emissions of lanthanide-doped metal oxides and salts. A review. Methods Appl Fluoresc 2016;4(4):23. 42001.

[11] Klick CC, Schulman JH. Luminescence in solids. Solid State Phys 1957;5:97–172.

[12] Riseberg LA, Moos HW. Multiphonon orbit-lattice relaxation of excited states of rare-earth ions in crystals. Phys Rev 1968;174(2):429–38.

[13] Riseberg LA, Weber MJ III. Relaxation phenomena in rare-earth luminescence. Prog Opt 1977;14:89–159.

[14] Moos HW. Spectroscopic relaxation processes of rare earth ions in crystals. J Lumin 1970;1–2:106–21.

[15] Nikolić MG, Jovanović DJ, Dramićanin MD. Temperature dependence of emission and lifetime in Eu^{3+}- and Dy^{3+}-doped $GdVO_4$. Appl Opt 2013;52(8):1716.

[16] Voron'ko YK, Sobol' AA, Shukshin VE, Zagumennyĭ AI, Zavartsev YD, Kutovoĭ SA. Raman spectroscopic study of structural disordering in YVO_4, $GdVO_4$, and $CaWO_4$ crystals. Phys Solid State 2009;51(9):1886–93.

[17] Blasse G. On the Eu^{3+} fluorescence of mixed metal oxides. IV. The photoluminescent efficiency of Eu^{3+}-activated oxides. J Chem Phys 1966;45(7):2356–60.

[18] Fonger WH, Struck CW. Eu^{+3} 5D resonance quenching to the charge-transfer states in Y_2O_2S, La_2O_2S, and LaOCl. J Chem Phys 1970;52(12):6364–72.

[19] Struck CW, Fonger WH. Quantum-mechanical treatment of Eu^{+3} 4f→4f and 4f $\Delta E / hv$ charge-transfer-state transitions in Y_2O_2S and La_2O_2S. J Chem Phys 1976;64(4):1784–90.
[20] Englman R, Jortner J. The energy gap law for radiationless transitions in large molecules. Mol Phys 1970;18(2):145–64.
[21] Li Y-Q, Li X-Y, Shindi AAF, Zou Z-X, Liu Q, Lin L-R, Li N. Synchronous fluorescence spectroscopy and its applications in clinical analysis and food safety evaluation. In: Geddes CD, editor. Reviews in fluorescence 2010. New York, NY: Springer New York; 2012. p. 95–117.
[22] Lenhardt L, Zeković I, Dramićanin T, Dramićanin MD, Bro R. Determination of the botanical origin of honey by front-face synchronous fluorescence spectroscopy. Appl Spectrosc 2014;68(5):557–63.
[23] Lenhardt L, Dramićanin MD. PARAFAC: a tool for the analysis of phosphor mixture luminescence. J Lumin 2016;170:136–40.
[24] Wang X, Wolfbeis OS, Meier RJ. Luminescent probes and sensors for temperature. Chem Soc Rev 2013;42(19):7834–69.
[25] Williams ATR, Winfield SA, Miller JN. Relative fluorescence quantum yields using a computer-controlled luminescence spectrometer. Analyst 1983;108(1290):1067–71.
[26] Bigger SW, Craig RA, Ghiggino KP, Scheirs J. Fluorescence anisotropy measurements in undergraduate teaching. J Chem Educ 1993;70(9):A234–9.
[27] Lakowicz JR. Fluorescence Anisotropy. In: Lakowicz JR, editor. Principles of Fluorescence Spectroscopy. Boston, MA: Springer US; 2006. p. 353–82.
[28] Duarte FJ. Tunable lasers handbook. Academic Press; 1995. p. 476.
[29] Alfano RR, editor. The supercontinuum laser source. New York: Springer-Verlag; 2006. p. 523.
[30] Brau CA. Free-electron lasers. Science 1988;239(4844):1115–21.
[31] Abramczyk H. Lasers. In: Abramczyk H, editor. Introduction to laser spectroscopy. Amsterdam: Elsevier Science; 2005. p. 59–106.
[32] Wikipedia. Commercial laser lines. Available from: https://en.wikipedia.org/wiki/Laser#/media/File:Commercial_laser_lines.svg.
[33] Instrumentation for fluorescence spectroscopy. In: Lakowicz JR, editor. Principles of Fluorescence Spectroscopy. Boston, MA: Springer US; 2006. p. 27–61.

Chapter 4

Schemes for Temperature Read-Out From Luminescence

4.1 INTRODUCTION

Luminescence of any substance is strongly affected by temperature. Luminescence changes its appearance with temperature variations and, generally, lessens in intensity with an increase in temperature until it eventually vanishes. Some substances which are nonluminescent at room temperature may present strong luminescence at low, cryogenic temperatures; some others, such as Dy^{3+} doped $Y_3Al_5O_{10}$, on the other hand, are luminescent even at very high temperatures (>1000 K). Changes of temperature can modify one or many features of luminescence, such as intensities of emission bands, shapes of emission spectra, or emission decays, producing a rich set of phenomena. Many of these can be exploited for temperature measurements; they can serve as principles of temperature measurement if the physics behind them is understood and correctly interpreted. Until now, a great number of measurement methods have been established based on temperature causing changes in luminescence features. The most important ones will be explained in this chapter.

Luminescence thermometry methods can be classified as time-integrated (steady-state) or time-resolved with respect to the temporal character of the observation of luminescence, Fig. 4.1. Time-integrated methods are executed with an illumination that has constant intensity over the time period of detection of luminescence and signal integration. These methods are founded on the changes in temperature caused by excitation and emission band energies and bandwidths, emission intensities, ratios of intensities of different emission bands, and polarization/anisotropy. Time-resolved methods are conducted with a pulsed illumination and an observation of the emission progression over the period between pulses. Similar information may be obtained using harmonically modulated illumination and signal analysis in the frequency domain. The lifetime of the material's excited state provides information about the material's temperature and can be evaluated from the decay of emission originating from a particular excited state. The initial rise of emission after excitation, most frequently observed with rare earth doped materials, is also a temperature sensitive feature of luminescence.

It is important to note that methods based on the changes of the ratio of intensities of emission bands, and the changes of excited-state lifetimes are used considerably more frequently than other luminescence thermometry methods. Except

Luminescence Thermometry. http://dx.doi.org/10.1016/B978-0-08-102029-6.00004-X

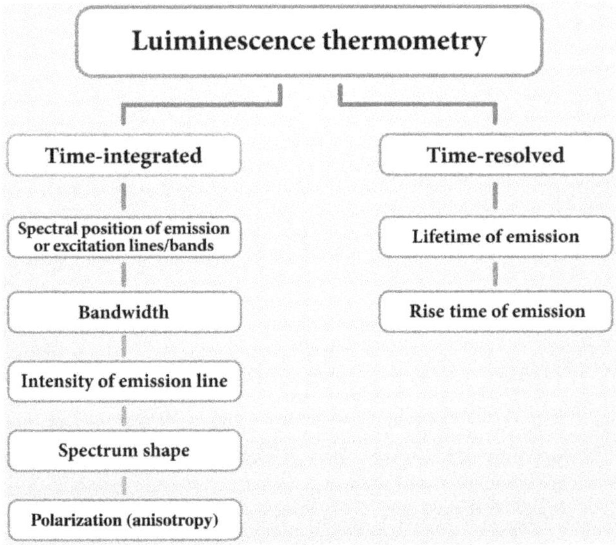

FIGURE 4.1 **Different types of temperature read-outs from luminescence.**

for the method based on anisotropy changes, all methods may exploit properties of a variety of materials, enabling measurements in different temperature ranges and environments; materials used as luminescence thermometry probes, and some selected applications, will be detailed in Chapters 6–8 of this book.

4.2 TEMPERATURE SENSING FROM EXCITATION AND EMISSION BAND POSITIONS AND BANDWIDTHS

Emission and excitation bands in a luminescence spectrum are defined by band position (a wavelength for which the emission intensity is maximal) and bandwidth (a difference in energy or wavelength for which intensity of emission has half of its maximal value). In general, the spectral positions of emission and excitation bands are temperature dependent in any material. These dependences arise from several processes, such as the changes in energy of electronic levels, dilatations of crystal lattice, changes in refractive index, and other phenomena. Therefore, measurement procedures can be developed to determine temperature from the luminescence bandshifts, provided that these shifts are sufficiently large to facilitate accurate measurements. This condition cannot be met for most materials. For example, alternations of transition energies of lanthanide ions are smaller than 0.1 cm^{-1}/K and come from changes in the crystal field that act upon lanthanide ions. The effect has been evidenced by Kusama et al. [1] from the line-shift of the Eu^{3+} $^5D_0 \rightarrow {}^7F_0$ emission (in Y_2O_2S) that occurs when temperature changes from −16°C to 72°C, Fig. 4.2A. The finding was recently confirmed by observation of Nd^{3+} (in LaF_3) 863 nm emission peak shifts to longer wavelengths with a temperature increase at the 0.007 nm/K rate [2].

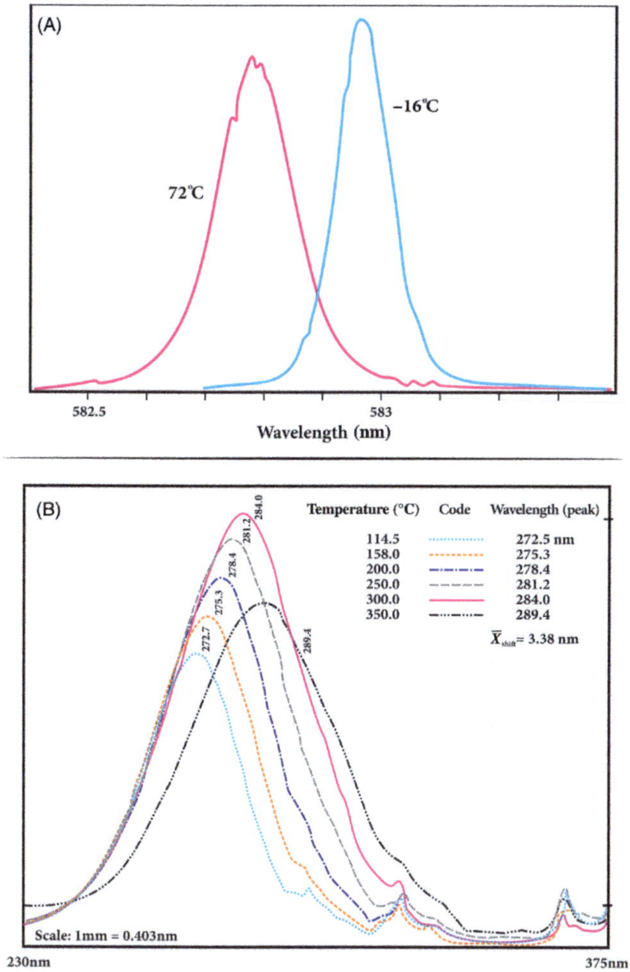

FIGURE 4.2 (A) $^5D_0 \rightarrow {}^7F_0$ emission of Y_2O_2S: Eu^{3+} measured at $-16°C$ and 72°C. (B) Excitation spectra of Eu^{3+} doped Y_2O_3 measured at different temperatures. *((A) adapted from Kusama H, Sovers OJ, Yoshioka T. Line shift method for phosphor temperature measurements. Jpn J Appl Phys 1976;15(12):2349–58, Copyright (2016) The Japan Society of Applied Physics and (B) adapted from Allison SW, Gillies GT. Remote thermometry with thermographic phosphors: Instrumentation and applications. Rev Sci Instrum 1997;68(7):2615–50, with the permission of AIP Publishing).*

Excitation spectra of materials doped with trivalent lanthanide ions are composed of a strong and broad absorption band, a so-called charge transfer (CT) band, and several narrow bands. The latter are a consequence of electronic transitions within dopant ions and have almost negligible temperature dependence, similar to the case of corresponding bands in emission spectra. On the other hand, the wavelength of CT band maximum has a notable temperature dependence which may reach 0.6 nm/K in the case of Eu^{3+} doped Y_2O_3, as shown in Fig. 4.2B [3], and can be used for reading temperatures.

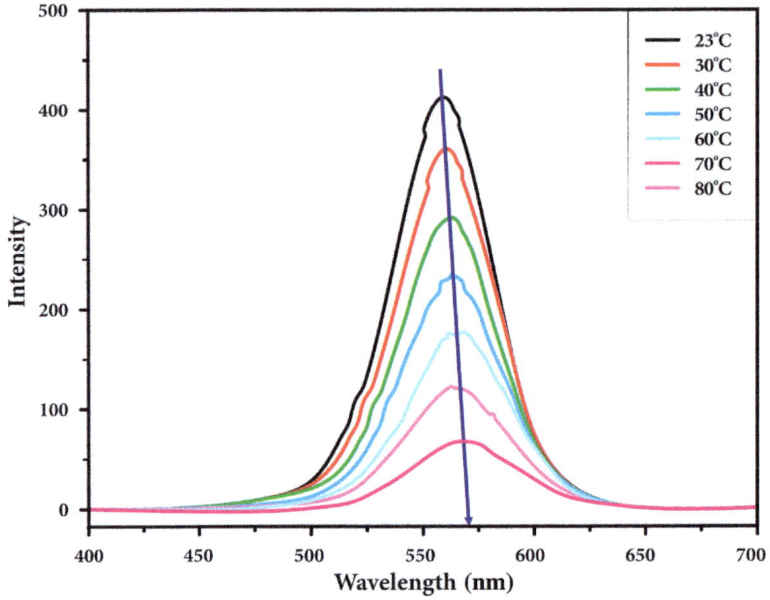

FIGURE 4.3 Changes of temperature with emission band maxima of CdTe QDs and double hydroxide monolayers. *(Adapted from Liang R, Tian R, Shi W, Liu Z, Yan D, Wei M, et al. A temperature sensor based on CdTe quantum dots–layered double hydroxide ultrathin films via layer-by-layer assembly. Chem Commun 2013;49(10):969–71 with permission of The Royal Society of Chemistry).*

Emission bands of quantum dots (QDs) are the exception to a material's general behavior of small bandshifts with temperature. With an increase in temperature the maximum emission of QDs may be either blue-shifted (to higher energies/shorter wavelengths) or red-shifted (to lower energies/longer wavelengths), typically by 0.1–0.2 nm/K. For example, the emission band maxima of ultrathin films, based on alternate assembly of CdTe QDs and double hydroxide monolayers, is red-shifted (0.193 nm/°C) with an increase in temperature [4], Fig. 4.3. One should note that band maxima can be measured with high accuracy, better than the accuracy of emission intensity measurements. For this reason, temperature sensitivities of QD band maxima are sufficient for realization thermometry with a good temperature resolution.

Emission bands broaden with an increase in temperature, and, in most cases, bandwidths linearly depend on temperature. The change is rather small for lanthanide emissions (small for the realization of temperature measurements); an approximate 50 cm^{-1} bandwidth increase has been recorded for the temperature change from 70 K to 700 K in the case of $^5D_0 \rightarrow {}^7F_2$ emission of Eu^{3+} doped Y_2O_3 particles [5], Fig. 4.4A. Somewhat larger bandwidth changes occurs with QDs. The emission bandwidth of CdSe QDs coated with ZnS (average size of particles between 7 and 12 nm) increases at a rate of 0.28 nm/°C from 24.4–43.6°C [6], Fig 4.4B.

(A)

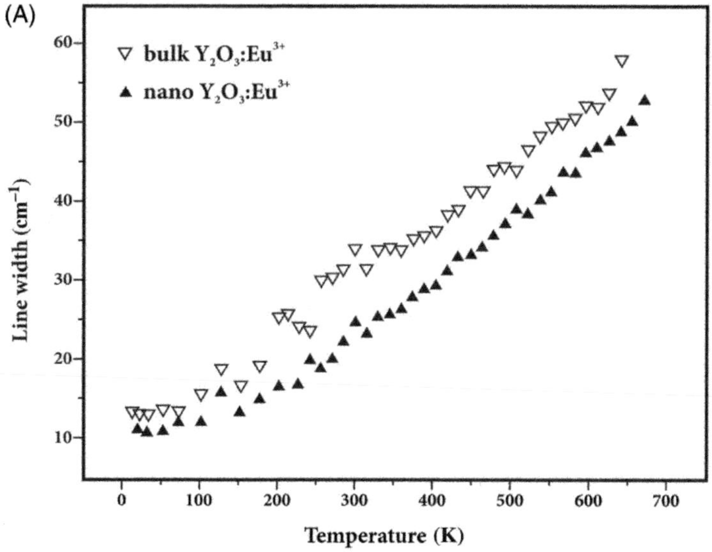

(B)

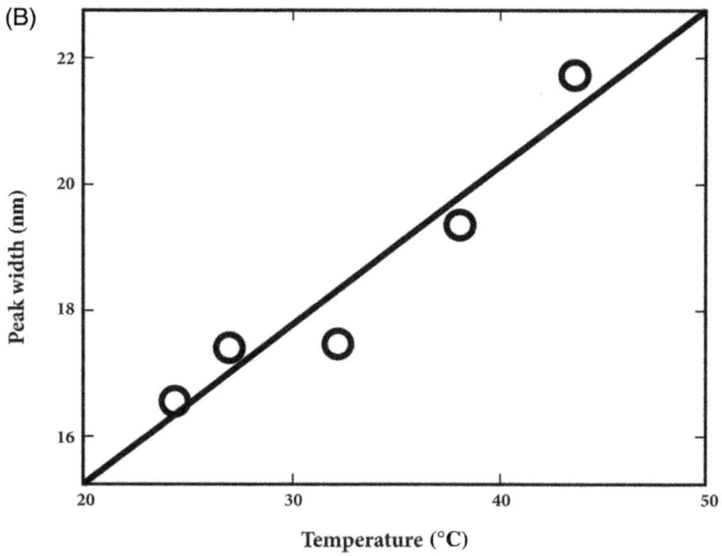

FIGURE 4.4 Temperature dependences of bandwidths of (A) $^5D_0 \rightarrow {}^7F_2$ emission of Y_2O_3: Eu^{3+} and (B) emission of CdSe QDs coated with ZnS. *((A) adapted from Peng H, Song H, Chen B, Wang J, Lu S, Kong X, et al. Temperature dependence of luminescent spectra and dynamics in nanocrystalline Y_2O_3:Eu^{3+}. J Chem Phys 2003;118(7):3277–82, with the permission of AIP Publishing and (B) adapted with permission from Li S, Zhang K, Yang J-M, Lin L, Yang H. Single quantum dots as local temperature markers. Nano Lett 2007;7(10):3102–5, copyright (2007) American Chemical Society).*

Measurement of temperature from emission bandwidths is in principle impractical. Changes of bandwidths are relatively small, and their estimation is time consuming. For these reasons, no application of this temperature read-out scheme has been presented yet.

4.3 SENSING TEMPERATURE FROM THE INTENSITY OF A SINGLE EMISSION BAND

The intensity of emission is quenched by temperature in all materials. Several processes are responsible for this phenomenon, and they are detailed in Chapter 3, Luminescence: the Basics, Methods, and Instrumentation. A decrease in intensity of dominant bands in emission spectra with an increase in temperature can be easily observed, as shown in Fig. 4.5 (except for Fig. 4.5D), and for this reason, early efforts in the development of luminescence thermometry were focused on the use of changes in emission band intensity. Up to now, a significant number of materials of almost all types have been tested for temperature readings as a result of changes in the intensity of a single emission band. These materials include lanthanide and transition metal doped metal oxides and salts, semiconductors, QDs, organic dyes, polymers, and hybrid materials. The changes can be quite high; a 2%/K change in emission intensity is regularly evidenced with QDs and organic dyes. Fig. 4.5 exemplifies the changes of emission band intensity with temperature for (1) QDs [7], (2) organic dyes [8], (3) transition metal doped phosphors [9], and (4) luminescent polymer [10].

Each material type shows some interesting phenomena. For example, the emission intensity of QDs habitually changes with temperature in a linear way. Organic materials show intense changes; however, they may be used only over a narrow temperature range. Emission intensity dependence on the temperature of these materials differs when the temperature is increased or decreased (it shows hysteresis). A negative quenching effect, i.e., an increase in emission intensity with temperature, is characteristic of some luminescent polymers, see Fig. 4.5D. Detailed elaboration of the effects of temperature on the luminescence of different materials is given in Chapters 6, 7 and 8 of this book.

The critical setback in taking temperature readings from a single emission band comes from the significant influence of measurement conditions on the measured emission intensity. Fluctuations in excitation and instabilities in detection are the primary cause of large measurement errors and so must be properly compensated. Also, the accuracy of imaging may be considerably compromised with unequal probe distribution and varying experimental geometries. Referenced sensing may alleviate this problem, but only to a modest degree. Therefore, this read-out scheme is rarely used in current luminescence thermometry applications.

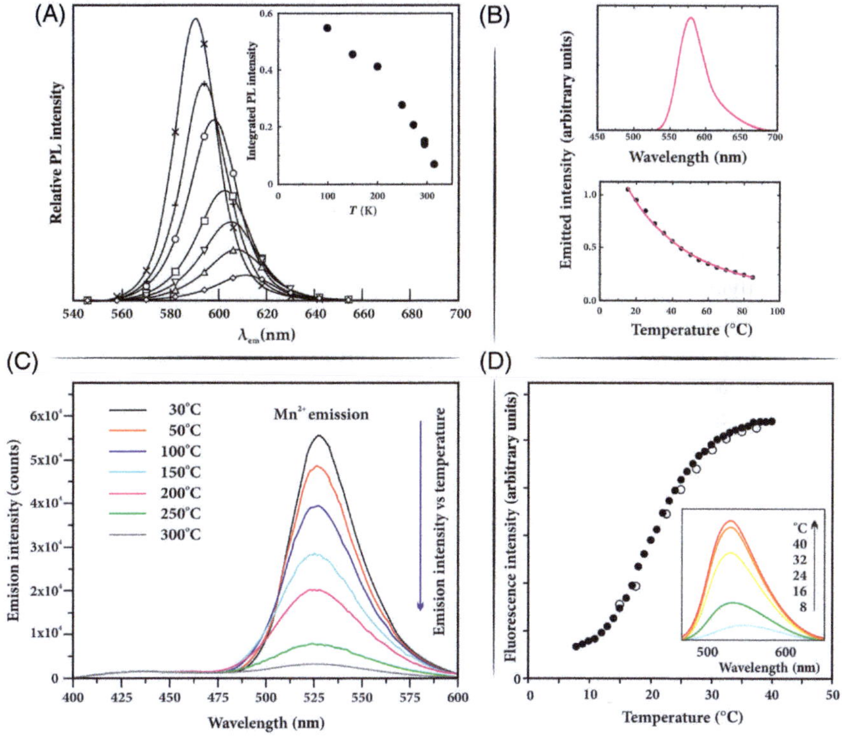

FIGURE 4.5 Temperature dependent emission intensity of (A) 5 nm CdSe QDs with ZnS overlayer, (B) rhodamine B solution, (C) Mn^{2+} doped Zn_2SiO_4 green phosphor, and (D) poly(DBD-*co*-DMAPAM-*co*-NTBAM) in buffer (pH 7). *Full circle*, heating; *Circle*, cooling. *((A) adapted from Walker GW, Sundar VC, Rudzinski CM, Wun AW, Bawendi MG, Nocera DG. Quantum-dot optical temperature probes. Appl Phys Lett 2003;83(17):3555. 20, with permission of AIP Publishing, (B) adapted from Jaque D, Vetrone F. Luminescence nanothermometry. Nanoscale 2012;4(15):4301–26, with permission of The Royal Society of Chemistry, (C) adapted from Lojpur V, Nikolić MG, Jovanović D, Medić M, Antić Ž, Dramićanin MD. Luminescence thermometry with Zn_2SiO_4:Mn^{2+} powder. Appl Phys Lett 2013;103(14):141912 (3 pp), with permission of AIP Publishing, and (D) adapted with permission from Uchiyama S, Kawai N, de Silva AP, Iwai K. Fluorescent polymeric AND logic gate with temperature and pH as inputs. J Am Chem Soc 2004;126(10):3032–3, copyright (2004) American Chemical Society).*

4.4 RATIOMETRIC TEMPERATURE SENSING

The most frequently used temperature read-out scheme in the current practice of luminescence thermometry is one which is founded on the determination of the ratio of intensities of different emission bands of a luminescent material [11]. The method is of paramount importance since it is self-referencing, demonstrates high sensitivity, and can be used for imaging. It alleviates problems caused by changes in measurement conditions by relying on measurements of the ratio of absolute intensities. It can be realized using a variety of probes of different types of material, giving this method great versatility. Ratiometric intensity methods may exploit emissions from a single emission center in the probe or two emission centers.

Ratiometric methods based on a single emission center customarily utilize rare earth doped phosphors, but other materials have also been used, such as Ruby (the ratio of two R-lines). Regarding rare earth doped materials, ratiometric reading of temperature most frequently exploits intensities of emissions originating from two closely separated excited states of lanthanide ions (the so-called Boltzmann-type ratiometric method), and, to a lesser extent, the ratio of intensities of emissions originating from electron transitions that end at different Stark sublevels. The method can be developed for both downshifting and upconversion emissions. An example of a Boltzmann-type ratiometric temperature reading is the ratio of upconversion emissions of Er^{3+} doped $GdVO_4$ nanoparticles coming from electronic transitions from the $^2H_{11/2}$ and $^4S_{3/2}$ levels to the $^4I_{15/2}$ ground level [12], whose temperature-dependent emission is shown in Fig. 4.6A.

One can consider two excited energy states of lanthanide ions thermally coupled when the energy difference between them is less or equal to 2000 cm^{-1}. Such an energy difference allows the promotion of electrons to a higher energy state from a lower one using thermal energy [13]. In this case, both states share the electronic population according to Boltzmann's distribution:

$$N_H = N_L \cdot \exp\left(-\Delta E / k_B T\right), \tag{4.1}$$

where N_H and N_L are the number of electrons in the higher and lower excited state, respectively, ΔE is the energy difference between these states, k_B is Boltzmann's constant, and T is absolute temperature. Then, the luminescence intensity ratio (*LIR*) of emissions from higher (I_H) and lower (I_L) excited states can be approximated by [14]:

$$\begin{aligned} LIR(T) &= \frac{I_H(T)}{I_L(T)} \\ &= c \cdot \frac{g_H A_H h\nu_H}{g_L A_L h\nu_L} \cdot \exp\left(-\Delta E / k_B T\right) = B \cdot \exp\left(-\Delta E / k_B T\right), \end{aligned} \tag{4.2}$$

where h is Plank's constant, g is the degeneracy of the excited state, A is the spontaneous emission rate, ν is the emission frequency, and H and L denote "higher" and "lower" energy states. The natural logarithm of Eq. (4.2) shows that log(*LIR*) has a linear dependence on inverse temperature, see Fig. 4.6B:

$$\log\left(LIR\right) = Log\left(B\right) - \frac{\Delta E}{k_B} \cdot \frac{1}{T}. \tag{4.3}$$

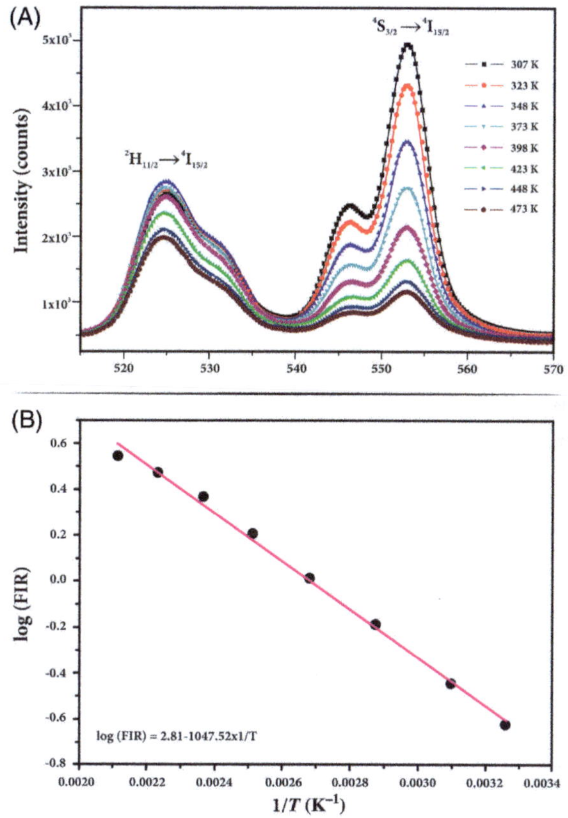

FIGURE 4.6 Boltzmann-type ratiometric intensity reading of temperature: (A) upconversion emission spectra of Er^{3+} doped $GdVO_4$ nanoparticles at different temperatures and (B) linear dependence of the logarithm of the ratio of emission transitions from the $^2H_{11/2}$ and $^4S_{3/2}$ levels to the $^4I_{15/2}$ ground level of Er^{3+} ions on inverse temperature, see Eq. (4.3) *(Adapted from Gavrilović TV, Jovanović DJ, Lojpur V, Dramićanin MD. Multifunctional Eu^{3+} - and Er^{3+}/Yb^{3+} -doped $GdVO_4$ nanoparticles synthesized by reverse micelle method. Sci Rep 2014;4:4209 (9pp)).*

So, the values of log(B) and $\Delta E/k_B$ may be derived from the line-slope and intercept with the y-axes of the log(LIR) versus the $1/T$ linear fit, respectively.

At low temperatures, the higher energy state will not be populated because the electrons will not have enough energy to bridge the energy gap, and because the nonradiative relaxation rate from the higher to lower energy state is very large for the closely spaced states. Therefore, the *LIR* read-out scheme has a low-temperature boundary that is proportional to ΔE, so that the lower the ΔE, the lower the temperature from which *LIR* can be used. As temperature increases, the higher energy state becomes populated, and hence, the emission from this state gradually increases in intensity at the expense of the population of the lower state. However, both emissions lessen in intensity because of temperature quenching (Chapter 3: Luminescence: the

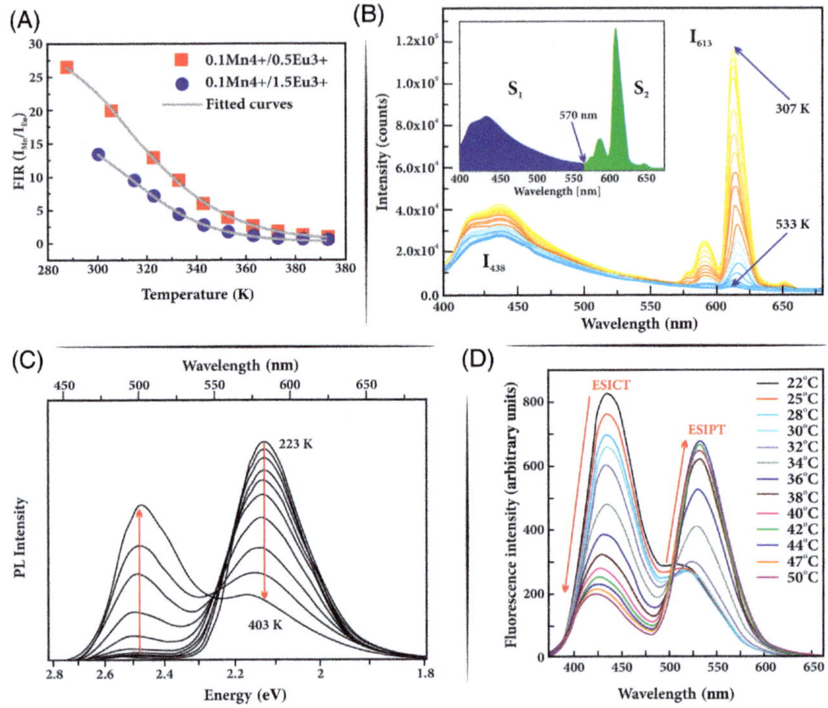

FIGURE 4.7 Temperature reading from the ratio of emission intensities originating from different emission centres. (A) *LIR* of $Y_3Al_5O_{12}$ doped with Mn^{4+} and Eu^{3+}. (B) Luminescence spectra of TiO_2: Eu^{3+} nanopowder measured at different temperatures. (C) Emission of CdSe: Mn^{2+} at different temperatures. (D) Temperature-dependent emission spectra of poly-NIPAM labeled with 3-hydroxyflavone. *((A) adapted from Chen D, Liu S, Zhou Y, Wan Z, Huang P, Ji Z. Dual-activator luminescence of RE/TM:$Y_3Al_5O_{12}$ (RE = Eu^{3+}, Tb^{3+}, Dy^{3+}; TM = Mn^{4+}, Cr^{3+}) phosphors for self-referencing optical thermometry. J Mater Chem C 2016;4(38):9044–51, with permission from The Royal Society of Chemistry, (B) adapted from Nikolić MG, Antić Ž, Ćulubrk S, Nedeljković JM, Dramićanin MD. Temperature sensing with Eu^{3+} doped TiO_2 nanoparticles. Sens Actuators B Chem 2014;201:46–50, copyright (2014), with permission from Elsevier, (C) adapted with permission from Vlaskin VA, Janssen N, van Rijssel J, Beaulac R, Gamelin DR. Tunable dual emission in doped semiconductor nanocrystals. Nano Lett 2010;10(9):3670–4, copyright (2010) American Chemical Society, and (D) adapted from Chen C-Y, Chen C-T. A PNIPAM-based fluorescent nano-thermometer with ratiometric readout. Chem Commun 2011;47(3):994–6 with permission from The Royal Society of Chemistry).*

Basics, Methods, and Instrumentation) until one of them, or both, become undetectable. The upper temperature boundary of *LIR* applicability thus depends mainly on the phonon spectrum of the host material and the type of lanthanide ion involved. One should note that measurement uncertainties become higher near these limits. One should also note that with the *LIR* method it is possible to use any emissions providing they originate from two thermally coupled excited levels, Fig. 4.7.

Luminescence thermometry probes having more than one emission center comprise a great versatility of materials. Combinations may include QDs doped with transition metal ions, materials doped with two rare earth ions, materials doped with rare earth ions and transition metal ions [15], two luminescent polymers, etc. In recent times the ratio between the emission of a lanthanide ion and the emission of defect states of the host material is found to be very sensitive to temperature [16,17]. Combination of two ratiometric methods, one based on a single emission center and the second based on two, can be used to extend the measurement range [18].

4.5 TEMPERATURE SENSING USING DECAY TIME

The temporal dependence of an emission generally shows three distinct behaviors, as shown in Fig. 4.8. In the first time period, the emission resembles the time dependence of the excitation pulse. In the second, the emission intensity rises due to the accumulation of electrons in the excited level, and then, in the third period, the emission decreases due to processes that were discussed in Chapter 3, Luminescence: the Basics, Methods, and Instrumentation. The first two periods are relatively short by comparison with the third, and in routine analyses they are frequently not measured (so called "dead time" [21]). However, the initial rise of an emission may be very useful for the determination of temperature, which will be discussed in the next section.

Time domain methods and frequency domain methods are commonly used for measurements of emission decay times; the first method utilizes pulsed excitation sources while the latter uses continuous wave light sources with modulated intensity. So far, only a few reports have been published on temperature

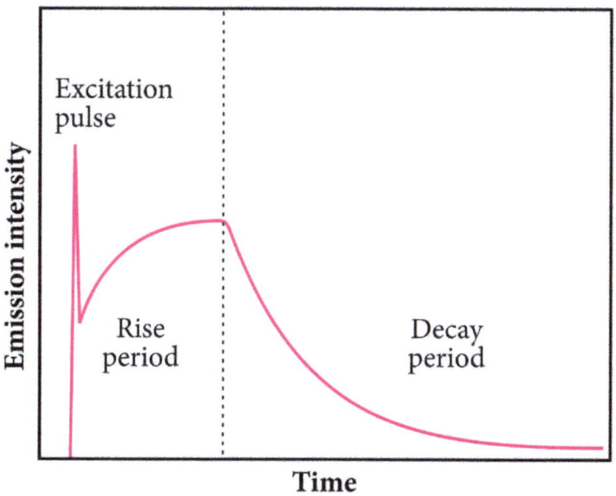

FIGURE 4.8 General shape of the emission temporal behavior.

sensing from frequency domain measurements of emission decay times. Reports considering the use of the time domain method for temperature read-outs from decay times, on the other hand, are extensive.

Self-referencing is the main feature of lifetime-based thermometry. The method is robust, since lifetime values are not affected by the concentration and distribution of probe material and are not influenced by fluctuations in excitation power. The method requires monitoring of just one emission band and for this reason it is suitable for thermal imaging. On the other hand, the temporal resolution of temperature measurements is smaller than from intensity-based thermometry. The main problem occurs with measurement range. When lanthanide doped materials are utilized as probes, at low temperatures, the changes of lifetimes are negligible or very small, see Fig. 4.9A. In most cases they cannot be used in a physiologically relevant temperature range (from room temperature to 60°C), thus, they cannot be used for most biomedical applications. Principally, these materials are suitable and used for high-temperature thermometry applications. Transition metal doped materials show similar behavior to lanthanide doped materials, with the exception that in some cases the temperature region in which the lifetime strongly changes may also cover a physiologically relevant temperature range. Lifetimes of QD emissions usually decrease with an increase in temperature, but the changes are rather small (Fig. 4.9B). Lifetime values and the extent of change notably depend on the size of the QDs [23]. There is also a particle size effect on the excited state lifetimes of rare earth doped materials. Consequently, the decay time of an emission of any phosphor sample is actually a composite value of individual decay times arising from the distribution of particle sizes. One should also note that in the case of phosphors decay time strongly depends on the concentration of luminescence centers.

4.6 TEMPERATURE SENSING BASED ON RISE TIME

The rise of emission intensity shortly after an excitation pulse occurs due to an accumulation of electrons in the excited level of the lowest energy in rare earth ions. The phenomenon is a consequence of the fast transitions to this level from higher energy levels and energy migration from the nearby excited ions; both processes are much faster than the radiative transition to the ground state from this level. The number of electrons in the emitting level N can be approximated as [24]:

$$N(t) = N_0 + N_1 \cdot \left[1 - \exp(-t/\tau_r)\right], \tag{4.4}$$

where N_0 and N_1 are numbers of electrons directly excited within the ion and accumulated by the transition from neighboring ions, respectively, while τ_r is a rise time constant. By analogy to the decay time constant, the rise time may be considered as the time passed until the emission intensity achieves $1-1/e$;

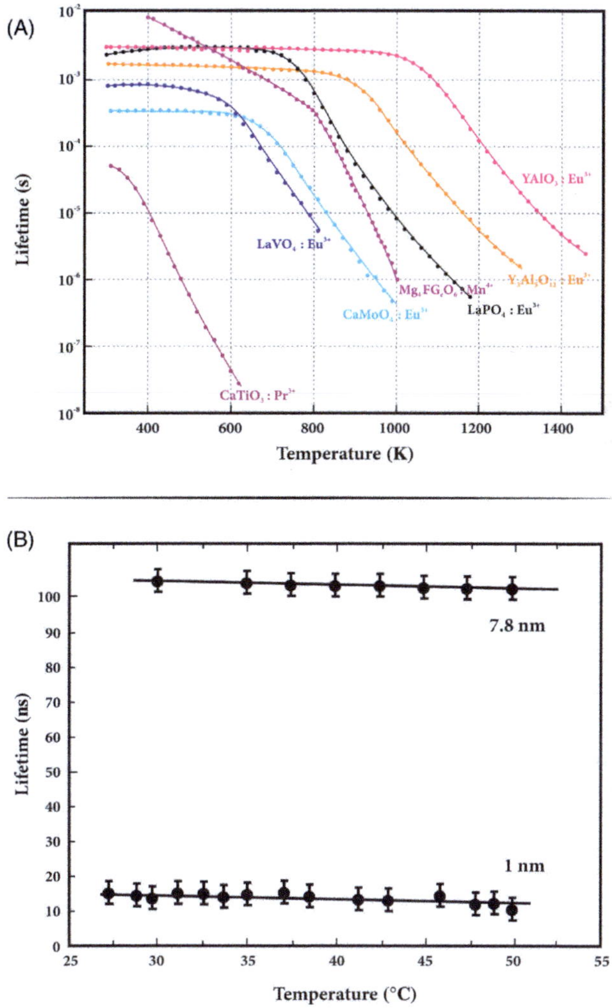

FIGURE 4.9 Changes of lifetime values with temperature for (A) several rare earth and transition metal doped materials and (B) 1 and 7.8 nm CdTe QDs. *((A) adapted from Brübach J, Pflitsch C, Dreizler A, Atakan B. On surface temperature measurements with thermographic phosphors: a review. Prog Energy Combust Sci 2013;39(1):37–60, copyright (2013), with permission from Elsevier and (B) adapted from Haro-González P, Martínez-Maestro L, Martín IR, García-Solé J, Jaque D. High-sensitivity fluorescence lifetime thermal sensing based on CdTe quantum dots. Small 2012;8(17):2652– 8, with permission from John Wiley and Sons).*

$e \cong 2.71828$ (the base of the natural logarithm) of its maximal value. Then, the temporal dependence of the emission intensity may be presented by:

$$I(t) = \{A + B \cdot [1 - \exp(-t/\tau_r)]\} \cdot \exp(-t/\tau_d), \tag{4.5}$$

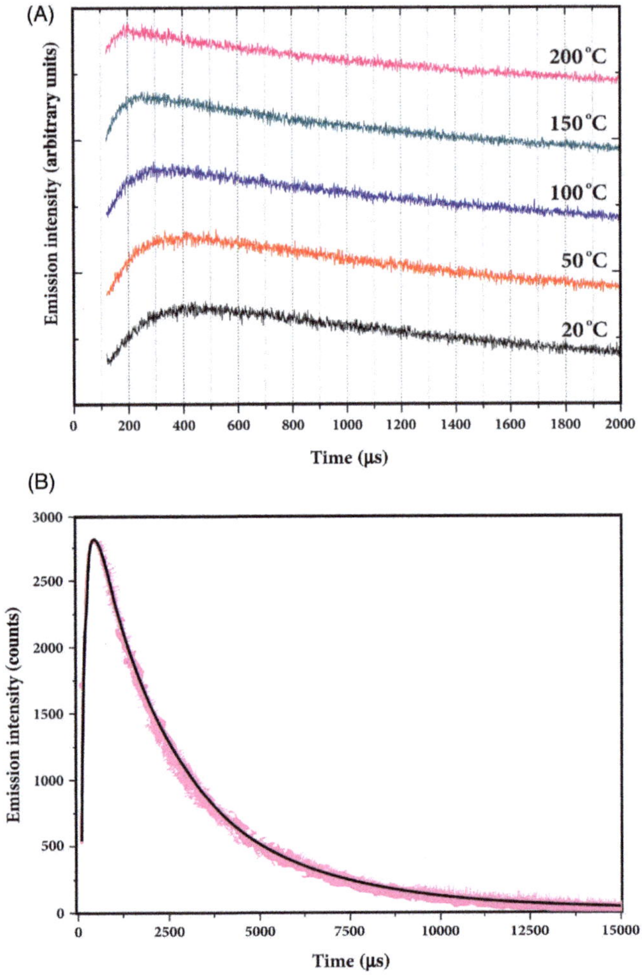

FIGURE 4.10 (A) Temporal behavior of Eu^{3+} emission in SrY_2O_4 at different temperatures. (B) An example of fitting using Eq. (4.5). *(Adapted from Lojpur V, Antić Z, Dramićanin MD. Temperature sensing from the emission rise times of Eu^{3+} in SrY_2O_4. Phys Chem Chem Phys 2014;16(46):25636–41, with permission from The Royal Society of Chemistry).*

where τ_d is the decay time constant and *A* and *B* are intensity parameters, considering that emission intensity is proportional to the number of electrons in the excited state level.

Temporal dependence of Eu^{3+} emission in SrY_2O_4 at different temperatures is presented in Fig. 4.10A [25]. Fig. 4.10B shows the example of experimental data fitting Eq. (4.5).

The rise time value of the Eu^{3+} emission in the 0.5 at% doped sample reduced (linearly) from 138 to 13 μs when the temperature was raised from 20

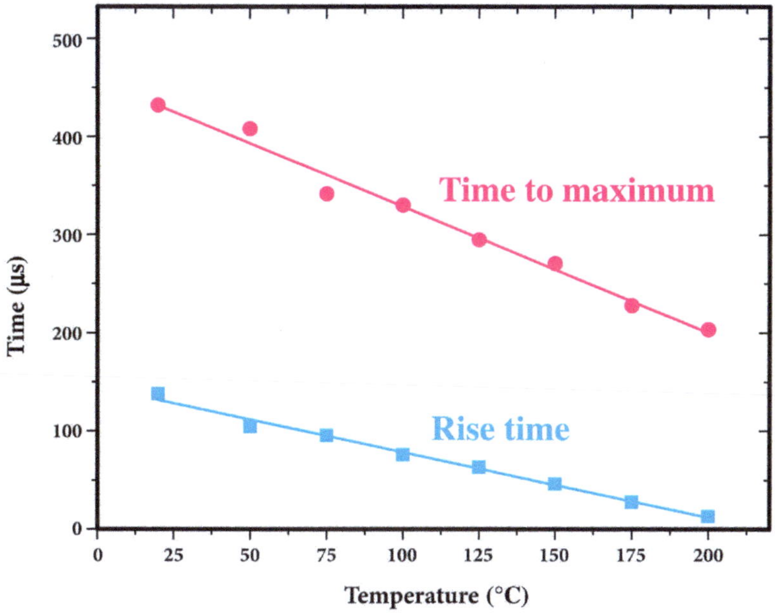

FIGURE 4.11 Temperature dependence of rise time and time-to-max values of Eu^{3+} emission in $SrAl_2O_4$. *(Adapted from Lojpur V, Antić Z, Dramićanin MD. Temperature sensing from the emission rise times of Eu^{3+} in SrY_2O_4. Phys Chem Chem Phys 2014;16(46):25636–41 with permission from The Royal Society of Chemistry).*

to 200°C, as shown in Fig. 4.11. This provided quite a high relative sensitivity of 5.53%K^{-1}at 473 K. A fitting procedure provides both rise time and decay time values thus enabling two temperature read-outs. These may complement each other in the way that rise times may be used in low-temperature operating ranges where decay times are quite insensitive to temperature changes (in the case of lanthanide doped materials), while decay times can indicate temperature in high-temperature regions. However, Eq. (4.5) is rather complicated—the evaluation of rise time values using the fitting procedure is a time-consuming process. Therefore, temperature monitoring has low temporal resolution and is impractical for imaging applications. Temperature reading can be carried out in a much simpler way by using the time at which the emission reaches a maximal value (time-to-max value) as an indication of temperature, as first proposed by Khalid and Kontis [26]. Fig. 4.11 shows that a time-to-max value of Eu^{3+} emission in $SrAl_2O_4$ also decreases linearly with temperature. The temperature reading is of lower sensitivity than a reading from rise time values, but the time-to-max values can be more precisely determined. Smaller measurement uncertainties compensate lower sensitivities to provide similar, or even better, temperature resolution compared to rise time–based temperature readings. Also, fast determination of temperature from time-to-max values facilitates the use of this method for thermal imaging.

4.7 TEMPERATURE SENSING FROM LUMINESCENCE POLARIZATION (ANISOTROPY)

Intensity of luminescence is strongly dependent on its polarization in anisotropic media. The degree of anisotropy r (see Eq. 3.12 in Chapter 3) is temperature dependent since temperature influences the Brownian rotational motion of the particles. Rotation accelerates with an increase in temperature, and the higher rotation of particles during their fluorescence lifetime causes a larger number of reemitted photons to lose memory of the incident light polarization [27]. Consequently, the degree of anisotropy of fluorescence decreases with rise in temperature. There are just a few examples of temperature readings being taken using the degree of anisotropy of fluorescence. One of them describes the mapping of intracellular temperature using green fluorescent protein [27], where it has been found that the degree of anisotropy of a green fluorescent protein in a phosphate saline buffer decreases with temperature at a rate of 0.1%/K, see Fig. 4.12. Other examples include the well-known fluorescein molecule for which a 3%/K decrease in degree of anisotropy has been recorded.

It is important to note that the degree of anisotropy presents the ratio of fluorescence intensities, so it is not affected by fluctuations of measurements of absolute intensities. Therefore, the process of temperature reading from degree of anisotropy is self-referenced, and with an appropriate calibration provides an absolute temperature measurement. The method is characterized by good temperature sensitivity, but low acquisition speed and a quite complicated experimental set-up for imaging purposes.

4.8 THE GENERIC LUMINESCENCE THERMOMETRY SYSTEM

The generic luminescence thermometry system is comprised of six elements: (a) light source, (b) wavelength selector on either excitation or emission light paths or on both, (c) sample chamber with controlled temperature, (d) sample holder/positioner, (e) temperature controller, (f) reference thermometer, (g) filter, (h) detector, and (i) data acquisition system. There are schematically presented in Fig. 4.13.

Different types of light sources may be used for the excitation of luminescence. Lasers are a frequent choice. They produce intense monochromatic light, so there is no need for a wavelength-selecting element on the excitation path. For time-resolved measurements it is necessary to use pulsed lasers or continuous wave lasers coupled with intensity modulators such as optical choppers or acousto-optical modulators. However, lasers are expensive and in lot of cases can be replaced without significant loss in performance with laser diodes or light-emitting diodes (LEDs) which are more affordable devices. Lamp sources are still used, mainly in the investigation of the thermometric performance of

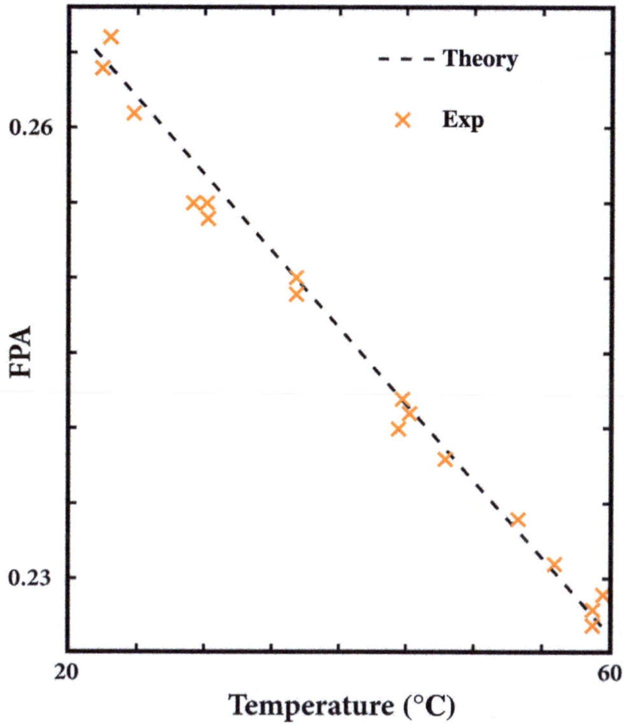

FIGURE 4.12 Temperature dependence of the degree of anisotropy of green fluorescent protein in a phosphate saline buffer. *(Adapted with permission from Donner JS, Thompson SA, Kreuzer MP, Baffou G, Quidant R. Mapping intracellular temperature using green fluorescent protein. Nano Lett 2012;12(4):2107–11, copyright (2010), American Chemical Society).*

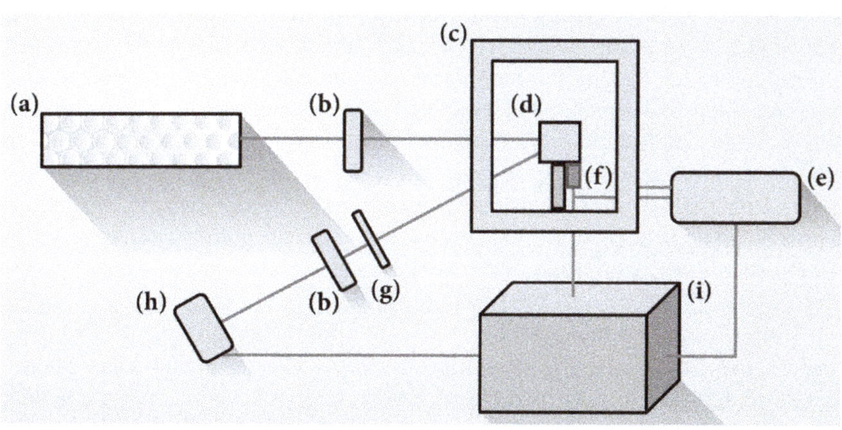

FIGURE 4.13 Generic luminescence thermometry system—a schematic.

different materials. Their polychromatic light must be dispersed on a dispersing element from which a light beam of a selected wavelength is directed on a sample. Lamp sources are also used for temperature read-outs from excitation spectra. Tunable and supercontinuum lasers, able to produce emissions over the given spectral interval, may also be used for the same purpose; however, these are rather expensive devices. Light dispersion can be accomplished with prisms and diffraction gratings; the latter is dominantly used in the modern spectrofluorometers.

The luminescence properties of sample should be measured in a controlled environment. Usually, a sample is placed in a chamber where the temperature can be precisely controlled and measured with a reference thermometer.

The detector can be either single-channel or multichannel. A single-channel detector, usually photomultiplier tube (PMT) or semiconductor, detects the intensity of one wavelength at a time. Multichannel detectors, such as charge-coupled device (CCD) cameras, record the intensity of emission over the range of wavelengths simultaneously. Also, modern instruments include some other important components, such as polarizers, filters, and optical fiber connectors.

Laboratory grade systems can be rather complex and robust devices capable of multiple temperature read-out measurements. However, for many applications, and for the convenience and mobility of measurements, systems may be designed as miniaturized devices.

4.9 STATIC (TIME-INTEGRATED) VERSUS TIME-RESOLVED METHODS: PROS AND CONS

Regarding the complexity and the cost of equipment, time-integrated measurements are favorable to time-resolved ones since they are simpler and cheaper. Also, time-integrated methods provide sufficient data for the majority of applications, so they are more regularly practiced than time-resolved ones. However, one should note that time-integrated measurements do not account for all the information imparted by luminescence since they are simply an average of the time-resolved phenomena [11]. Actually, all time-integrated temperature read-out methods can also be carried out using pulsed excitation and delayed detection. What is more, by using delayed detection some problems that compromise time-integrated measurements may be overcome, such as strong background luminescence. In this way measurements of higher accuracy may be achieved. One must to be careful, though, to use delays in detection much shorter than the lifetimes of any of the involved emissions.

Both time-integrated and time-resolved approaches provide self-referencing read-out methods, *LIR* and lifetime, respectively. In the early stage of development of luminescence thermometry, temperature was regularly determined from emission decay times, mostly for high-temperature applications. In recent times, the research focus has shifted to time-integrated methods, especially to the *LIR* method, and to physiologically relevant temperature ranges. Both approaches present some advantages over one another. The *LIR* technique is faster, simpler, and requires less sophisticated instrumentation than the lifetime method, and therefore it has received more attention lately. In addition, *LIR* methods present larger relative sensitivities than lifetime methods. On the other hand, *LIR* requires monitoring of two distinctively separated emissions to achieve self-referencing, while measurements of lifetime are performed on one emission band only. Also, measurements of temporal emission changes are much more accurate than emission intensity measurements. Reduced uncertainties compensate for lower sensitivities so that time-resolved methods may provide better temperature resolution than *LIR* methods [16]. Lifetime methods suffer from small operating ranges, and the low-temperature boundary can be at a relatively high temperature in a lot of cases. This obstacle may be overcome by reading temperature from rise times. The high-temperature boundary, on the other hand, is generally at higher temperatures in lifetime methods than in *LIR* methods. However, the performance of thermometry that uses materials with a single emission center, in both time-resolved and time-integrated approaches, is mostly governed by the intrinsic luminescence properties of the dopant lanthanide ion with a small contribution coming from the host material's characteristics. The relative sensitivity of *LIR*, which uses emissions from the thermally coupled excited levels of lanthanide ions, is directly proportional to the energy gap between these levels. The largest energy gap is in the Eu^{3+} ion at about 1750 cm^{-1}, meaning that relative sensitivity varies with temperature at $251{,}800/T^2$ [$\%K^{-1}$], and that the maximum achievable theoretical relative sensitivity of the technique is around 2.8 $\%K^{-1}$ at 300 K for different hosts. Bearing this in mind and considering the large number of publications that have been reported on *LIR* thermometry with different ion/host combinations, it is not reasonable to expect that further research focused simply on the scrutiny of performance on different combinations of single lanthanide luminescence centers and hosts will not bring any progress in the field. However, this *LIR* method can be further developed by improving its temperature resolution which is rather low—rarely better than 1 K. This can be accomplished by enhancing measurement signals which would yield larger absolute sensitivities (if the signal-to-noise ratio is maintained) and/or by improving signal-to-noise and measurement procedures to lower measurement uncertainties.

Further improvement in *LIR* relative sensitivities may be expected with the use of multiple emission centers, as discussed in Section 3.2. These centers may be all lanthanide ions or a combination of lanthanide ions and other types of emission centers. Such probes may also provide multiple read-outs of temperature which, when combined, give wider operating ranges with excellent relative sensitivity and resolution. We should also note that the investigations into thermometry that use near-infrared emissions have just begun and that they have potentially valuable applications, especially in biomedicine.

REFERENCES

[1] Kusama H, Sovers OJ, Yoshioka T. Line shift method for phosphor temperature measurements. Jpn J Appl Phys 1976;15(12):2349–58.

[2] Rocha U, Jacinto da Silva C, Ferreira Silva W, Guedes I, Benayas A, Martínez Maestro L, et al. Subtissue thermal sensing based on neodymium-doped LaF_3 nanoparticles. ACS Nano 2013;7(2):1188–99.

[3] Allison SW, Gillies GT. Remote thermometry with thermographic phosphors: instrumentation and applications. Rev Sci Instrum 1997;68(7):2615–50.

[4] Liang R, Tian R, Shi W, Liu Z, Yan D, Wei M, et al. A temperature sensor based on CdTe quantum dots–layered double hydroxide ultrathin films via layer-by-layer assembly. Chem Commun 2013;49(10):969–71.

[5] Peng H, Song H, Chen B, Wang J, Lu S, Kong X, et al. Temperature dependence of luminescent spectra and dynamics in nanocrystalline Y_2O_3:Eu^{3+}. J Chem Phys 2003;118(7):3277–82.

[6] Li S, Zhang K, Yang J-M, Lin L, Yang H. Single quantum dots as local temperature markers. Nano Lett 2007;7(10):3102–5.

[7] Walker GW, Sundar VC, Rudzinski CM, Wun AW, Bawendi MG, Nocera DG. Quantum-dot optical temperature probes. Appl Phys Lett 2003;83(17):3555.

[8] Jaque D, Vetrone F. Luminescence nanothermometry. Nanoscale 2012;4(I15):4301–26.

[9] Lojpur V, Nikolić MG, Jovanović D, Medić M, Antić Ž, Dramićanin MD. Luminescence thermometry with Zn_2SiO_4:Mn^{2+} powder. Appl Phys Lett 2013;103(14):141912–3.

[10] Uchiyama S, Kawai N, de Silva AP, Iwai K. Fluorescent polymeric AND logic gate with temperature and pH as inputs. J Am Chem Soc 2004;126(10):3032–3.

[11] Dramićanin MD. Sensing temperature via downshifting emissions of lanthanide-doped metal oxides and salts. A review. Methods Appl Fluoresc 2016;4(4):42001–23.

[12] Gavrilović TV, Jovanović DJ, Lojpur V, Dramićanin MD. Multifunctional Eu^{3+}- and Er^{3+}/Yb^{3+}-doped $GdVO_4$ nanoparticles synthesized by reverse micelle method. Sci Rep 2014;4:4209–19.

[13] Quintanilla M, Benayas A, Naccache R, Vetrone F, Quintanilla M, Benayas A, et al. Luminescent nanothermometry with lanthanide-doped nanoparticles. In: Carlos LD, Palacio F, editors. Thermometry at the nanoscale: techniques and selected applications. Royal Society of Chemistry; 2015. p. 124–66.

[14] Wade SA, Collins SF, Baxter GW. Fluorescence intensity ratio technique for optical fiber point temperature sensing. J Appl Phys 2003;94(8):4743–56.

[15] Chen D, Liu S, Zhou Y, Wan Z, Huang P, Ji Z. Dual-activator luminescence of RE/TM: $Y_3Al_5O_{12}$ (RE=Eu^{3+}, Tb^{3+}, Dy^{3+}; TM=Mn^{4+}, Cr^{3+}) phosphors for self-referencing optical thermometry. J Mater Chem C 2016;4(38):9044–51.

[16] Nikolić MG, Antić Ž, Ćulubrk S, Nedeljković JM, Dramićanin MD. Temperature sensing with Eu^{3+} doped TiO_2 nanoparticles. Sens Actuators B Chem 2014;201:46–50.

[17] Dramićanin MD, Antić Ž, Ćulubrk S, Ahrenkiel SP, Nedeljković JM. Self-referenced luminescence thermometry with Sm^{3+} doped TiO_2 nanoparticles. Nanotechnology 2014;25(48):485501–6.

[18] Ćulubrk S, Lojpur V, Ahrenkiel SP, Nedeljković JM, Dramićanin MD. Non-contact thermometry with Dy^{3+} doped $Gd_2Ti_2O_7$ nano-powders. J Lumin 2016;170:395–400.

[19] Vlaskin VA, Janssen N, van Rijssel J, Beaulac R, Gamelin DR. Tunable dual emission in doped semiconductor nanocrystals. Nano Lett 2010;10(9):3670–4.

[20] Chen C-Y, Chen C-T. A PNIPAM-based fluorescent nanothermometer with ratiometric readout. Chem Commun 2011;47(3):994–6.

[21] Dowell LJ. Fluorescence Thermometry. Appl Mech Rev 1992;45(7):253–60.
[22] Brübach J, Pflitsch C, Dreizler A, Atakan B. On surface temperature measurements with thermographic phosphors: a review. Prog Energy Combust Sci 2013;39(1):37–60.
[23] Haro-González P, Martínez-Maestro L, Martín IR, García-Solé J, Jaque D. High-sensitivity fluorescence lifetime thermal sensing based on CdTe quantum dots. Small 2012;8(17):2652–8.
[24] Ranson RM, Evangelou E, Thomas CB. Modeling the fluorescent lifetime of Y_2O_3: Eu. Appl Phys Lett 1998;72(21):2663–4.
[25] Lojpur V, Antić Z, Dramićanin MD. Temperature sensing from the emission rise times of Eu^{3+} in SrY_2O_4. Phys Chem Chem Phys 2014;16(46):25636–41.
[26] Khalid AH, Kontis K. 2D surface thermal imaging using rise-time analysis from laser-induced luminescence phosphor thermometry. Meas Sci Technol 2009;20(2):25305.
[27] Donner JS, Thompson SA, Kreuzer MP, Baffou G, Quidant R. Mapping intracellular temperature using green fluorescent protein. Nano Lett 2012;12(4):2107–11.

Chapter 5

Methods of Analysis for Luminescence Thermometry Measurements

5.1 INTRODUCTION

Luminescence thermometry requires measurements of different luminescence features depending on the chosen temperature read-out scheme (Chapter 4). Most commonly, these are emission spectra and emission decays, but they can also be polarization/anisotropy or just readings of emission intensities at specific wavelengths. Measurement data consist of signals of interest, but they also contain baseline offset and noise, and sometimes even measurement artifacts. They are all entwined by instrumental function. Since measurements need to be performed over the wide range of temperatures, the accumulation of data may be substantial. A large amount of data does not present a problem these days as current technologies expedite archiving and processing. However, a bottleneck which does exist is the ability to interpret and use data. A basic knowledge of signal processing and numerical analysis is, therefore, a must if one wishes to practice luminescence thermometry. Moreover, real-time applications demand automated signal preprocessing and processing. At present, one can employ a great number of computer-based numerical methods to reduce signal noise, remove baseline offset, undertake deconvolution, and compensate for instrumental artifacts. Data-processing techniques can convert data into more useful forms, but also, can detect and measure peak characteristics (position, maximal intensity, and width) and extrapolate data beyond the range of measurement, etc. Reliable and fast data processing is also needed to test hypotheses and optimize measurement approaches.

The content of this chapter aims to introduce readers to basic concepts of numerical analysis methods which are often needed for processing luminescence thermometry data. The chapter deals with baseline offset and noise, and the ways to remove or compensate for them when looking at experimental data. A brief outline on numerical differentiation and integration is also given. Next, a numerical quantification of fundamental spectral features is presented. This includes peak detection, quantification of peak height, width, and area. A method for numerical resolution enhancement is exemplified for the case of small

Luminescence Thermometry. http://dx.doi.org/10.1016/B978-0-08-102029-6.00005-1

peaks superimposed on one dominant peak. The chapter also deals with the methods for calculating excited state lifetime from data obtained with either time-resolved or phase-modulated techniques. Finally, the chapter concludes with consideration given to the evaluation of figures of merit of luminescence thermometers.

It is impossible to cover all digital signal processing techniques and statistics, consequently this chapter does not cover wavelet transforms, pattern recognition, or factor analysis. The intention is to introduce the reader to basic concepts; for more information and advanced topics references are given in the chapter to suitable literature. One should note that for simplicity of presentation, much of text implicitly considers "spectrum" and "data" as the luminescence intensity versus radiation wavelength. However, these considerations can equally treat other data, such as data presented in wavenumbers.

5.2 REMOVAL OF BASELINE OFFSET

Observed luminescence spectra are composed of the desirable "true" luminescence contribution and unwanted contributions, of which the most critical are the baseline offset and noise. The first two contributions also entail an instrument function (also termed the blurring function). Baseline offset and noise cause downgrading of measurement accuracy and precision, and effectively reduce the dynamic range of measurement. Hence, depending on their magnitude and the specificity of application, their removal may be required. In the measured spectrum, the baseline offset, frequently termed as background, appears as a low-frequency component, while the noise appears as a high-frequency component. The "true" signal usually has a frequency between these two.

Baseline offset may be independent of wavelength, but, generally, has a spectral shape. Its removal can be accomplished either by modification of measurement conditions (instrumental approach), by preprocessing measured data (noninstrumental approach), or by combining both of these methodologies. Luminescence measurement is, in fact, a single-beam technique, contrary to some other spectroscopic techniques, and as such, it is significantly influenced by experimental conditions. The best approach is to fix measurement instruments and conditions so that no significant baseline offset occurs. This can be accomplished by changing the wavelength, and even the type, of the excitation source, by introducing appropriate filters, modifying signal collection geometry and data acquisition, etc. In luminescence measurements, the baseline offset value is mainly influenced by the dark current of the detection device, direct current (DC) bias of electronic components, and stray light. The dark current can be measured with a closed optical shutter, or in the absence of excitation. Subsequently, the dark current value can be subtracted from the measured spectrum. Compensation for stray light is a more complex issue. Stray light is caused by multiple reflections within a spectrofluorimeter and is difficult for quantitative estimations. In cases in which stray light does not fluctuate in intensity, it can be compensated for by repeating measurements with and without the sample, and then subtracting data obtained without the sample from data obtained with the sample at each spectral point. The contribution of black-body radiation, i.e., the incandescence of samples at elevated temperatures, may be resolved in the same manner. Black-body radiation has a notable spectral shape, being of lower intensity at shorter wavelengths (see Chapter 6), and may considerably influence luminescence thermometry measurements in which samples are frequently exposed to high temperatures.

There are, however, circumstances when an instrumental approach to baseline offset removal is not feasible. These can be a consequence of the restrictions imposed by the application, such as an inability to alter/improve instrumental conditions or spend time repeating measurements. Some circumstances are related to fast changes in measurement conditions or changes within the sample that is investigated (for example in situ environmental and biomedical applications) which introduces a drift of the baseline offset between repeated measurements. One should also note that rationing of two spectra increases the noise by factor of $2^{1/2}$.

In principle, it is difficult to derive a theoretical description of the baseline contribution and then to use it to separate the baseline from the experimental spectra. For this reason, data preprocessing methods (i.e., noninstrumental methods) rely on numerical, statistical, and artificial intelligence methods for the estimation and removal of baseline offset. Baseline corrections can be done manually or automatically, or as a combination of the two. Manual correction starts from the visual inspection of a measured spectrum. The simplest way of correcting the baseline is to identify the smallest value of measured signal and to subtract it from all data points. This "one-point" correction sometimes leads to a surprisingly good improvement of results. Better correction can be achieved with an approximation of baseline offset below peaks with a smooth function. First, the trace of the baseline should be observed in spectral regions between peaks. In the so-called three-point method (Fig. 5.1), two data points near a peak should be selected to represent the ends of a signal peak, one with a lower wavelength (λ_1) and one with a higher wavelength (λ_2) than peak wavelengths. The intensities of the signal should be noted, I_1 and I_2, respectively. The straight line that connects these points becomes the baseline (Fig. 5.1A) and is further used for calculation of the "true" peak area or peak height. For example, the measured intensity of emission I_M at peak maximum (λ_M) is composed of the "true" component (I_T) and baseline offset (I_B) component (Fig. 5.1), $I_M = I_T + I_B$. The task is then to derive I_B, so the "true" peak intensity can be calculated as $I_T = I_M - I_B$. Using the three-point method (Fig. 5.1A) the baseline offset at λ_M is:

$$I_B = I_1 + (\lambda_M - \lambda_1) \cdot \frac{I_2 - I_1}{(\lambda_2 - \lambda_1)} \tag{5.1}$$

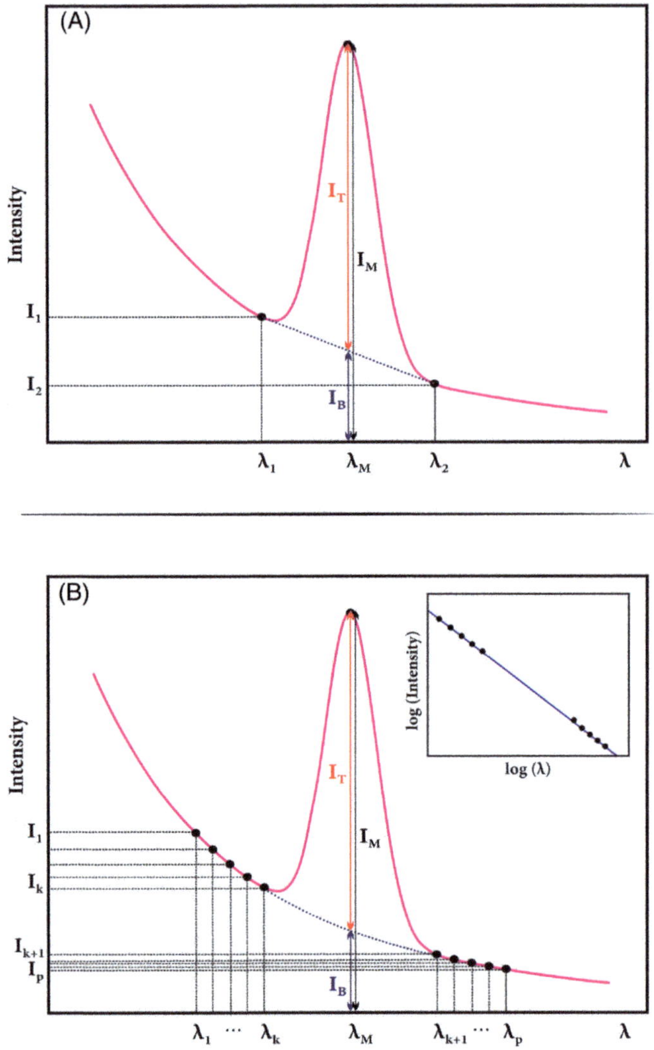

FIGURE 5.1 Approximation of baseline offset using (A) a three-point method (i.e., linear segment) and (B) a power function. Inset shows logarithm of intensity versus logarithm of wavelength for baseline data used for power function fitting, see Eq. (5.3).

If the straight line does not properly fit the real baseline, that is, if a more precise baseline approximation is needed, one may use some smooth, monotonic function to approximate the baseline. A very good approximation can be achieved with a power function:

$$I_B(\lambda) = C \cdot \lambda^n, \tag{5.2}$$

where C is a constant and n can be both a negative (baseline decreases with wavelength) or positive (baseline increases with wavelength) real number. In this case, several data points should be selected in spectral regions before and after a signal peak (Fig. 5.1B) and fitted to Eq. (5.2). Alternatively, a natural logarithm of data values can be plotted and fitted to a linear equation:

$$\log\left[I_B(\lambda)\right] = \log(C) + n \cdot \log(\lambda), \tag{5.3}$$

which is obtained as the logarithm of Eq. (5.2). Then, n is obtained from the line slope and log(C) is an intersect of the line with the y-axis. The "true" peak height is:

$$I_T = I_M - C \cdot \lambda_M^n, \tag{5.4}$$

When integrated emissions are needed, a correction is applied to the area below the measured emission intensity S_M by subtracting the value of the area below the baseline S_B. Using the three-point correction, the "true" integrated emission S_T can be derived as:

$$S_T = S_M - S_B = \int_{\lambda_1}^{\lambda_2} I_M(\lambda) \cdot d\lambda - \frac{1}{2}(I_1 + I_2)(\lambda_2 - \lambda_1). \tag{5.5}$$

An approximation of the baseline with power function S_T can be calculated from the following expression:

$$\begin{aligned} S_T = S_M - S_B &= \int_{\lambda_k}^{\lambda_{k+1}} I_M(\lambda) \cdot d\lambda - \frac{C}{n+1}\left(\lambda_{k+1}^{n+1} - \lambda_k^{n+1}\right) && \textit{for } n \neq -1, \\ &= \int_{\lambda k}^{\lambda k+1} I_M(\lambda) \cdot d\lambda - C \cdot \log \frac{\lambda_{k+1}^{n+1}}{\lambda_k^{n+1}} && \textit{for } n = -1, \textit{ and} \\ C &= \left[\frac{I_k I_{k+1}}{(\lambda_k \lambda_{k+1})^n}\right]^{\frac{1}{2}}. \end{aligned} \tag{5.6}$$

Further improvement in the precision of baseline approximation may be achieved using more complex functional forms such as polynomials [1] and splines. Manual baseline removal produces reasonable results, but it is a tedious process. Fortunately, there are many free software programs or commercial software programs which may be used for this task.

There are also many methods specifically designed for automation of baseline removal. Certain methods do not require any *a priori* information, while others need some information regarding the noise, baseline, instrument function, or signal [2]. For baseline estimation they rely on the maximum entropy approach [3], robust local regression [4], noise median method [5], first derivative method [6], artificial neural networks [2], signal removal methods [7], spectral shift methods [6], Fourier transform method [6,8], wavelet transform method, etc. Approaches in these methods have quite different theoretical underpinnings; therefore, it would be beyond the scope of this book to go into operational details. An excellent literature review on these methods can be found in Ref. [2].

The following analysis of the influence of baseline offset on the luminescence intensity ratio (*LIR*) temperature read-out (see Chapter 4) may be indicative of potential errors in measurements and interpretation of results. Let us consider the case of two separated emission peaks which have measured peak intensities of I_{M1} and I_{M2}, and baseline intensity values of I_{B1} and I_{B2}. The "true" *LIR* is:

$$LIR_T = \frac{I_{T2}}{I_{T1}}, \tag{5.7}$$

where I_{T1} and I_{T2} are "true" values of peak intensities. Let us simplify the case by considering that baseline intensity values do not change between measurement runs at different temperatures (i.e., there is no baseline drift), and that I_{T1} is the intensity of "an ideal" reference peak, meaning that it also does not change with temperature (I_{T1} = const.). Then, the measured *LIR* is:

$$\begin{aligned} LIR_M(T) &= \frac{I_{M2}(T)}{I_{M1}} = \frac{I_{T2}(T)+I_{B2}}{I_{T1}+I_{B1}} = \frac{\frac{I_{T2}(T)}{I_{T1}}+\frac{I_{B2}}{I_{T1}}}{1+\frac{I_{B1}}{I_{T1}}} \\ &= \frac{1}{1+\frac{I_{B1}}{I_{T1}}}\cdot LIR_T(T)+\frac{\frac{I_{B2}}{I_{T1}}}{1+\frac{I_{B1}}{I_{T1}}} \\ &= C_1 \cdot LIR_T(T)+\frac{C_2}{C_1}. \end{aligned} \tag{5.8}$$

where C_1 and C_2 are constants. Even for this simplified case one can note that baseline affects the measured *LIR* value, and also may affect figures of merit of thermometry (see Chapter 3) which are calculated from *LIR* versus temperature dependence. For example, the Boltzmann-type *LIR* (see Chapter 4) that is commonly used with lanthanide doped thermographic phosphors (see Chapter 6) theoretically predicts the "true" *LIR* in the following form:

$$LIR_T(T) = B \cdot \exp(-\Delta E / k_B T), \tag{5.9}$$

where B is a constant that depends on the radiative properties of the material, ΔE is the energy gap between emitting levels, and k_B is the Boltzmann constant

(Chapter 6). However, if baseline offset is not removed, the measured intensity ratio becomes:

$$LIR_M(T) = C_1 \cdot B \cdot \exp(-\Delta E / k_B T) + \frac{C_2}{C_1}. \quad (5.10)$$

It is not difficult to observe that the fitting of *LIR* would provide an erroneous *B* value, and that an additional constant term (C_2/C_1) is present. The additional constant term in *LIR* is frequently used to account for the effect of the peak overlap [9], so the presence of the baseline may mislead the interpretation of results.

5.3 DEALING WITH NOISE

Noise can be recognized as high-frequency fluctuations of the measured signal intensity around the level of the "true" signal intensity plus the baseline offset multiplied by instrument function. The amplitude of the fluctuation, i.e., the noise level, is different at different emission wavelengths and causes imprecision in intensity measurements. In luminescence thermometry, as in other measurement methods, the deterioration of measurement accuracy caused by noise is largest for low signal intensities which usually occur near measurement range boundaries. Signals equal to, or smaller than, the noise level are almost impossible to distinguish from the noise in quantitative measurements.

Assuming the properties of an investigated specimen remain constant, the primary sources of noise in a measurement are the measured optical signal itself, the detector, and the data acquisition electronics. The number of photons counted in a single measurement has an intrinsic statistical uncertainty called photon noise (also termed as shot noise or Poisson noise) since measurements of optical signals are in fact measurements of stochastic quantum events [10]. Photon noise cannot be eliminated or reduced. However, since the maximum variance in the number of counted photons is proportional to the square root of the total number of detected photons, by increasing the signal magnitude signal-to-noise ratio will be improved. The performance of a detector (and other electronic components) is affected by the thermal noise (also called dark noise) produced by the thermally facilitated stochastic generation of electrons along with the read noise introduced during processes of amplifying and quantifying the electronic signal. While the thermal noise can be reduced by cooling the detector and electronics, the read noise is an inherent property of the detector being used, and can be lowered only by technical and design improvement of that device. The main contribution to read noise comes from Johnson (white noise) and flicker noise ($1/f$ noise).

Noise cannot be simply subtracted from the measured spectrum or emission decay curve since it is not a constant. When necessary, the noise effect on the measurement can be reduced by modifying the measurement conditions and procedures, i.e., experimentally or postexperimentally (data processing) by smoothing or digitally filtering data. Modifications can be made to enhance the signal level (e.g., by collecting more photons) by increasing the excitation power and/or by more opening slits, since at higher signal levels the ratio of signal to photon noise is larger. Of course, the signal can be increased using instruments with more efficient optics and detectors. By lowering ambient temperature or by detector cooling the thermal noise level will be reduced. The signal-to-noise ratio can be improved by increasing the integration time of detection or by coadding several individual integrations. Both signal and noise enlarge with longer integration times, but noise less so than signal. Frequently, the noise is reduced by reading data repeatedly and then finding the mean of the signal (the so-called ensemble average method). In this way, random noise is reduced by the square root of the number of reads (Fig. 5.2).

As mentioned earlier, noise may be numerically suppressed by the method of data processing, called smoothing. Smoothing is, in fact, a form of digital low-pass filtering which blocks the high-frequency signal components to give an emphasis on the lower frequency ones (signal and baseline offset). Over time, a significant number of smoothing algorithms and procedures have been developed; the most popular ones are the moving average (also termed the running mean) and the Savitzky–Golay algorithms. The goal of these algorithms is to modify signal data values on account of the values of neighboring data in such way that high-frequency variations are eliminated, under the assumption that the "true" signal is smooth.

The simplest form of the moving average algorithm is the rectangular smooth (*rs*) (also called the unweighted sliding-average smooth); it simply substitutes each point in the signal with the average of that point and $(n - 1)$ neighboring points, where n is an odd integer called the smooth width. Neighboring data can be positioned after output data (forward smoothing), before (backward smoothing), or symmetrically around the output data. The last scheme is the most common, and for equally spaced data y_i can be written in the following form:

$$y_i^{rs} = \frac{1}{2n+1}\sum_{j=-n}^{n} y_{i+j}. \tag{5.11}$$

Forward smoothing is obtained by summation from $j = 0$ to $2n$, and backward smoothing by summation from $j = -2n$ to 0. However, only the symmetrical scheme does not alter the x-axis positions of spectral features (such as the spectral positions of peaks), and this explains why n is always an odd integer. Better noise suppression is achieved with a large smooth width since the noise reduction is proportional to the square root of n, like in the experimental ensemble average method. However, with large smooth widths, the smoothed spectrum or decay curve may be distorted. This problem is alleviated using the triangular smooth (*ts*), which preserves the area under the peaks and is implemented with a weighted smoothing function:

$$y_i^{ts} = \sum_{j=-n}^{n} w_j \cdot y_{i+j} \bigg/ \sum_{j=-n}^{n} w_j, \tag{5.12}$$

where w_j is the weight of the jth data in the smoothing window (Table 5.1).

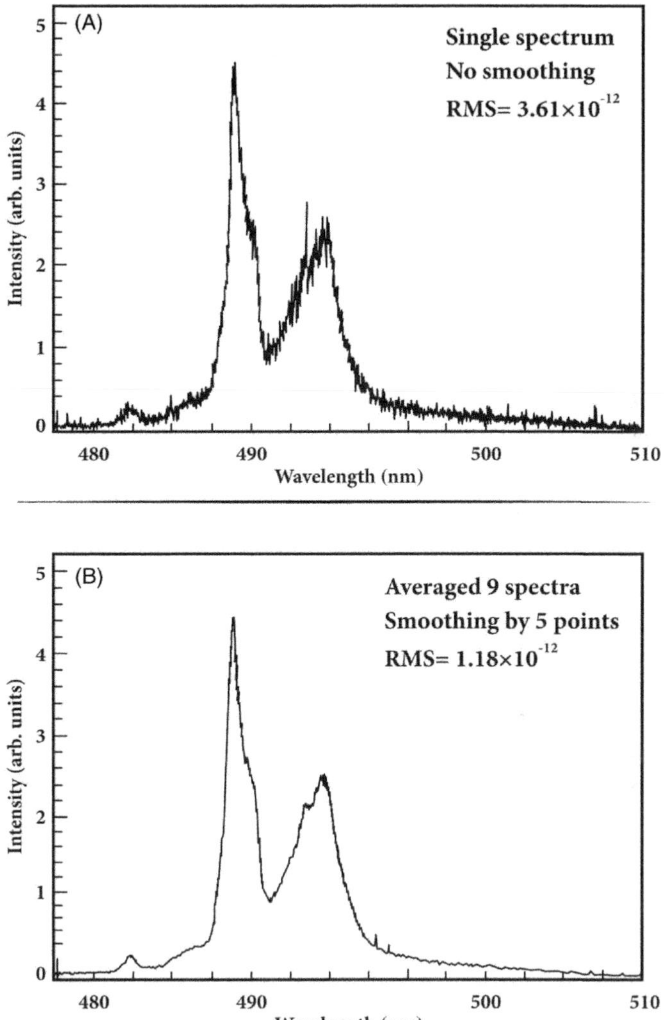

FIGURE 5.2 Emission spectra of Tb^{3+} doped Y_2O_2S excited with high-energy UV radiation: (A) spectrum obtained after one scan with no postmeasurement data smoothing, and (B) spectrum obtained as an average of 9 measurements after 5-point moving average smoothing. *RMS*, root mean square. *(Adapted from Yoshioka T, Ogawa M. Measurements of luminescence of phosphors. In: Yen WM, William M, Shionoya S, Yamamoto H, editors. Phosphor handbook. Boca Raton: CRC Press/ Taylor and Francis; 2007. p. 843–86, with permission from Taylor & Francis Group, LLC).*

One should be aware that moving average smoothing is quite sensitive to outliers and presents certain problems at both ends of the data. But more importantly, it alters the peak height and width even though spectral peak position and peak area are preserved. A Savitzky–Golay smooting (*sg*) [12] provides less noise reduction, but it does not change the true peak height or width. The

TABLE 5.1 Values of Normalized Weighting Coefficients for Triangular Smoothing for Smooth Widths of 3, 5, 7, 9, 11, and 13 points

Position	Weighting coefficients w_j for triangular smoothing: smoothing width/number of points n					
j	3	5	7	9	11	13
−6						1
−5					1	2
−4				1	2	3
−3			1	2	3	4
−2		1	2	3	4	5
−1	1	2	3	4	5	6
0	2	3	4	5	6	7
+1	1	2	3	4	5	6
+2		1	2	3	4	5
+3			1	2	3	4
+4				1	2	3
+5					1	2
+6						1
$\sum_j w_j$	4	9	16	25	36	49

idea of the method is to approximate the underlying function within the moving window not by a constant (whose estimate is the average), but by a polynomial of higher order which is the least-square fitted to data in the moving window. The procedure uses an equation of the same form as Eq. (5.12):

$$y_i^{sgs} = \sum_{j=-n}^{n} c_j \cdot y_{i+j} \bigg/ \sum_{j=-n}^{n} c_j, \tag{5.13}$$

but uses coefficients obtained from fitting c_j instead of normalized weighting coefficients (see Tables 5.2 and 5.3). In fact, there are particular sets of coefficients that inherently realize polynomial least-squares fitting inside a moving window. These sets differ for different polynomial orders and number of points, and depend also on the positioning of data in the moving window with respect to output point. Two sets are given in Table 5.2 (2nd and 3rd order polynomials) and Table 5.3 (4th and 5th order polynomials), both for the symmetrical data scheme. One should note that Savitzky–Golay coefficients can be easily calculated for the selected polynomial order and data frame. Also note that a

TABLE 5.2 Coefficients for Savitzky–Golay Smoothing Using Quadratic or Cubic Polynomial and Smooth Widths of 5, 7, 9, 11, and 13 Points

Position	Normalized weighting coefficients for triangular smoothing: smoothing width/number of points n				
j	5	7	9	11	13
−6					−11
−5				−36	0
−4			−21	9	9
−3		−2	14	44	16
−2	−3	3	39	69	21
−1	12	6	54	84	24
0	17	7	59	89	25
+1	12	6	54	84	24
+2	−3	3	39	69	21
+3		−2	14	44	16
+4			−21	9	9
+5				−36	0
+6					−11
$\sum_j c_j$	35	21	231	429	143

Savitzky–Golay method can be used for differentiation of data; sets of coefficients can be found in many textbooks as well as in Ref. [12].

Apart from these two popular smoothing algorithms, noise reduction may be achieved by many other algorithms/methods, such as exponential smoothing [13], smoothing spline [14], kernel smoothing [15], Fourier transfer [16], wavelets [17], etc.

5.4 DIFFERENTIATION AND INTEGRATION OF SPECTRAL DATA

Differentiation of spectral and time-resolved luminescence data is frequently needed for further data processing and analysis. It can be used for the evaluation of the spectral position of the peak maximum, for the detection of small peaks obscured by stronger ones, for artificial resolution enhancement, etc. In luminescence thermometry, differentiation of data is needed for calculations of temperature sensitivities and some other figures of merit of thermometry (Chapter 3), as will be demonstrated later in this chapter.

The simplest numerical differentiation uses the finite difference algorithms based on the forward, backward, and central difference forms. For the first

TABLE 5.3 Coefficients for Savitzky–Golay Smoothing Using Quartic and Quintic Polynomial and Smooth Widths of 7, 9, 11 and 13 Points

Position	Normalized weighting coefficients for triangular smoothing: smoothing width/number of points *n*			
j	**7**	**9**	**11**	**13**
−6				110
−5			18	−198
−4		15	−45	−160
−3	5	−55	−10	110
−2	−30	30	60	390
−1	75	135	120	600
0	131	179	143	677
+1	75	135	120	600
+2	−30	30	60	390
+3	5	−55	−10	110
+4		15	−45	−160
+5			18	−198
+6				110
$\sum_j c_j$	231	429	429	2431

derivative of I with respect to λ (assuming equidistant data) these forms respectively provide the following equations:

$$
\begin{aligned}
I'_j(\lambda) &= \frac{I_{j+1} - I_j}{\lambda_{j+1} - \lambda_j} = \frac{I_{j+1} - I_j}{\Delta\lambda}, \qquad (for 1 < j < N-1) \\
I'_j(\lambda) &= \frac{I_j - I_{j-1}}{\lambda_j - \lambda_{j-1}} = \frac{I_j - I_{j-1}}{\Delta\lambda}, \qquad (for 2 < j < N) \\
I'_j(\lambda) &= \frac{I_{j+1} - I_{j-1}}{\lambda_{j+1} - \lambda_{j-1}} = \frac{I_{j+1} - I_{j-1}}{2\Delta\lambda}, \qquad (for 2 < j < N-1)
\end{aligned}
\tag{5.14}
$$

where N is total number of data points. One should note that forward and backward differences reduce the number of points in the data set by one (at the beginning and end of the data set respectively), and central differences by two (one at each end of the data set). Second and higher derivative orders can be calculated by successive first derivate calculations. Alternatively, higher derivate orders can be also calculated with higher order finite differences:

$$
I'_j(\lambda) = \frac{I_{j+1} - I_{j-1}}{\lambda_{j+1} - \lambda_{j-1}} = \frac{\frac{1}{2} I_{j+1} - \frac{1}{2} I_{j-1}}{\Delta\lambda}, \quad (for 2 < j < N-1) \tag{5.15}
$$

$$I_j''(\lambda) = \frac{I_{j-1} - 2I_j + I_{j+1}}{\Delta\lambda^2}. \quad (for\ 2 < j < N-1) \tag{5.16}$$

One should observe that two consecutive first-order derivations reduce the number of points by four, while the direct use of Eq. (5.16) reduces it by two. The given equations contain coefficients derived for the equidistant wavelength spacing and the basic accuracy of differentiation. It is possible though to derive coefficients for any order of derivative approximated to any order of accuracy, and on an arbitrary grid in one dimension [18]. For the centered approximations on equidistant grids the coefficients c_j for second-, fourth-, and sixth-order accuracy (the higher the order the better the accuracy of approximation) are given in Table 5.4 [18] and the differentiation formula is:

$$I'^{m}(\lambda) = \sum_{j=-(q-1)}^{q-1} c_j \cdot I_j / (\Delta\lambda)^m, \tag{5.17}$$

where m is differentiation order and q is the order of accuracy.

A finite difference algorithm is sufficient for differentiation of luminescence thermometry data in the majority of cases. However, one should be aware of a large number of algorithms and differentiation procedures for

TABLE 5.4 Coefficients *cj* for the Derivation of the *m*th-Order Central Difference (Eq. (5.17)) for *q* = 2, 4, and 6 Orders of Approximation Accuracy (Equidistant Grid)

		j								
m	*q*	−4	−3	−2	−1	0	1	2	3	4
1	2				−1/2	0	1/2			
	4			1/12	−2/3	0	2/3	−1/12		
	6		−1/60	3/20	−3/4	0	3/4	−3/20	1/60	
2	2				1	−2	1			
	4			−1/12	4/3	−5/2	4/3	−1/12		
	6		1/90	−3/20	3/2	−49/18	3/2	−3/20	1/90	
3	2			−1/2	1	0	−1	1/2		
	4		1/8	−1	13/8	0	−13/8	1	−1/8	
	6	−7/240	3/10	−169/120	61/30	0	−61/30	169/120	−3/10	7/240
4	2			1	−4	6	−4	1		
	4		−1/6	2	−13/2	28/3	−13/2	2	−1/6	
	6	7/240	−2/5	169/60	−122/15	91/8	−122/15	169/60	−2/5	7/240

discrete data, some of which are much more advanced than the one described here. Their description can be found in many treatises on numerical analysis, for example, in Refs. [19–23]. What is important here is to understand that for the effective use of differentiation in luminescence thermometry, as in any quantitative application, it is imperative to use sufficient smoothing along with differentiation. This is because differentiation reduces the signal-to-noise ratio. As the rule of thumb, two ordinary smoothing (rectangular) runs should be used with a first derivative, and three with a second; it makes no difference if the smoothing is done before or after the derivation. Then, the signal-to-noise ratio of the derivative will be even better than that of the unsmoothed original. The Savitzky–Golay smoothing, discussed in the Section 5.3, can also be utilized for differentiation. The advantage of its use is a simultaneous process of differentiation and smoothing.

When processing luminescence spectral data, integration is used to measure the area under a curve plotted on a spectrum. The precise determination of the integrated peak areas is key to achieving quality ratiometric luminescence thermometry. The simplest way to calculate the area below an intensity curve is to use the left, right, or midpoint Riemann sum of intensities over the wavelength region of interest. Suppose this region consist of N data. Then the right Riemann sum provides the area below curve S from the following equation:

$$S^R = \Delta\lambda \cdot \sum_{j=2}^{N} I_J, \tag{5.18}$$

where $\Delta\lambda$ is the distance between data points (assuming equidistant data points). The error that accumulates over the entire calculation is of order $\Delta\lambda$. Integration may be performed using the trapezoidal rule:

$$S^T = \frac{\Delta\lambda}{2} \cdot \sum_{j=2}^{N} \left(I_{J-1} + I_J \right), \tag{5.19}$$

which has a smaller error than the Riemann sum (rectangular rule), of order $(\Delta\lambda)^2$. Further improvement in accuracy may be obtained using Milne's rule, Simpson's rule, and Boole's rule. Of course, many other integration methods are also available, for details see Refs. [19–23].

5.5 QUANTIFICATION OF THE FEATURES OF LUMINESCENCE SPECTRA

Before we start discussing the methods of quantification of spectral features, it is useful to mention basic conversion relations between different units (nanometers, wavenumbers, and electronvolts) in which one can express the emission intensity, emission bandwidth, wavelength, and energy.

1. From cm^{-1} or eV to nm:

$$\lambda[nm] = \frac{10^7}{\tilde{\upsilon}[cm^{-1}]} = \frac{1239.84}{E[eV]},$$
$$I(\lambda) \sim \tilde{\upsilon}^2 I(\tilde{\upsilon}) \sim E^2 I(E),$$
$$\Delta\lambda[nm] at \lambda_0 \cong \frac{10^7 \cdot \Delta\tilde{\upsilon}[cm^{-1}]}{\tilde{\upsilon}_0^2} = \frac{1239.84 \cdot \Delta E[eV]}{E_0^2} \tag{5.20}$$

For example, a peak with a maximum at 20,000 cm^{-1} having a bandwidth of 20 cm^{-1} is a peak at 500 nm with a bandwidth of 0.5 nm.

2. From eV or nm to cm^{-1}:

$$\tilde{\upsilon}[cm^{-1}] = \frac{10^4 \cdot E[eV]}{1.23984} = \frac{10^7}{\lambda[nm]},$$
$$I(\tilde{\upsilon}) \sim I(E) \sim \lambda^2 I(\lambda),$$
$$\Delta\tilde{\upsilon}[cm^{-1}] at\ \tilde{\upsilon}_0 = \frac{10^4 \cdot \Delta E[eV]}{1.23984} \cong \frac{10^7 \cdot \Delta\lambda[eV]}{\lambda_0^2}. \tag{5.21}$$

3. From cm^{-1} or nm to eV:

$$E[eV] = 1.23984 \cdot 10^{-4} \tilde{\upsilon}[cm^{-1}] = \frac{1239.84}{\lambda[nm]},$$
$$I(E) \sim I(\tilde{\upsilon}) \sim \lambda^2 I(\lambda),$$
$$\Delta E[eV] at E_0 = 1.23984 \cdot 10^{-4} \Delta\tilde{\upsilon}[cm^{-1}] \cong \frac{1239.84 \cdot \Delta\lambda[eV]}{\lambda_0^2}. \tag{5.22}$$

One should note that conversion from wavelength to energy (cm^{-1} or eV) changes spectra so that they look mirror-like, but more "dense" at high-wavelength spectral regions (Fig. 5.3). In addition, peak positions and peak shapes may slightly change since conversion of intensity includes multiplication of intensity by squared wavelength which increases intensity values differently over the spectral range.

Peak spectral positions can be numerically evaluated from I–λ data sets by simple comparison of the I value at each point with values on both sides of that point (the λ value at which value of intensity is larger than values in adjacent points is the position of a peak maximum). However, data should be smoothed before processing to avoid false-peak detection which may result from the noise. Also, it is common practice to introduce the amplitude threshold value, i.e., the minimal value of intensity for which the wavelength, having higher intensity than adjacent ones, could be considered as a peak maximum. A better approach

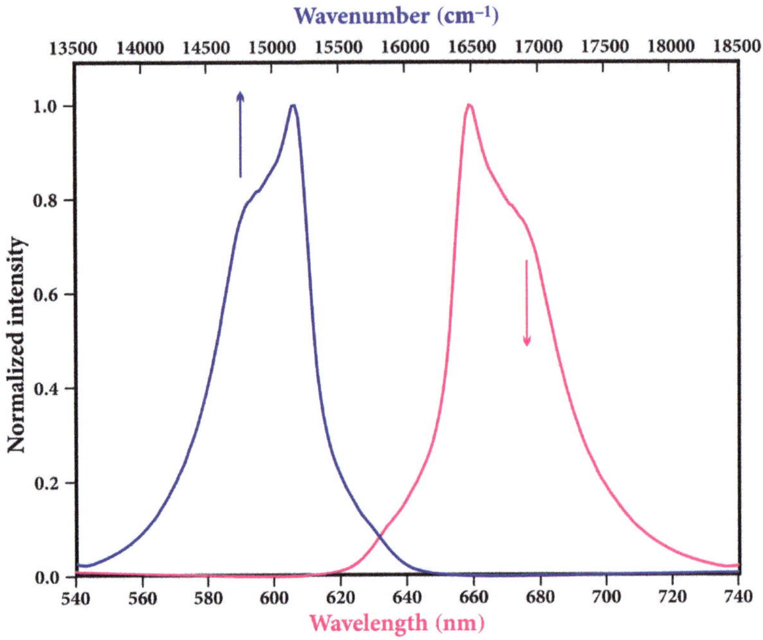

FIGURE 5.3 **Intensity of Mn^{4+} doped Mg_2TiO_4 as a function of wavelength (bottom axes) and as a function of wavenumber (upper axes).**

is to use the first derivative of data and to find the so-called downward going, zero-crossing points, i.e., wavelengths at which the first derivative of intensity changes sign from positive to negative (which is equivalent to intensity reaches the maximum) (Fig. 5.4).

Extensive smoothing of the first derivate is needed to minimize false-peak detection due to zero-crossing induced by noise. After smoothing, the algorithm should be set to detect only those zero-crossings whose slope exceeds the so-called slope threshold (a certain predetermined minimum slope value) and amplitude threshold. Then, *intensities of peaks* can be taken from data that correspond to zero-crossing wavelengths after baseline removal. The method could be further extended with *the least squares fitting* of a section of the original unsmoothed spectrum in the small region around the zero-crossing wavelength. With this extension, it would be possible *to derive peak position, intensity, and peak width in one procedure* after baseline removal. For ease of fitting, the peak can be approximated by a parabola around its maximum (I_{max}, λ_p), because the Taylor's series expansion of the majority of peak functional forms, for example, Gaussian and Lorentzian:

$$I^{Gauss}(\lambda) = I_{max} \cdot \exp\left[-c \cdot (\lambda - \lambda_p)^2\right],$$
$$I^{Lorntz}(\lambda) = I_{max} \cdot \frac{1/c}{1/c + (\lambda - \lambda_p)^2}, \tag{5.23}$$

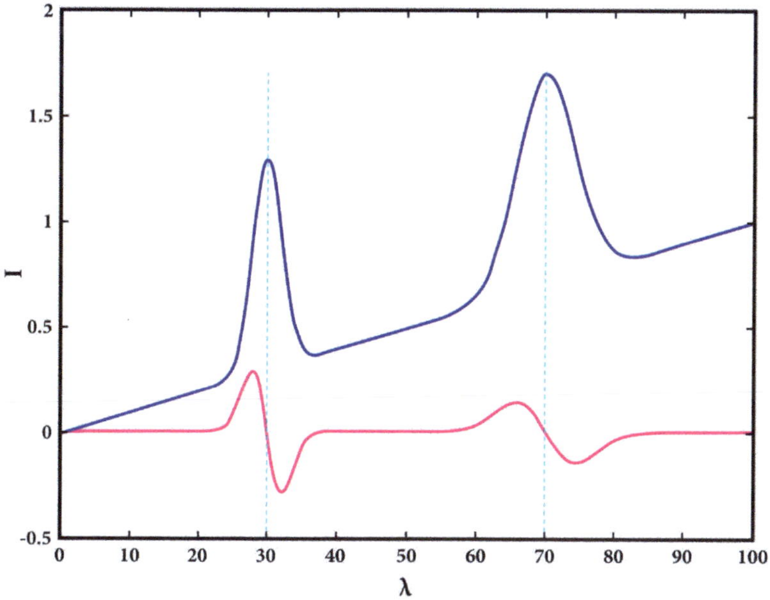

FIGURE 5.4 Zero-crossings of the first derivative of a spectrum [bottom curve in grey *(red in the web version)*] corresponds to peaks maxima positions [upper curve in black *(blue in the web version)*]. *(Adapted from O'Haver TC. A pragmatic introduction to signal processing: with applications in scientific measurement. CreateSpace Independent Publishing Platform; 2016. p. 146).*

respectively, is of the same parabolic form:

$$I(\lambda)\Big|_{\lambda \text{around } \lambda_p} = \mathrm{I}_{max} \cdot \left[1 - c \cdot \left(\lambda - \lambda_p\right)^2\right], \tag{5.24}$$

where c is a constant that determines the peak shape attributes. Peak width is specified for a certain portion of the peak maximum intensity; usually it is specified as half the intensity and termed peak width at half height, $w_{1/2}$. With known peak position and maximal intensity, the peak width at any height can be found as the difference of wavelengths for which intensity reaches a specified portion of the maximal intensity at both sides of the peak position. One can use the same comparison procedure as the one used for the determination of the peak position. The peak area S can be approximately calculated once the peak height (peak intensity, I_{max}) and width are derived:

$$S \doteq \mathrm{I}_{\max} \times w_{1/2} \text{or } S \doteq \mathrm{I}_{\max} \times \frac{w_{0.85} + w_{0.15}}{2}, \tag{5.25}$$

where $w_{0.85}$ and $w_{0.15}$ are peak widths calculated for the wavelengths at which the signal reaches 85% and 15% of the maximal peak intensity, respectively. The last equation shows that peak width at half height may be deduced from the previously determined peak area and height as: $w_{1/2} \cong S / I_{max}$. There are several methods for graphical integration of peak area, such as the so-called perpendicular drop, triangulation, and tangent skim; these are particularly useful for deducing areas of overlapping peaks. Numerically, the spectral parameters of overlapping peaks can be determined using fractional derivatives [25] and fitting. Generally speaking, the most convenient way for peak area calculation is numerical integration for resolved peaks, and peak fitting for overlapping peaks.

The *Emission barycenter* represents an average energy of transition and is not necessarily equal to the emission peak position. It can be derived from the emission spectrum as a "center of mass" of the peak (band) that corresponds to the transition of interest. Since it is related to energy, the spectrum should be represented as intensity of emission versus energy (or wavenumbers, which is equal) and, then, barycenter wavenumber ($\tilde{\upsilon}_{BC}$) or energy (E_{BC}) can be calculated using the following integration over the peak spectral region:

$$\tilde{\upsilon}_{BC} = \frac{\int_{\tilde{\upsilon}_1}^{\tilde{\upsilon}_2} \tilde{\upsilon} \cdot I(\tilde{\upsilon}) d\tilde{\upsilon}}{\int_{\tilde{\upsilon}_1}^{\tilde{\upsilon}_2} I(\tilde{\upsilon}) d\tilde{\upsilon}}, \quad E_{BC} = \frac{\int_{E_1}^{E_2} E \cdot I(E) dE}{\int_{E_1}^{E_2} I(E) dE}. \tag{5.26}$$

For this purpose, some of the earlier mention numerical integration techniques can be employed.

In some cases peaks are poorly resolved, so it is difficult, or even impossible, for one to precisely determine their positions in a spectrum. If peaks are artificially narrowed their overlap will be reduced or eliminated. This can be achieved with *numerical resolution enhancement* [26,27] also called *peak sharpening*. The most convenient method is generating a new spectrum (I^{enh}) as the weighted sum of the original spectrum and it's second and fourth derivatives:

$$I^{enh}(\lambda) = I(\lambda) - w_1 \frac{d^2 I(\lambda)}{d\lambda^2} + w_2 \frac{d^4 I(\lambda)}{d\lambda^4}, \tag{5.27}$$

where w_1 and w_2 are weighting factors which are arbitrarily chosen but the optimum values of which depend on the width, shape, and digitization interval of the spectrum. The spectrum in Fig. 5.5A shows a broad band with several shoulders on the right side of the band maximum which indicate small peaks superimposed on a broader one. Obviously, it is difficult to determine the exact position of these small peaks. Spectral enhancement using only the original spectrum subtracted with the weighted second derivate ($w_2 = 0$; Eq. (5.27)) produces a "sharpened" spectrum given in Fig. 5.5B. Now, the small peaks are satisfactorily resolved for the evaluation of their spectral position.

One can vary weighting factors in several iterations and observe the changes in the "sharpened" spectrum which will be reflected in resolution, signal-to-noise ratio, and baseline flatness. However, it is reasonable to start with w_1 = peak

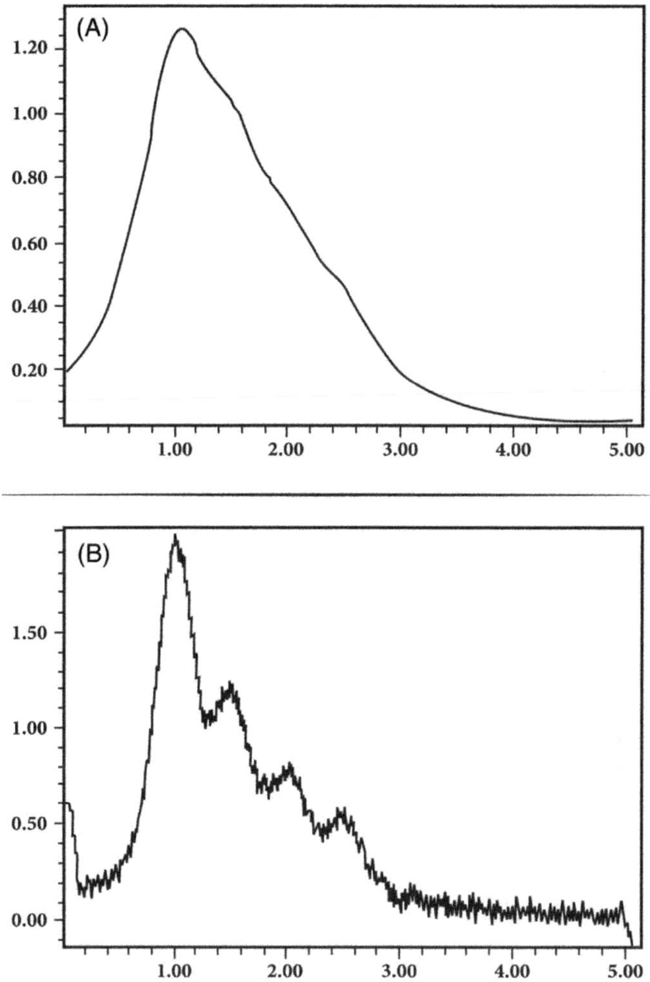

FIGURE 5.5 Numerical spectral enhancement ("peak sharpening"): (A) the original spectrum and (B) the "sharpened spectrum" obtained by subtraction of the weighted second derivate from the original spectrum. *(Adapted from O'Haver TC. A pragmatic introduction to signal processing: with applications in scientific measurement. CreateSpace Independent Publishing Platform; 2016. p. 146).*

width^2/25 and w_2 = peak width^4/800 for Gaussian shape peaks, and w_1 = peak width^2/6 and w_2 = peak width^4/700 for Lorentzian shape peaks (peak width is the full-width at half maximum of the peaks, expressed in number of data points) [24].

An interesting property of this procedure is that it does not change the total peak area and peak position. However, it changes peak intensities. Note that the numerical resolution enhancement deteriorates signal-to-noise ratio, simply because it adds derivative components which, as we discussed earlier, have a lower signal-to-noise ratio than the original signal.

5.6 EVALUATION OF LIFETIMES FROM TIME-RESOLVED MEASUREMENTS

As discussed throughout this book, the value of the excited state lifetime is a strong indication of temperature. It is exploited for temperature measurements with probes prepared from almost all types of materials whose lifetimes span from below nanoseconds to several milliseconds. The time domain and frequency domain (also called phase modulated) methods are standard techniques for measurements of time-resolved luminescence (see Chapter 3), and both provide data from which excited state lifetimes can be derived. Time domain measurements provide emission intensity (or a complete spectrum), convoluted by the instrument function, as a function of time measured after the excitation pulse. The phase-modulated technique uses modulated light at variable frequency and gives the harmonic response of a sample. The harmonic response is the Fourier transform of the impulse response. Thus, these methods provide theoretically equal data. As in other types of measurements, these data consist not only of a signal of interest but also of noise and baseline (most frequently DC offset). Thus, data should be first preprocessed by methods of baseline and noise removal, discussed earlier in this chapter. A more complex issue is the evaluation of emission decay parameters from data since these parameters are mathematically correlated, i.e., the change of the value of one parameter can be compensated by the change of another, to afford the same mathematical response. This problem is an inherent feature of analyses, which cannot be avoided, so one needs to be careful in data processing and the interpretation of results. To make things more complicated, emission decays may arise from more than one emission center or may have complex profiles. The lifetime τ is the inverse value of the sum of the rates which depopulate the excited state (see Chapter 3). If we assume a conventional single exponential decay of an emission intensity ($I(t)$; see full magenta line in Fig. 5.6A):

$$I(t) = I_0 \cdot \exp(-t/\tau), \tag{5.28}$$

where I_0 is the emission intensity at $t = 0$, then τ is the time for which the emission intensity decreases to 1/e of its initial value (I_0). The lifetime can be resolved from the slope of a plot of log $I(t)$ versus t (see blue dotted line in Fig. 5.6A) or by fitting the data to Eq. (5.28). The fastest method to estimate the lifetime value of a single exponential emission decay is to measure emission intensities at two different time intervals and to calculate the value as:

$$\tau = \frac{t_2 - t_1}{\log(I_1/I_2)}, \tag{5.29}$$

where I_2 and I_1 are emission intensities measured at t_2 and t_1, respectively. However, if the emission decay is not a single exponential the evaluation of lifetime by these procedures will provide a false result.

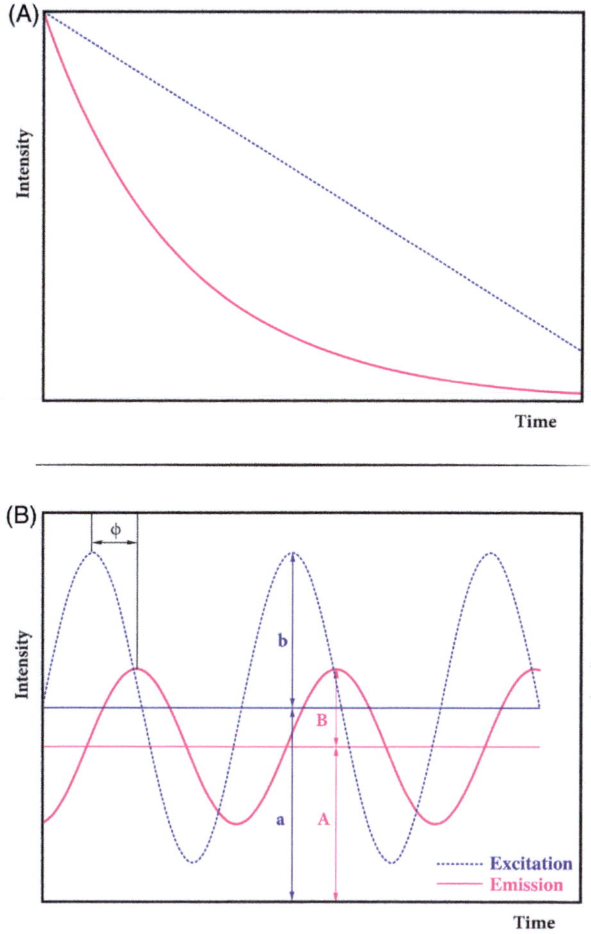

FIGURE 5.6 Profiles of emission decay data used in (A) the time domain method and (B) the phase modulated method.

Emission decay data utilized in a phase modulated method are presented in Fig. 5.6B. The modulated excitation radiation (blue dotted line in Fig. 5.6B) causes a modulated emission (full magenta line in Figure 5.6B) of the same frequency f, but with a phase delay called the phase angle (ϕ). The phase delay can be derived from the positions of peak maximums or from the zero-crossing times of the excitation and emission signals by some of the numerical procedures previously described in this chapter. Then, the lifetime value can be obtained using the following equation:

$$\tau = \frac{1}{2\pi f}\tan(\phi). \tag{5.30}$$

Alternatively, or additionally, the lifetime value can be evaluated from the modulation factor:

$$m = \frac{B / A}{b / a}, \tag{5.31}$$

where A and a are the average intensities, and B and b are the amplitudes of the harmonic parts of the emission and excitation signal, respectively. Once modulation factor m is derived from the data, the lifetime can be calculated as:

$$\tau = \frac{1}{2\pi f}\left[\frac{1}{m^2} - 1\right]. \tag{5.32}$$

Both Eqs. (5.31) and (5.32) should, in principle, give the same result. It is common practice to derive these values for different modulation frequencies and then to calculate their average or perform data fit over the frequency region. As in the case of a time domain method, when emission decay is not a single exponential the procedure will yield an erroneous lifetime value.

Analysis of complex intensity decay forms, such as multiexponential forms, are considerably more difficult. With double exponential forms, one may try to apply the so-called component stripping method, if lifetimes of components are expected to differ considerably. This approach relies on fitting one component at longer times of emission decay and then fitting a complete decay profile while keeping parameters of this component fixed. However, one should be very cautious since fitting parameters are strongly correlated, and different parameter combinations may provide the same decay profile. An example of this situation is considered in Refs. [28,29], with two double exponential decays:

$$\begin{aligned} I_1(t) &= 7500 \cdot \exp(-t / 5.5) + 2500 \cdot \exp(-t / 8.0), \\ I_2(t) &= 2500 \cdot \exp(-t / 4.5) + 4500 \cdot \exp(-t / 6.7), \end{aligned} \tag{5.33}$$

which apparently look entirely different, but yield almost identical emission decay curves, as shown in Fig. 5.7. The plot of intensity decays on a linear scale shows that they are indistinguishable at all times, while on a logarithmic scale one notices some minor differences at 30–50 ns [29].

For data obtained from time domain techniques and phase modulated techniques, the most common approach for the determination of lifetimes is nonlinear least squares data fitting on some common intensity decay law (Table 5.5). The quality of fit is judged from an error analysis, but also from a comparison of calculated values with expected values considering the underling physical and chemical processes in the investigated sample. The absence of such analysis may lead to a fallacy in the interpretation of luminescence decay parameters [30].

The selection of a particular intensity decay law comes from the specifics of the investigated material. The simple single exponential decay is used when

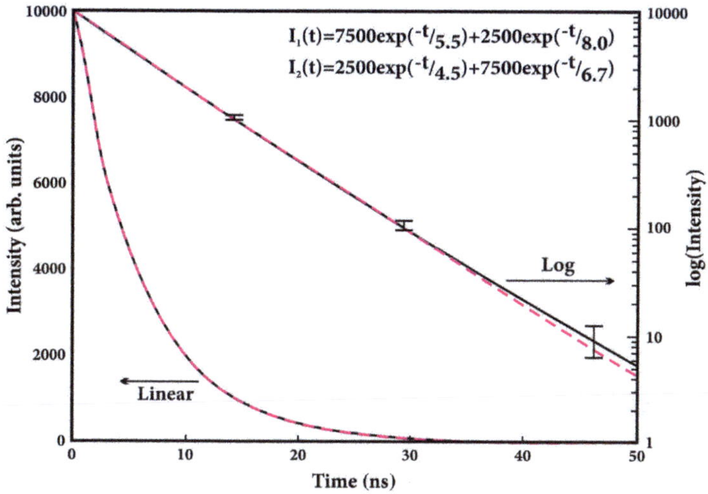

FIGURE 5.7 Comparison of two intensity decays given in Eq. (5.33): on a linear scale (left) and logarithmic scale (right). The *error bars* represent Poisson noise *(Adapted from Lakowicz JR. Time-domain lifetime measurements. In: Lakowicz JR, editor. Principles of fluorescence spectroscopy. Boston, MA: Springer US; 1999. p. 95–140, with permission from Springer).*

TABLE 5.5 Mathematical Expressions for Intensity Decay Laws Commonly Used for the Derivation of Excited State Lifetimes

Name	Expression, $I(t)$
Single exponential	$A \cdot \exp(-t/\tau)$
Multiexponential	$\sum_{i=1}^{N} A_i \cdot \exp(-t/\tau_i)$
Transient	$A \cdot \exp\left(-t/\tau - 2Bt^{1/2}\right)$
Inokuti–Hirayama	$A \cdot \exp\left(-t/\tau - \alpha \cdot (t/\tau)^{3/Q}\right)$
Stretched exponential	$A \cdot \exp(-t/\tau)^{\beta}$
Lifetime distributions	$\int_{\tau=0}^{\infty} A(\tau) \cdot \exp(-t/\tau) d\tau$

material presents a single emission center residing in a homogeneous and static environment. Even in such a case, some deviations from a single exponential law may occur, for example, from interactions between emission centers. Multiexponential decay law is most commonly used with a mixture of emission centers or with one emission center residing in different environments; in both situations, the one lifetime value τ_i and the one fractional intensity A_i are assigned to each emission center, or emission center in each environment, respectively.

Transient intensity decay law is used to model emission decays influenced by collisional quenching. The B parameter may be considered as an estimate of the quencher concentration and diffusion coefficient. It can also be used to model the effects of the resonance energy transfer; in this case, the transient law expression may have different powers of time in the exponent (1/6, 1/3, or 2). The Inokuti–Hirayama model [31] considers the electric multipolar interactions in the temporal dependence of the emission of lanthanide doped phosphors. In this model α is a phenomenological parameter containing the energy transfer probability, and Q is an indication value of the electric multipolar interaction: dipole–dipole ($Q = 6$), dipole–quadrupole ($Q = 8$), and quadrupole–quadrupole ($Q = 10$). In polymer studies, a stretched exponential functional form may be used when one expects a distribution of relaxation times. The lifetime distribution is used in situations when a finite number of lifetimes is not expected, but instead, their continuous distribution $A(\tau)$ is more reasonable, such as for an emission center in a mixture of solvents, so that a range of environments exists [29]. Typically, Gaussian or Lorentzian shape lifetime distributions are used [32].

However, one should note that the excitation pulse is not instantaneous concerning the emission decay in the general case. Therefore, a time-resolved luminescence response from a sample is the convolution of the true excited state decay and the excitation source intensity function. Deconvolution is therefore needed to be included in the process of lifetime evaluation. Many methods have been proposed for such a purpose with different success—mainly depending on the quality of the experimental data and assumed intensity decay function. These techniques include: the nonlinear least squares, the maximum entropy method, Prony's method, the sine transforms, phase–plane methods, and the method-of-moments; their performances are thoroughly examined in Refs. [33,34].

Finally, one should be aware of the difficulties in evaluating excited state lifetimes. What is of importance for luminescence thermometry is that the same procedure and assumed intensity decay law are used for derivation of lifetimes over the complete temperature range. As discussed in Chapter 4, temperature sensing from lifetimes is based on the change of lifetime value with temperature, so the potential error in lifetime determination may be considered as a systematic error which does not have to seriously compromise thermometry. Then, it is convenient to present lifetime temperature dependence as a ratio of the dependence and lifetime value at a certain temperature, usually one from the beginning of the measurement range.

5.7 CALCULATING THE PERFORMANCE OF A LUMINESCENCE THERMOMETER

As for other types of thermometers, the performance of luminescence thermometers can be judged from their figures of merit. The general consideration of the figures of merit in thermometry are presented in Chapter 3 along with specificities related to luminescence thermometry. The figures of merit of luminescence thermometry are also detailed in recent review articles [9, 35]. Of importance here, is their derivation from luminescence thermometry data.

The most frequently reported figures of merit in the literature on luminescence thermometry are sensitivity (S) and relative sensitivity (S_R). Sensitivity is defined as a quotient of the change in an indication (Q), for example, peak position or lifetime, and the change in the temperature value. Relative sensitivity is the ratio of sensitivity and indication values. They can be calculated from the following equations:

$$\begin{aligned} S &= \left|\frac{dQ}{dT}\right|, \\ S_R &= \frac{S}{Q} = \frac{1}{Q}\left|\frac{dQ}{dT}\right|. \end{aligned} \tag{5.34}$$

Two approaches are used for the calculation. In the first, the functional dependence of Q on T is known. It may be linear, Boltzmann-type, or polynomial, for example. The data are fitted to the selected functional form, and the obtained parameters are used for the analytical calculation of the first derivate to provide continuous functions of sensitivity and relative sensitivity over the temperature region of interest. With this approach, it is possible to extrapolate values of sensitivities beyond the temperature range limits. The second method is based on numerical differentiation of the experimental data. In this procedure, as discussed earlier, heavy data smoothing is necessary. Numerical differentiations of unsmoothed data may give unrealistic values of sensitivities. Two issues ought to always be considered: that the temperature difference between data points is not large, and that data may be lost at both edges of the temperature interval if the basic numerical integration algorithms are used. Commonly, it is not advisable to use sensitivity data obtained in this way for extrapolation beyond the measurement range. Temperature resolution, or uncertainty in temperature (ΔT_{min}), is the smallest change in temperature that causes a perceptible change in the measured indication. At this point, let us analyze the simplest case in which temperature uncertainty arises only from an uncertainty in indication, ΔQ. The functional dependence can be calculated from the Taylor's series expansion of T versus Q:

$$\Delta T_{min} = \frac{dT}{dQ}\Delta Q + \frac{1}{2!}\frac{1d^2T}{dQ^2}\Delta Q^2 + \cdots + \frac{1}{n!}\frac{d^nT}{dQ^n}\Delta Q^n. \tag{5.35}$$

As in the previous elaboration, the knowledge of an analytical expression of the functional dependence of T versus Q considerably eases the job, since one can first fit experimental data to this dependence, and then perform analytical calculations of derivate, and finally uncertainty in temperature. Numerical procedures for higher order differentiations inevitably increase the noise, and may lead to erroneous conclusions in the analysis. Fortunately, in many situations, it is acceptable to use only the first term of Taylor's expansion [36]. If uncertainty in the indication is given in estimated standard deviation of measurements s, then the simplified form of the expression for the temperature resolution is:

$$\Delta T_{min} = \frac{s}{S}. \tag{5.36}$$

In addition to differentiation, for the calculation of the temperature resolution, a statistical calculation of the standard deviation is needed. Calculations of other figures of merit are also based on the differentiations and basic statistical calculations; therefore, they will not be exemplified.

REFERENCES

[1] Gan F, Ruan G, Mo J. Baseline correction by improved iterative polynomial fitting with automatic threshold. Chemometr Intell Lab Syst 2006;82(1–2):59–65.

[2] Schulze G, Jirasek A, Yu MML, Lim A, Turner RFB, Blades MW. Investigation of selected baseline removal techniques as candidates for automated implementation. Appl Spectrosc 2005;59(5):545–74.

[3] Phillips AJ, Hamilton PA. Improved detection limits in fourier transform spectroscopy from a maximum entropy approach to baseline estimation. Anal Chem 1996;68(22):4020–5.

[4] Ruckstuhl AF, Jacobson MP, Field RW, Dodd JA. Baseline subtraction using robust local regression estimation. J Quant Spectrosc Radiat Transfer 2001;68(2):179–93.

[5] Friedrichs MS. A model-free algorithm for the removal of baseline artifacts. J Biomol NMR 1995;5(2):147–53.

[6] Mosier-Boss PA, Lieberman SH, Newbery R. Fluorescence rejection in raman spectroscopy by shifted-spectra, edge detection, and fft filtering techniques. Appl Spectrosc 1995;49(5):630–8.

[7] Grosswendt B. Automatische analyse von γ-spektren hoher auflösung. Nucl Instr Methods 1971;93(3):461–72.

[8] Atakan AK, Blass WE, Jennings DE. Elimination of baseline variations from a recorded spectrum by ultra-low frequency filtering. Appl Spectrosc 1980;34(3):369–72.

[9] Brites CDS, Millán A, Carlos LD. Lanthanides in luminescent thermometry. In: Bünzli J-C, Pecharsky VK, editors. Handbook on the physics and chemistry of rare earths. Elsevier; 2016. p. 339–427. (Chapter 281).

[10] Waters JC. Accuracy and precision in quantitative fluorescence microscopy. J Cell Biol 2009;185(7):1135–48.

[11] Yoshioka T, Ogawa M. Measurements of luminescence of phosphors. In: Yen WM, William M, Shionoya S, Yamamoto H, editors. Phosphor handbook. Boca Raton: CRC Press/Taylor and Francis; 2007. p. 843–86.

[12] Savitzky A, Golay MJE. Smoothing and differentiation of data by simplified least squares procedures. Anal Chem 1964;36(8):1627–39.
[13] Gardner ES. Exponential smoothing: the state of the art. J Forecast 1985;4(1):1–28.
[14] Reinsch CH. Smoothing by spline functions. Numer Math 1967;10(3):177–83.
[15] Wand MP, Jones MC. Kernel smoothing. Boca Raton: Chapman and Hall/CRC; 1995. p. 224.
[16] Angelini C, Canditiis DDe. Fourier frequency adaptive regularization for smoothing data. J Comput Appl Math 2000;115(1–2):35–50.
[17] Mallat S, Hwang WL. Singularity detection and processing with wavelets. IEEE Trans Inf Theory 1992;38(2):617–43.
[18] Fornberg B. Generation of finite difference formulas on arbitrarily spaced grids. Math Comput 1988;51(184):699–1699.
[19] Hamming RW. Numerical methods for scientists and engineers. 2nd ed. New York, NY, USA: Dover Publications, Inc.; 1986.
[20] Burden RL, Faires JD. Numerical analysis. Brooks/Cole Publishing Company; 1997. (Mathematics Series).
[21] Quarteroni A, Saleri F, Gervasio P. Numerical differentiation and integration. In: Quarteroni A, Saleri F, Gervasio P, editors. Scientific computing with matlab and octave. Berlin, Heidelberg: Springer Berlin Heidelberg; 2010. p. 107–28.
[22] Teodorescu P, Stănescu N-D, Pandrea N. Numerical differentiation and integration. Numerical analysis with applications in mechanics and engineering. John Wiley & Sons, Inc.; 2013. pp. 377-449.
[23] Wiki contributors W. Introduction to numerical methods/numerical differentiation. 2015. Available from: https://en.wikibooks.org/w/index.php?title(Introduction_to_Numerical_Methods/Numerical_Differentiation&oldid=2754046.
[24] O'Haver TC. A pragmatic introduction to signal processing: with applications in scientific measurement. CreateSpace Independent Publishing Platform; 2016. p. 146.
[25] Kharintsev SS, Salakhov MK. A simple method to extract spectral parameters using fractional derivative spectrometry. Spectrochim Acta Part A Mol Biomol Spectrosc 2004;60(8–9):2125–33.
[26] Anderssen RS, Hegland M. Derivative spectroscopy: an enhanced role for numerical differentiation. J Integr Equations Appl 2010;22(3):355–67.
[27] Hegland M, Anderssen RS. Resolution enhancement of spectra using differentiation. Inverse Probl 2005;21(3):915.
[28] Grinvald A, Steinberg IZ. On the analysis of fluorescence decay kinetics by the method of least-squares. Anal Biochem 1974;59(2):583–98.
[29] Lakowicz JR. Time-domain lifetime measurements. In: Lakowicz JR, editor. Principles of fluorescence spectroscopy. Boston, MA: Springer US; 1999. p. 95–140.
[30] James DR, Ware WR. A fallacy in the interpretation of fluorescence decay parameters. Chem Phys Lett 120(4–5), 1985, 455–9.
[31] Inokuti M, Hirayama F. Influence of energy transfer by the exchange mechanism on donor luminescence. J Chem Phys 1965;43(6):1978–89.
[32] Lakowicz JR, Cherek H, Gryczynski I, Joshi N, Johnson ML. Analysis of fluorescence decay kinetics measured in the frequency domain using distributions of decay times. Biophys Chem 1987;28(1):35–50.
[33] Irvin JA, Quickenden TI, Sangster DF. Criterion of goodness of fit for deconvolution calculations. Rev Sci Instrum 1981;52(2):191–4.

[34] O'Connor DV, Ware WR, Andre JC. Deconvolution of fluorescence decay curves. a critical comparison of techniques. J Phys Chem 1979;83(10):1333–43.
[35] Dramićanin MD. Sensing temperature via downshifting emissions of lanthanide-doped metal oxides and salts. A review. Methods Appl Fluoresc 2016;4(4):42001. 23 pp.
[36] Baker SN, McCleskey TM, Baker GA. An ionic liquid-based optical thermometer. In: Rogers RD, Seddon KR, editors. Ionic liquids iiib: fundamentals, progress, challenges, and opportunities. American Chemical Society; 2005. p. 14–171. (ACS Symposium Series; vol. 902).

Chapter 6

Lanthanide and Transition Metal Ion Doped Materials for Luminescence Temperature Sensing

6.1 INTRODUCTION

Luminescence of lanthanide and transition metal ions incorporated into insulating and wide band-gap inorganic materials has found numerous applications. It is indispensably used in lighting (fluorescent lamps and light-emitting diodes (LEDs)), displays (electroluminescent, plasma, field-emission, and cathode-ray tubes), solid-state lasers (such as ruby laser and Nd:YAG), detectors (such as scintillators and X-ray screens), persistent luminescence paints, and for the prevention of counterfeiting. With the development of nanoscience and nanotechnology, new application possibilities have emerged, especially in bioimaging, photovoltaics, and physical and chemical sensing. Such a large variety of valuable uses is a consequence of the fascinating features of the luminescence of rare earth and transition metal ions. Their emissions cover much of the ultraviolet-visible-near-infrared spectral region, have strong intensities, and usually long-lasting temporal behavior.

Because of the characteristics mentioned above, availability of well-developed synthesis procedures, and excellent chemical and temperature stabilities, these materials were among the first tested for temperature sensing by luminescence. They are also the class of materials which up to now has received, by far, the greatest attention in luminescence thermometry, and are utilized in almost all forms: micro and nanopowders, crystals, composites, glasses, glass-ceramics, thin films, and coatings. All temperature read-out schemes described in Chapter 4, have been successfully employed with them. Lanthanide and transition metal ion doped materials offer the largest temperature ranges of measurements and the best repeatability among all materials, as well as showing good sensitivities.

This chapter describes the physical phenomena that govern luminescence of lanthanide and transition metal ion doped materials, and highlights their characteristics which are relevant for luminescence thermometry. Common routes for preparations of single crystals, powders, thin films, and coatings of

Luminescence Thermometry. http://dx.doi.org/10.1016/B978-0-08-102029-6.00006-3

these materials are concisely presented in Section 6.4. This chapter concludes with an overview of their use in downshifting and upconversion excitation-emission schemes.

6.2 CHARACTERISTICS OF LANTHANIDE AND TRANSITION METAL ION LUMINESCENCE CENTERS

The presence of rare earth and transition metal ions as optically active centers in insulators and wide band-gap semiconductors give rise to many interesting optical features. As shown in Chapter 3, these centers consist of dopant ions introduced into the crystal and their first neighboring ions (ligands). Both types of centers have energy levels within the energy gap of the host material between which optical transitions may appear. However, rare earth and transition metal ions considerably differ in their electronic configurations and interactions with ligands, so spectroscopic features of their centers also notably differ. The basic description of the spectroscopic properties of these centers, along with the description of physical interactions that are responsible for them, is presented in this section.

The transition metals of the iron group are elements from the fourth period of the periodic table (from ^{22}Ti to ^{30}Zn). When introduced into solids, the outer 4s electrons are striped from the elements to form bonds while some 3d electrons are also lost or gained. In this way "so-called" 3d transition ion centers are created in solids having the $1s^22s^22p^63s^23p^63d^n$ ($0 < n < 10$) electronic configuration. Incompletely filled d shells provide many energy levels between which optical transitions may occur. The crystal field potential of surrounding ions strongly interacts with 3d electrons since they are positioned outside the iron core and have relatively large orbitals. Therefore, spectroscopic properties of transition metal ion optical centers, such as intensities of absorption and emission bands, their spectral positions and widths (they can be both narrow and broad), are principally a consequence of the d^n electron configuration of an ion and the crystal field potential of ligands. Accordingly, the groups of ions of the same $3d^n$ electron configuration in Table 6.1 demonstrate similar optical behavior. Elements from palladium (in the fifth period of the periodic table) and platinum (in the sixth period of the periodic table) groups form similar series of elements to the iron group. In these series, 4d and 5d shells are filled, respectively. However, their optical properties are not of practical interest in terms of the scope of this book.

In a crystalline environment, interaction of 3d electrons with the crystal field of surrounding ligands is much stronger ($\sim 10^4$ cm^{-1}) than the spin-orbit interaction ($\sim 10^2$ cm^{-1}). Hence, spectroscopic properties are usually described by the crystal field theory, considering, when necessary, spin-orbit interaction as a perturbation to the obtained solution.

To understand the energy level structure of transition metal ion centers we will introduce here the crystal field parameters 10Dq, A, B, and C, without going

TABLE 6.1 Electronic Configurations of 3d Transition Metal Ions

Configuration	Ions							
$3d^1$	Ti^{3+}	V^{4+}	Cr^{5+}	Mn^{6+}				
$3d^2$	Ti^{2+}	V^{3+}	Cr^{4+}	Mn^{5+}				
$3d^3$		V^{2+}	Cr^{3+}	Mn^{4+}				
$3d^4$			Cr^{2+}	Mn^{3+}				
$3d^5$			Cr^{+}	Mn^{2+}	Fe^{3+}			
$3d^6$				Mn^{+}	Fe^{2+}	Co^{3+}		
$3d^7$						Co^{2+}	Ni^{3+}	
$3d^8$						Co^{+}	Ni^{2+}	
$3d^9$							Ni^{+}	Cu^{2+}
$3d^{10}$								Cu^{+}

into the details of crystal field theory and its applications—which can be found in many books and publications, for example, in [1]. Now, 10Dq (or Δ) is the crystal field strength (also called the ligand field splitting parameter) which has dimension of energy, and is in fact a combined parameter, which incorporates the properties of both the central ion and its environment. A, B, and C are so-called Racah parameters (named after the Italian–Israeli physicist Giulio Racah, who first described them). A is an average total interelectron repulsion and it is constant among the d electron configuration. B and C correspond with individual d electron repulsions. B is the most important Racah parameter while C is needed only in selected cases. Levels of the free ions are determined by electron–electron interaction and labeled ^{2S+1}L states (usually called L–S terms), where S is the total spin and L is the total angular momentum. The energy difference between these states can be given in terms of Racah parameters. When an ion is placed into a crystal field, orbitals of 3d electrons are split, which consequently creates many more electron configuration states than in the case of the free ion. Yukito Tanabe and Satoru Sugano (Japanese physicists) have calculated the energies of these states for all d^n configurations and presented them in the form of diagrams in which the x-axis is expressed in terms of Dq/B and the y-axis is in terms of energy E divided by Racah's parameter B. Today, these diagrams are called Tanabe–Sugano diagrams and considerably ease the interpretation of optical spectra of 3d ions in various host materials. In these diagrams one line represents one electronic state, including both "allowed" and "spin-forbidden" (i.e., electrons change spin when moving between levels) electronic states. Each state is given its symmetry label (e.g., A_{1g}, T_{2g}) and term symbols (e.g., ^3P, ^1S) are listed in order of increasing energy. The ground state is of zero energy. The relative order of energies is determined using Hund's rules. Splitting of free-ion

TABLE 6.2 Splitting of Free-Ion Terms of d^n Configuration in an Octahedral Field

Free-ion Term	Degeneracy	Terms in Oh
S	1	A_{1g}
P	3	T_{1g}
D	5	$E_g + T_{2g}$
F	7	$A_{2g} + T_{1g} + T_{2g}$
G	9	$A_{1g} + E_g + T_{1g} + T_{2g}$
H	11	$E_g + T_{1g} + T_{1g} + T_{2g}$
I	13	$A_{1g} + A_{2g} + E_g + T_{1g} + T_{2g} + T_{2g}$

terms of d^n configuration in an octahedral (Oh) field is given in Table 6.2. A vertical line is drawn in some Tanabe–Sugano diagrams (for d^4, d^5, d^6, and d^7 configurations) at a specific Dq/B value where discontinuity in the slopes of the excited states' energy levels occurs. Luminescence centers to the left of this line are high-spin, while those to the right are low-spin.

The Tanabe–Sugano diagram for the d^5 configuration is presented in Fig. 6.1 as an example. It can be used to explain Mn^{2+} emission which is known to occur in more than 500 inorganic compounds. Mn^{2+} has a sextet (^6S) ground

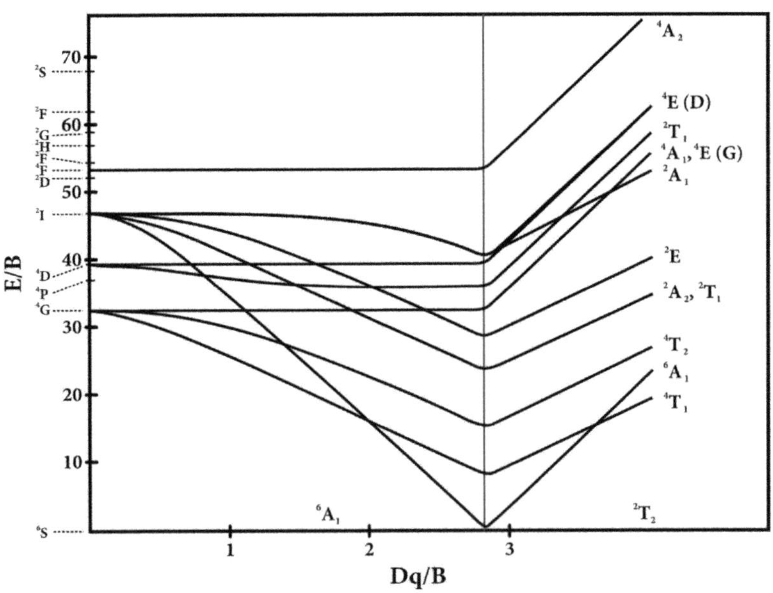

FIGURE 6.1 **Tanabe–Sugano diagram for the d^5 electronic configuration.**

state, and in the weak crystal field (Dq/B $\approx$ 1) its emission comes from the $^4T_1(^4G) \rightarrow ^6A_1(^6S)$ transition. By observing the slopes of curves on the diagram, one can expect that emission wavelengths increase (decrease in energy) with an increase of crystal field strength. This happens for example in the case of Mn^{2+} doped ZnF_2 and MgF_2. The ionic radii of Mg^{2+} is smaller than that of Zn^{2+} producing shorter Mn-F distances, so the crystal field acting on Mn^{2+} is stronger in MgF_2, and consequently the emission is redshifted (toward longer wavelengths). Similar behavior is observed with Zn_2SiO_4 and Zn_2GeO_4.

However, the decrease of Mn^{2+}-ligand distances does not necessarily lead to an increase in emission wavelength. Changes in Mn^{2+} coordination, such as from octahedral to tetrahedral, produce a considerable change in Dq. The tetrahedral Dq value is smaller and the energy of transitions in Mn^{2+} increases creating a blueshift (toward shorter wavelengths). Octahedral Mn^{2+} coordination in $CaSiO_3$ thus has a red emission while the analogous transition for Mn^{2+} in Zn_2SiO_4 yields a green one. Fe^{3+} has the same electronic configuration as Mn^{2+}, so its emissions can be interpreted from the same Tanabe–Sugano diagram (Fig. 6.1). However, the larger valence of Fe^{3+} than Mn^{2+} brings a much stronger crystal field, reducing the $^4T_1(^4G) \rightarrow ^6A_1(^6S)$ transition and shifting the emission into the deep-red spectral region. Energies of transitions are also affected by the *nephelauxetic effect.* Quantitatively, this effect is observed through an increase of transition energies and reduction of B and C Racah parameters, and may be regarded to increase the covalency between the cation and ligands. The reduction of B and C parameters differs for 3d ions (and also for different ligands) in the following order:

$$Mn^{2+} < Ni^{2+} < Cr^{3+} < Fe^{3+} < Co^{3+}.$$

For this reason, the stronger crystal field (Dq) of Fe^{3+} compared with Mn^{2+} is to a small extent compensated by the reduction of the B parameter, however, the Dq/B of Fe^{3+} ($\sim$1.2) is still larger than of Mn^{2+} ($<$1) and thereby is able to produce a considerable redshift of emission even in its tetrahedral coordination. Finally, it is important to note here that for both ions luminescent transitions are forbidden and correspondingly the emission decay times are rather long; information which is of interest to luminescence thermometry.

Similar analyses of optical transitions from Tanabe–Sugano diagrams can be done for 3d ions in other electronic configurations. Among them, interest exists for the Cr^{3+} and Mn^{4+} ($3d^3$ configuration) which emit in deep-red and near-infrared spectral regions. Cr^{3+} is important because it is used in lasers. In fact, the first solid-state laser was built in 1960 from ruby (Al_2O_3:Cr^{3+}). It is also important for the development of persistent luminescence materials. Recently, red persistent luminescence of Cr^{3+} doped $MgGa_2O_4$ and $ZnGa_2O_4$ for in vivo bioimaging has been reported [2]. Transition metal ion luminescence becomes more important when consideringthe continuous decrease in the availability of rare earths. Currently, a huge research is ongoing to find

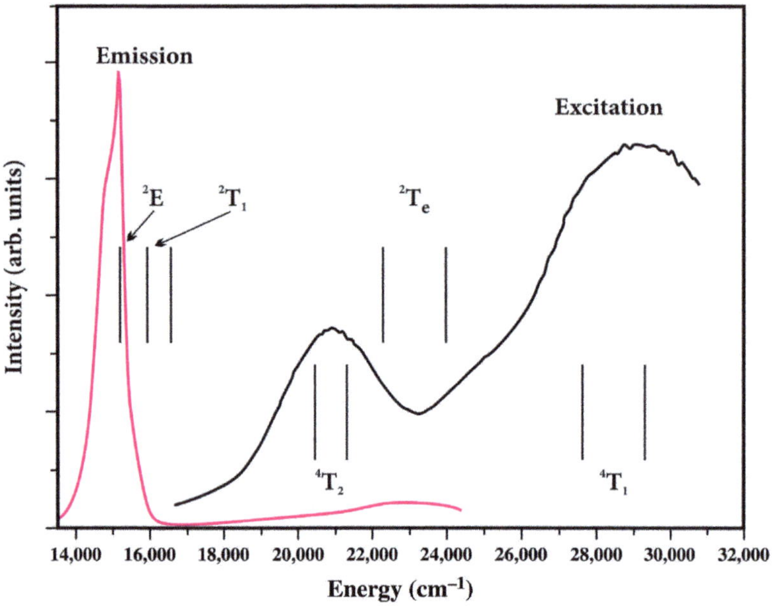

FIGURE 6.2 Experimental excitation (*black line*) and emission (gray (*red in the web version*) *line*) spectra of Mn^{4+} in inverse-spinel Mg_2TiO_4 with calculated energy levels. *(Source: Reproduced from Medić MM, Brik MG, Dražić G, Antić ZM, Lojpur VM, Dramićanin MD. Deep-red emitting Mn^{4+} doped Mg_2TiO_4 nanoparticles. J Phys Chem C 2015;119(1):724–30, copyright (2015) American Chemical Society).*

replacements for the well-established rare earth Eu^{3+} activator in red phosphors. So far, emission of the Mn^{4+} ions is used in various applications, such as lighting, holographic recording, thermoluminescence dosimetry, and to improve the color-rendering index of phosphor converted, white light-emitting diodes (WLEDs) [3], Fig. 6.2.

Regarding new applications of luminescence thermometry in bioimaging, 3d ion doped nanomaterials with emission in the first biological window (NIR-I: 650–950 nm) and especially in the second and third windows (NIR-II: 1000–1350 nm, NIR-III: 1550–1870 nm) gain importance. Use of near-infrared light in the range of the second and third biological windows is increasing remarkably as it provides improvements in penetration depth into biological tissues, image contrast, and reduced phototoxicity and photobleaching over conventional illumination. Luminescence thermometry may be a valuable tool for temperature monitoring at intercellular and intracellular levels, as already shown in several typical applications, some of which will be detailed in Chapter 10. One finds that Ti^{2+}, Co^{2+}, Ni^{2+}, V^{2+}, and Cr^{4+} in some host materials can emit luminescence at wavelengths located in biological windows. For example, $^4T_2 \rightarrow ^4T_1$ transition of Co^{2+} ions in octahedral sites is interesting because of its long wavelength. The decay time of this transition in $KMgF_3$ changes from 3.3 ms at a temperature of 1.5 K to 2.5 μs at 550 K. Such a large change of decay time opens up the possibility for measurements of temperature with high-temperature resolution. Similar observations have been recorded for Ni^{2+} doped $KMgF_3$.

The rare earth elements comprise lanthanides, ^{21}Sc and ^{39}Y. Among many types of luminescence centers, those involving lanthanide ions have gained much attention and found important applications over the years. Lanthanides comprise 15 elements that belong to the sixth period of the periodic table, from ^{57}La to ^{71}Lu. Lanthanum and lutetium have a single valence electron in the d shell so they are generally labeled as group three elements, but they are frequently considered in discussions of lanthanide element properties. The filling of the 4f shell starts after lanthanum and completes with ytterbium, as shown in Table 6.3 which provides the electronic configurations of lanthanide elements. The 4f orbitals penetrate the [Xe] core and are isolated, and thus they do not participate in bonding.

Lanthanide luminescence centers in solids are formed from the triply or doubly charged lanthanide ions. Even though they are formed from the same elements, these two types of lanthanide centers provide distinct luminescence appearances, as will be shown later in the text. Considering that centers involving trivalent lanthanide ions are much more used than divalent ones, especially in luminescence thermometry, their optical properties will be given in more details.

TABLE 6.3 Electronic Configuration of Lanthanides

Element		Electronic configuration
^{57}La	Lanthanum	[Xe] $5d^1$ $6s^2$
^{58}Ce	Cerium	[Xe] $4f^1$ $5d^1$ $6s^2$
^{59}Pr	Praseodymium	[Xe] $4f^3$ $6s^2$
^{60}Nd	Neodymium	[Xe] $4f^4$ $6s^2$
^{61}Pm	Promethium	[Xe] $4f^5$ $6s^2$
^{62}Sm	Samarium	[Xe] $4f^6$ $6s^2$
^{63}Eu	Europium	[Xe] $4f^7$ $6s^2$
^{64}Gd	Gadolinium	[Xe] $4f^7$ $5d^1$ $6s^2$
^{65}Tb	Terbium	[Xe] $4f^9$ $6s^2$
^{66}Dy	Dysprosium	[Xe] $4f^{10}$ $6s^2$
^{67}Ho	Holmium	[Xe] $4f^{11}$ $6s^2$
^{68}Er	Erbium	[Xe] $4f^{12}$ $6s^2$
^{69}Tm	Thulium	[Xe] $4f^{13}$ $6s^2$
^{70}Yb	Ytterbium	[Xe] $4f^{14}$ $6s^2$
^{71}Lu	Lutetium	[Xe] $4f^{14}$ $5d^1$ $6s^2$

[Xe], Xenon core.

TABLE 6.4 Electronic Configurations of Trivalent Lanthanide Ions in the Ground State

Ion	$4f^n$	4f electrons							S	L	J	$^{2S+1}L_J$
Ce^{3+}	$4f^1$	↑							1/2	3	5/2	$^2F_{5/2}$
Pr^{3+}	$4f^2$	↑	↑						1	5	4	$^3H_4$
Nd^{3+}	$4f^3$	↑	↑	↑					3/2	6	9/2	$^4I_{9/2}$
Pm^{3+}	$4f^4$	↑	↑	↑	↑				2	6	4	$^5I_4$
Sm^{3+}	$4f^5$	↑	↑	↑	↑	↑			5/2	5	5/2	$^6H_{5/2}$
Eu^{3+}	$4f^6$	↑	↑	↑	↑	↑	↑		3	3	0	$^7F_0$
Gd^{3+}	$4f^7$	↑	↑	↑	↑	↑	↑	↑	7/2	0	7/2	$^8S_{7/2}$
Tb^{3+}	$4f^8$	↑↓	↑	↑	↑	↑	↑	↑	3	3	6	$^7F_6$
Dy^{3+}	$4f^9$	↑↓	↑↓	↑	↑	↑	↑	↑	5/2	5	15/2	$^6H_{15/2}$
Ho^{3+}	$4f^{10}$	↑↓	↑↓	↑↓	↑	↑	↑	↑	2	6	8	$^5I_8$
Er^{3+}	$4f^{11}$	↑↓	↑↓	↑↓	↑↓	↑	↑	↑	3/2	6	15/2	$^4I_{15/2}$
Tm^{3+}	$4f^{12}$	↑↓	↑↓	↑↓	↑↓	↑↓	↑	↑	1	5	6	$^3H_6$
Yb^{3+}	$4f^{13}$	↑↓	↑↓	↑↓	↑↓	↑↓	↑↓	↑	1/2	3	7/2	$^2F_{7/2}$

In triply charged lanthanide ions all 5d and 6s electrons are detached and the 4f shell is just partly occupied (having from 1 to 13 electrons, Table 6.4). The 4f electrons are responsible for the transitions which give ions exciting luminescence properties. Ions with no 4f electrons, that is, Sc^{3+}, Y^{3+}, and La^{3+}, or with a filled 4f shell, that is, Lu^{3+}, present no luminescence in the visible and near-infrared spectral region. The 4f electrons are shielded from the external electric fields by the outer $5s^2$ and $5p^6$ electrons. Electron transitions between 4f states are forbidden according to Laporte's selection rule, however, when lanthanide ions are surrounded by a nonsymmetric local environment the f–f transitions becomes partially allowed. For this reason, the f–f luminescence is specific for each lanthanide ion (due to the characteristic 4f energy levels of each ion), presents narrow-emission lines, and weakly depends on the crystalline environment of the host material. Electronic states are noted as $^{2S+1}L_J$ (where L represents S, P, D, F, G, H, I, K, L, M,... for $L = 0, 1, 2, 3, 4, 5, 6, 7, 8, 9, \ldots$, respectively; L is orbital angular momentum, S is spin angular momentum, and J is total angular momentum; $J = L–S$ when the number of 4f electrons is smaller than 7 and $J = L + S$ when the number of 4f electrons is larger than 7). Electronic configurations of trivalent lanthanide ions are given in the Table 6.4. The well-known Dieke diagram given in Fig. 6.3 presents the energy levels of 4f electrons of trivalent lanthanide ions experimentally evaluated in $LaCl_3$ crystals.

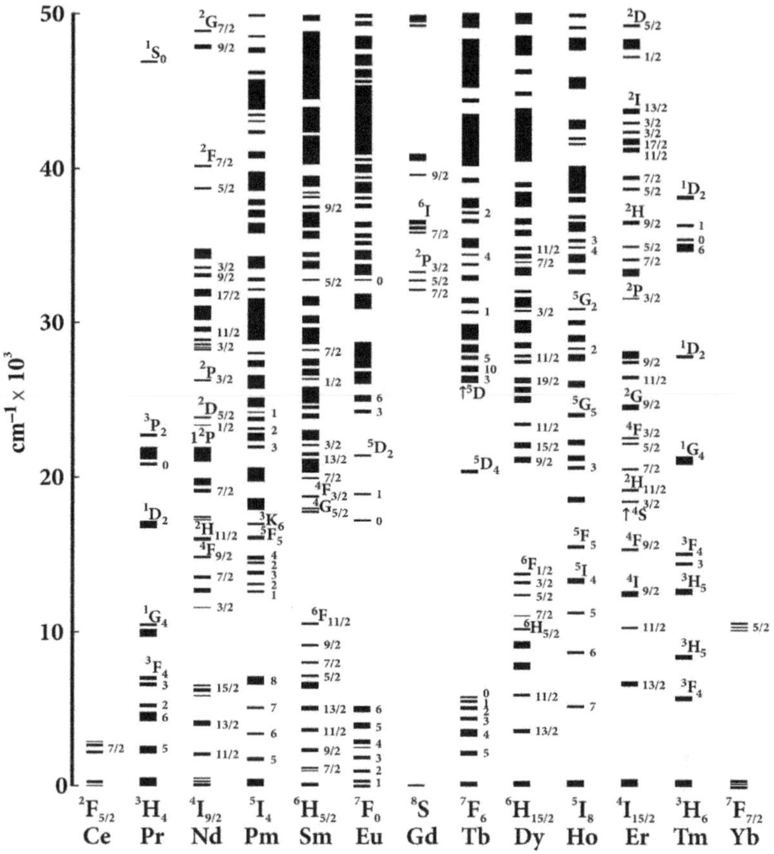

FIGURE 6.3 Dieke diagram. Energy levels of 4f electrons of trivalent lanthanide ions in $LaCl_3$ crystal.

Because of the shielding of 4f electrons their interaction with the crystal field of ligands ($\sim 10^2$ cm^{-1}) is considerably smaller than the spin-orbit interaction ($\sim 10^3$ cm^{-1}). The crystal field produces a small change of energy and causes splitting of levels into a number of sublevels (the so-called Stark effect). The maximum number of sublevels (Stark components) is $2J + 1$ when J is an integer number or $J + ½$ when J is a half-integer number. However, the appearance of splitting, and the exact number of sublevels, is governed by the symmetry of the crystal field surrounding the lanthanide ion and selection rules. Further sublevel splitting due to hyperfine interactions is barely detectable by means of optical spectroscopy methods. The influence of principal physical interactions on the development of the energy level structure of a trivalent europium ion when introduced into a solid is illustrated in Fig. 6.4.

Electronic transitions between 4f levels occur due to magnetic dipole or electric dipole interactions. Since these transitions occur between states of the same

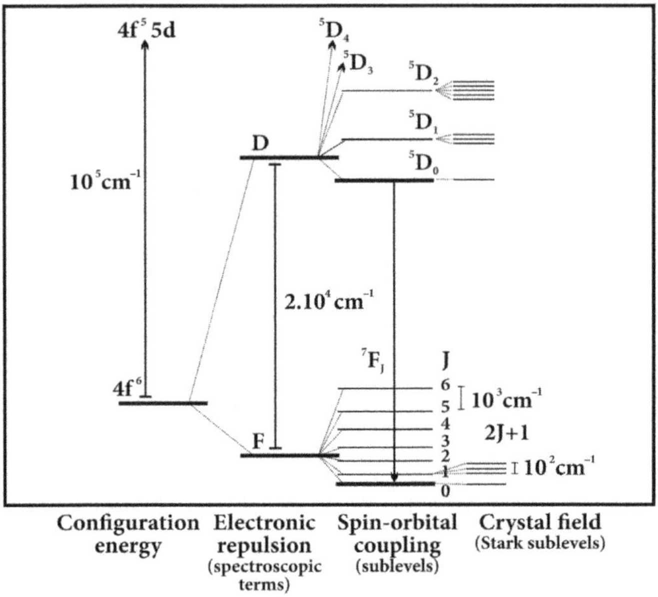

FIGURE 6.4 **Energy levels of trivalent europium ion centers in solids.**

parity only luminescence from magnetic dipole interaction is expected since this transition is parity allowed (selection rule: $\Delta J = 0, \pm 1$ except for $0 \rightarrow 0$). Electron dipole transitions in free 4f ions are parity forbidden, but when a lanthanide ion is located in the site which lacks inversion symmetry a mixing of opposite parity states of the $4f^{(n-1)}5d$ configuration into the $4f^n$ states can occur. Even a small admixture of even and odd parity states allows a small electric dipole component to the transition, and since the electric process is about five orders of magnitude stronger than the dipole one, it may become a principal component in the transition [4]. Selection rules for the electric dipole transitions are derived from the Judd–Offelt theory [5,6]:

1. $\Delta J \leq 6$, $\Delta S = 0$, and $\Delta L \leq 6$ (Russell–Saunders approximation), and
2. for lanthanide ions with an even number of electrons
   $J = 0 \leftrightarrow J' = 0$ is forbidden
   $J = 0 \leftrightarrow$ odd J' is weak
   $J = 0 \leftrightarrow J' = 2, 4, 6$ is strong.

Even with small crystal field energies, some mixing of states from different levels (J-mixing) occurs in addition to a splitting of levels. This affects the J-selection rules. For example, because of J-mixing, the $^5D_0 \leftrightarrow {}^7F_0$ transition of Eu^{3+} is frequently seen even if it is forbidden by the $J = 0 \leftrightarrow J' = 0$ selection rule. It is important to note here that luminescence decay times of emitting $4f^n$ levels are typically long, in the microsecond to millisecond range. Luminescence of trivalent lanthanide centers is also characterized by charge transfer states, which occur when electrons from ligands are transferred to a 4f orbital. This process results with a strong, broadband absorption since it is allowed.

The abundance of transitions between $4f^n$ energy levels in trivalent lanthanide ions provides luminescence emissions that cover huge energy range of about 40,000 cm^{-1}. In addition, in trivalent lanthanide ions radiative transitions may occur between the next higher energy configuration $4f^{n-1}5d^1$ (formed when one of the 4f electrons is transferred to a 5d orbital) and $4f^n$. These so-called 4f–5d transitions can be allowed (they can be both spin-allowed and spin-forbidden) with short luminescence decay times. Contrary to $4f^n$ levels, the positions of the $4f^{n-1}5d^1$ levels are intensely influenced by the crystal field interaction due to a strong interaction of the 5d-electron with neighboring ligands. Optical spectra of these transitions are characterized by wide emission and absorption bands, caused by the coupling of the 5d electron with the lattice phonons. Barely resolvable Stark splitting adds to the broadening of the emission bands [7].

The divalent lanthanide ions also have an outer electronic configuration of $4f^n$ and contain one more electron than the corresponding trivalent lanthanide ion. For example, the electronic configuration of Eu^{2+} is $4f^7$ and it is identical to that of Gd^{3+}. Here, higher lying energy configurations overlap some of the low lying $4f^n$ energy levels. Therefore, $4f^n \rightarrow 4f^{n-1}5d$ transitions may occur in addition to $4f^n$ ones in the energy range which is suitable for applications. They produce intense and broad absorption and emission bands since they are parity-allowed.

Among divalent lanthanide ions, Eu^{2+} is the most important one in terms of practical use. Its ground level is $^8S_{7/2}$ and emissions occur from the next highest levels from the $4f^65d$ states, since the lowest excited state of $4f^n$ levels is higher than the $4f^65d$ level in most materials (it is located at about 28,000 cm^{-1}). The wavelength of this emission extensively varies between different host materials. Red Eu^{2+} luminescence is observed in CaS, yellow in Ba_2SiO_5, green in $SrAl_2O_4$, blue in $BaAl_{12}O_{19}$ and BAM ($BaMgAl_{10}O_{17}$), near-ultraviolet in $(Sr,Mg)P_2O_7$, and violet in Ba(F,Br). In the weak crystal fields the lowest excited state of the $4f^7$ configuration (^6P_J) may have lower energy than the $4f^65d$ state. In such a case, narrow-band emissions occur in the ultraviolet spectral region (around 360 nm) due to f–f transitions (Eu^{2+} in $SrAlF_5$ or $BaAlF_5$ hosts). These days Eu^{2+} broadband emissions are investigated in many host materials for the development of phosphors for a new generation of ultraviolet excited WLEDs.

Some other divalent lanthanide ions may be also of interest. Sm^{2+} emits intensively around 700 nm. Its emission comes either from the low-lying $4f^55d$ level or from the $^5D_0$ ($4f^6$) level, depending on the host material. Emission at 2.4 μm from the $^5I_7 \rightarrow {}^5I_8$ transition of Dy^{2+} doped CaF_2 is used for a single frequency laser. Laser action is also achieved with emission from the $^2I_{5/2} \rightarrow {}^5I_{7/2}$ transition of Tm^{2+}.

6.3 DOWNSHIFTING, UPCONVERTING, SCINTILLATING, AND QUANTUM-CUTTING PHOSPHORS

Photoluminescence of lanthanide and transition metal ion centers is excited through absorption of radiation by the host lattice, by the absorption of ions themselves (which also emit radiation—so-called activator ions), or by the additionally introduced impurities (so-called sensitizers) which absorb the radiation and then transfer the energy to the activators. Sensitizers are most frequently also rare earth or transition metal ions. Materials that include these types of centers (but are not limited only to them) can be divided into four general types depending on the energy they absorb and emit, and the number of photons involved in these processes. These functional materials are downshifting, upconverting, scintillating, and quantum-cutting phosphors.

Downshifting phosphors absorb radiation of certain energies (usually high energy—vacuum ultraviolet or ultraviolet) and subsequently reemit photons of lower energies (higher wavelengths) than those absorbed. These are the oldest type of phosphors and ones with the most practical and valuable use, both presently and in the past. Due to detailed studies of their properties and extensive work on the procedures for their preparation, current downshifting phosphors have great internal quantum efficiencies (>90%, meaning that for 100 absorbed photons a material emits more than 90 photons). They are predominantly prepared from transition metals ion activators since the lanthanide ones are much more expensive and used for specialty applications (including luminescence thermometry). The thermographic properties of downshifting phosphors have been recently reviewed in Ref. [8].

Low-energy radiation (long wavelengths) can be converted to high-energy radiation (the radiation of shorter wavelengths) by some of the following processes: anti-Stokes Raman emission, two-photon excitation, second harmonic generation, cooperative luminescence, cooperative sensitization, and energy (sometimes called frequency) upconversion (which includes excited state absorption (ESA), energy transfer (ET) upconversion, and the sensitized ET upconversion processes). The last process, energy upconversion, is of the greatest importance for modern technology. Phosphors capable of conversion (upconversion) of low-energy photons to high-energy photons are called *upconversion phosphors*. They are usually used to convert infrared to visible light in displays, as infrared detectors, biolabels, security markers, lasers, solar cells, etc. Two or more low-energy photons are required to produce one photon of higher energy so the energy conservation rule is preserved. Materials doped with lanthanide ion centers are extensively used because these centers provide a great number of long-lived electronic states at various energies which facilitate the absorption of low-energy photons and their conversion into radiation with higher energy. Host materials are generally chosen from those having low maximal phonon energies, like fluorides (of about 200 cm^{-1}), to reduce nonradiative losses. However, upconversion emissions generally suffer from small internal quantum efficiencies (the largest efficiencies in fluorides is around 20%) and therefore are still inadequate for many applications. Consequently, considerable research is currently focused on both the development of novel, more efficient upconversion materials and efficiency improvements in existing ones.

Generally, two different types of upconversion can be recognized: the single-ion excited-state absorption process and the two-ion (sensitized) ET upconversion process. The later is much more exploited. With lanthanide ion centers, Yb^{3+} is commonly used as sensitizer ion since it has a relatively large absorption cross section at 980 nm, so materials can be efficiently excited with relatively cheap light sources. Er^{3+}, Ho^{3+}, and Tm^{3+} are the activator ions whose emissions are most frequently facilitated by ET from excited Yb^{3+} ions. Transition metal ion doped materials were among the first examples of upconversion; single crystals of Ni^{2+} doped MgF_2 showed green 500 nm emission under 752.5 nm excitation. Later, upconversion was demonstrated in other halides, mainly with Ni^{2+} and Mn^{2+}, but also with Cr^{3+} and Ti^{2+}. Presently, upconversion from transition metal ions has become more popular since, as mentioned in the previous sections, their energy states are strongly influenced by the local crystal field, allowing for chemical variation of their photophysical characteristics. Also, upconversion from mixed lanthanide/transition metal ion doped compounds is gaining popularity—operating either as lanthanide sensitized transition metal upconversion, or transition metal sensitized lanthanide upconversion. The former may be of particular interest for use in luminescence thermometry, and generally involves the Yb^{3+} ion as the sensitizer and Mn^{2+}, Cr^{3+}, and Mn^{4+} as the activator.

Quantum cutting (quantum splitting) is, phenomenologically speaking, the opposite process to upconversion in terms of the involved excitation and emission energies and the number of involved photons. *Quantum-cutting phosphors* are used to convert one high-energy photon into two or more photons of lower-energy. Such materials, in principle, may substantially improve the overall energy conversion of fluorescent lamps.

Quantum splitting of a high-energy photon can be obtained by several schemes [9]; two of the most important are presented in Fig. 6.5. In the first, between the ground (G) and the excited (E) states of the luminescent ion there is an intermediate state (I), Fig. 6.5A. Both the E and I states can deexcite radiatively, so after high energy excites an electron from its ground state to its excited state (G $\rightarrow$ E), two radiative transitions, E $\rightarrow$ I and I $\rightarrow$ G, occur. The activator ion produces two photons per absorbed one and, therefore, the internal quantum efficiency of the process equals 2. In the second scheme, Fig. 6.5B, two ions are involved, a sensitizer (S) and an activator (A). In this case, excitation comes through the sensitizer ion S. If the energy levels of the sensitizer can excite the emission of the activator ion A, by cross-relaxation via the radiative transition E2 $\rightarrow$ E1 process, then activator ion A produces one photon due to its deexcitation to the ground state. The second photon is generated when the energy localized on the E1 excited state of the sensitizer ion is

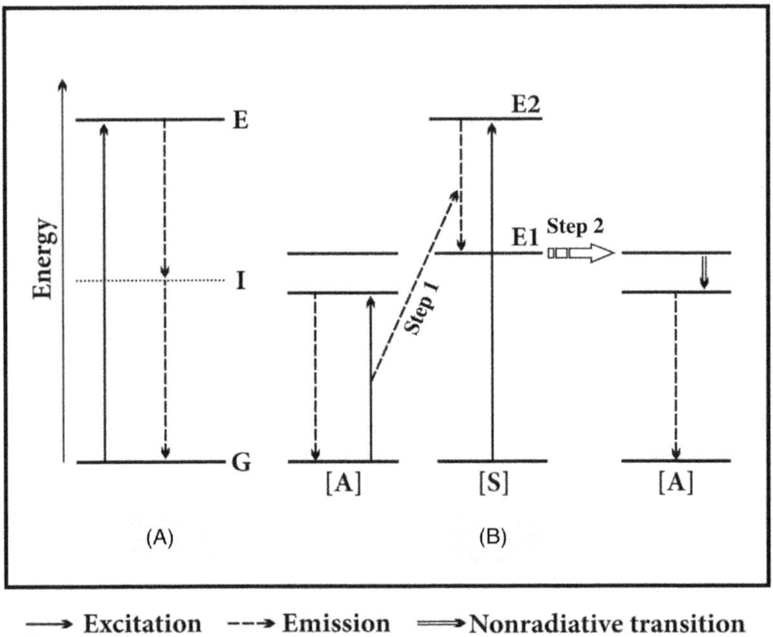

FIGURE 6.5 Two schemes of possible quantum cutting process: (A) quantum cutting through an intermediate state and (B) quantum cutting involving two ions (sensitizer and activator). *(Source: Reproduced from Srivastava AM, Ronda CR. Quantum-splitting systems. In: Luminescence: from theory to applications. Wiley-VCH Verlag GmbH & Co. KGaA; 2007. p. 75–103. with permission from Wiley).*

transferred by an ET process to the second activator ion, which then radiatively deexcites. Again, the internal quantum efficiency equals 2.

Up to now, quantum cutting has been demonstrated mostly with Pr^{3+} ions, and combinations of Gd^{3+}/Eu^{3+} and $Er^{3+}/Gd^{3+}/Tb^{3+}$ ions. It is important to note that generally internal quantum efficiencies exceeding 100% do not necessarily imply overall quantum efficiencies larger than 100%. From the consideration of their applicability for use in luminescence thermometry, quantum-cutting phosphors may be treated in the same way as downshifting phosphors; the same ascertainment stands for scintillating phosphors.

Scintillating phosphors are in fact downshifting phosphors designed to convert high-energy X-rays and gamma-rays into visible light. They are used for high-energy physics, medical imaging devices (such as computed tomography and positron emission tomography scans), industrial inspection scenarios, and security applications. In addition to general requirements for downshifting phosphors, these materials need to have some additional characteristics. They should have both a high physical density and a high stopping power for high-energy radiation, high yields, short decay times, and be a spectral match with a detector of visible light. Diverse luminescence centers are used to facilitate scintillation.

Regarding lanthanide luminescence centers, Ce^{3+} and Pr^{3+} are most often used, with Eu^{3+} (in a $(Y,Gd)_2O_3$ host) being used less often. It is interesting to note that luminescence thermometry was demonstrated with Eu^{3+} doped $(Y_{0.75}Gd_{0.25})_2O_3$ in the field of high-energy gamma-rays, where the energy from gamma-rays provided the excitation of emission [10]. Lanthanide luminescence centers are also used in some X-ray intensifying screens ($LaOBr:Tm^{3+}$ and $Gd_2O_2S:Tb^{3+}$).

6.4 WAYS TO PREPARE RARE EARTH AND TRANSITION METAL DOPED MATERIALS

6.4.1 Crystal Growth

For certain applications, such as scintillators and lasers, phosphors are almost exclusively prepared in the form of crystals. Also, detailed fundamental investigations of the properties of phosphors are preferably carried out on single crystals. Therefore, crystal is an indispensable form of phosphor. Crystal growth can be accomplished by a variety of methods, ranging from a small inexpensive technique to sophisticated expensive processes while preparation times can span from minutes to months. The importance of crystal growth for phosphor applications may be evidenced from the fact that scintillators present 12% of all grown crystals, and that solid-state lasers significantly rely on YAG ($Y_3Al_5O_{12}$) crystals.

General classification of crystal growth methods can be done in relation to the type of transport of crystal constituents during the crystallization process: solid-to-solid phase in solids (solid growth), liquid-to-solid phase transformation (liquid growth), and vapor-to-solid phase transformation (vapor growth). In this categorization, liquid growth comprises both melt and solution growth. A survey of the methods shows that the majority of single crystals are nowadays grown from melt (almost 80%); the estimated share of different crystal growth methods is given in Fig. 6.6.

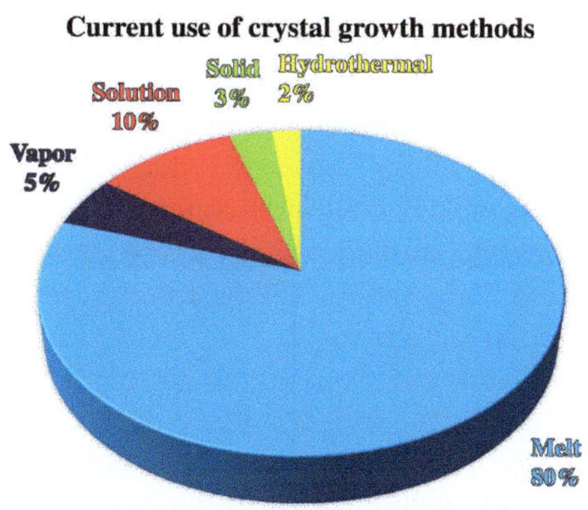

FIGURE 6.6 Estimated share of the different crystal growth methods.

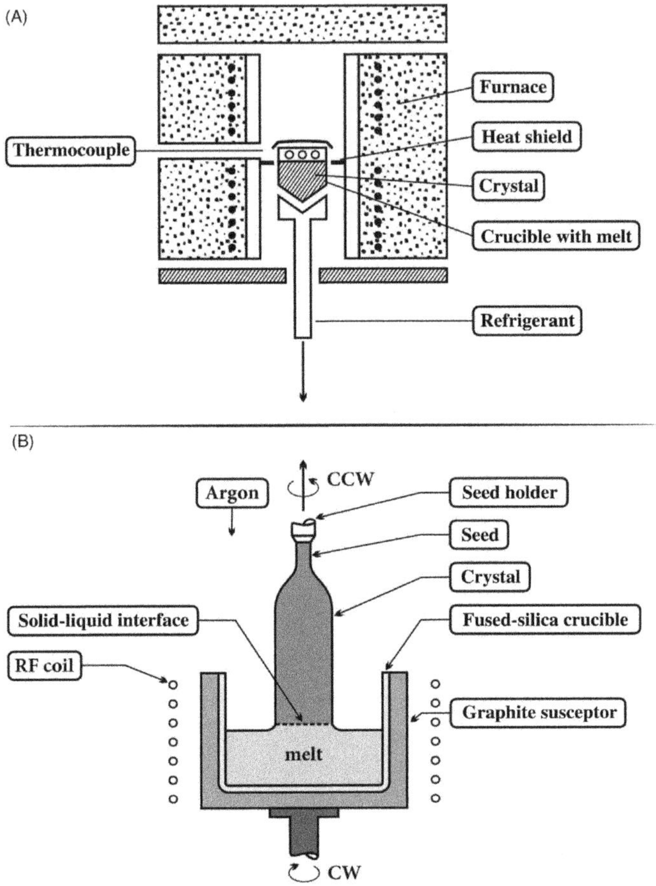

FIGURE 6.7 Schematic representations of (A) Bridgman and (B) Czochralski crystal growth processes.

The science of crystal growth is vast and rich in information, therefore, it is impractical to detail all the growth techniques. To meet the purpose of this book, only the most commonly used ones will be mentioned and briefly explained; all of them are based on the growth from melt. Materials can be grown in single crystals from their own melt if they melt congruently without decomposition at the melting point, and do not undergo any phase transformation between the melting point and room temperature. The best known methods are: Bridgman, Czochralski, and Verneuil techniques.

The Bridgman technique (the "pulling down" method) is a popular method of producing crystals and is mainly used where the Czochralski process proves too difficult. For example, it may be used for the production of scintillating BGO ($Bi_4Ge_3O_{12}$) crystals. The method involves heating a polycrystalline material above its melting point in a vertical cylindrical container, tapered conically with its point at the bottom, and slowly cooling it from the end where a seed crystal is located. Single crystal material is progressively formed along the length of the container. The process can be carried out horizontally or vertically—the latter being more common (Fig. 6.7A).

Crystals of many important oxide phosphors (such as $LaAlO_3$, YAG, and GGG) are grown by the Czochralski process (the "pulling up" method), Fig. 6.7B. This process is usually performed in an inert atmosphere, such as argon, and in an inert chamber, such as quartz. First, material is melted using resistance or induction heating with precise control over the temperature. Second, a seed crystal is lowered to touch the molten charge, with the temperature of the seed being kept lower than the temperature of the melt. With these temperature conditions, the melt solidifies on the seed, and the seed is pulled up with simultaneous rotation of the seed rod and the crucible.

The Verneuil process (the flame fusion process) has been known since 1902, and can produce synthetic rubies and sapphires, as well as rutile and strontium titanate. It is largely used in the manufacture of synthetic corundum and spinel gemstones. Despite some improvements in the method, the process remains virtually unchanged to this day. In this method, the starting material is finely powdered (to fine, dry, and micron-size powder) and placed in a container within a furnace which has an opening at the bottom. The powder is then shaken through the wire mesh and permitted to fall through the oxyhydrogen flame (with a temperature of at least 2000°C). As the powder passes through the flame, it melts into small droplets, which fall onto an earthen support rod placed below. The droplets gradually form a sinter cone on the rod, the tip of which is close enough to the core to remain liquid. The seed crystal forms at the tip, and, as more droplets fall, a single crystal starts to form.

6.4.2 Preparation of Powders

Powders are the most common form of luminescent temperature probes, utilized either as pressed pellets (with or without calcination) or as coatings after mixing with binders. Numerous routes have been developed over the years for the preparation of phosphor powders. Here, we will briefly explain several so-called soft-chemical methods. Physical methods for preparation of powders are based on similar physical principles to those used for the preparation of thin films.

6.4.2.1 Processes Involving Solids

The solid-state method, sometimes called the ceramic method, is a high-temperature method of the sintering reaction between solid-state precursors. The method is straightforward; temperature, pressure, and time of heating can be varied, and temperature should be at least two-thirds the melting point of one of the precursors for sintering to begin. Thoroughly powdered stoichiometric amounts of precursors are mixed and heated; when volatile precursors are used, some excess quantity is needed. Some of the processes can be additionally optimized by controlling the aspects of nucleation and diffusion rate of the ions, by using phase similarities between the reagents and the products or by choosing reactants that decompose prior to, or during, the reaction. Also, ball milling is sometimes used to additionally activate precursors. The treated mixture may sometimes be pulverized and reheated repeatedly; using higher temperatures or forming pellets before the next heating cycle starts [11].

The molten salt method is an attractive alternative to the solid-state method since it presents higher reaction rates and lower reaction temperatures. It can also be used for the preparation of small single crystals. The synthesis is based on salts with a low melting point, often with a eutectic composition of 0.5 NaCl–0.5 KCl, which are mixed with the reactants and heated above the melting point of the salts. In such an obtained melt, product particles are formed, so some of the characteristics of the product can be varied by optimizing the salt mixture, temperature, and duration of heating. After slow cooling to room temperature, the salts are washed away from the products [12].

Low-temperature, self-propagating synthesis is a highly cost-effective method for obtaining powders such as oxides. By combining precursor nitrates with sodium hydroxide, one can obtain the mixture of stable insoluble oxides and water-soluble nitrates. After washing, the final product of good crystallinity can be obtained with additional temperature treatments.

6.4.2.2 Processes Involving Liquids—Solution Based

Combustion methods represent several synthesis techniques which are based on the combustion of a highly organic-rich intermediate product prepared by solution-based precursors. The temperature of combustion and additional heat treatments can be varied to achieve the desired phase, morphology, and crystallinity of the product.

The in situ polymerizable complex method, often called the Pechini method, is one of the most used methods for the preparation of uniformly doped powders of various compounds. Precursors are water soluble salts from which cations are chelated with some water soluble polycarboxylic acid, for example, citric acid or ethylenediaminetetraacetic acid (EDTA). In this way, cations are homogenously ordered through the solution. When a polyhydroxy alcohol, such as ethylene glycol, is introduced into the solution at slightly elevated temperatures, networks are formed by polyesterification, and cations take fixed positions. After water removal, a highly viscous resin is formed and then burned, producing a porous powder [13].

The polymer complex method [14] uses water soluble salt precursors and water soluble oxygen-rich polymers, such as polyethylene glycol, to homogenously order cations through the coordination with free electron pairs from oxygen, while the polymer is also used as fuel for combustion. In that way, synthesis is faster than in the Pechini method as well as being more ecologically friendly since fewer organics are used in the procedure. After burning, a porous powder is formed which is the result high amounts of gas by-products escaping from the solid.

Sol–gel synthesis is a technique that uses metal alkoxides as precursors for the production of primarily silicate and titanate powders. The process involves acid hydrolysis of the alkoxides, gelation, drying, and finally combustion. The network arrangement, which is produced in the gelation stage, can be acid or base catalysed. The gel drying is accompanied by its substantial shrinkage and densification which results in a formation of a xerogel. The final product is achieved after the xerogel firing [15].

The precipitation method, along with coprecipitation, is the method of forming a product of low solubility in water or other solvents, followed by calcination treatment to form crystalline products. In this way, the reaction can start by exchange reaction, hydrolysis, oxidation, reduction, etc. Final products of various shapes and sizes can be tuned by controlling the key points of the precipitation reaction: nucleation, growth, Ostwald ripening, growth termination, and nanoparticle stabilization [15].

The solvothermal or hydrothermal method (when the solvent is water), is the method where a precursor solution or precipitates are placed in a stainless steel autoclave, and a reaction is conducted under conditions of elevated pressure and temperature [16]. Various components of precursor preparations, as well as conditions of autoclaving, may form different morphologies, from quantum dots (QDs) to a variety of two-dimensional and three-dimensional shapes. Nowadays, it is common to combine the hydrothermal technique with other techniques, such as a microwave synthesis.

Spray and flame pyrolysis. Both types of pyrolysis start with liquid feed, where precursors are dissolved in a solvent and sprayed into a heat zone by a carrying gas. With spray pyrolysis, ultrasound is used to prepare a fine aerosol that is carried into a tube furnace. The fine spherical powder particles or thin film coating is collected or formed at the exit of the tube. In flame pyrolysis, an oxidizing gas carries the aerosol into a flame zone where it is combusted. Both techniques are flexible in terms of choosing precursors, solvents, and processing conditions, giving the user control over particle size, composition, and morphology [17].

Reverse micelles, also called water-in-oil microemulsions, represent water droplets in nonpolar solvents which are stabilized by a layer of surfactant molecules. The micelles are an optically isotropic and thermodynamically stable mixture of water, oil, surfactant, and cosurfactant. They can be used as almost ideal nanoreactors for the preparation of many inorganic compounds. Two routes are possible for the synthesis of nanoparticles in reverse micelles: (1) the mixing of two separate reverse micelles with different reactants (precursors), and (2) the introduction of all precursors in one reverse micelle. In the first, a more practiced route, two reverse micelles are mixed after which the exchange of their intermicellar content takes place. Then, the contents react chemically, building the desired product which then precipitates in water droplets of inverse micelles.

This complex process is illustrated in Fig. 6.8. It is executed through the following stages: (1) the Brownian motion of inverse micelles, (2) collisions and

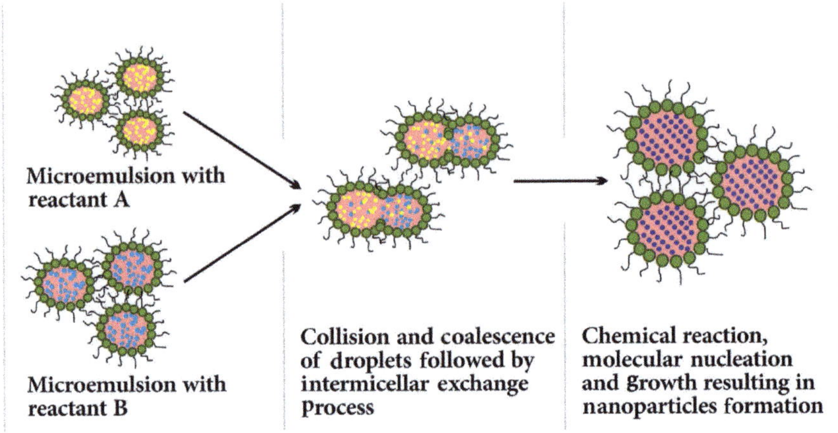

FIGURE 6.8 A simplified schematic of the reaction stages involved in the synthesis of nanoparticles by two reverse micelle methods. *(Source: Adapted from Dhand C, Dwivedi N, Loh XJ, Jie Ying AN, Verma NK, Beuerman RW, et al. Methods and strategies for the synthesis of diverse nanoparticles and their applications: a comprehensive overview. RSC Adv 2015;5(127):105003–37, with permission from The Royal Society of Chemistry).*

coalescence of droplets, (3) opening of layers of surfactant molecules, (4) diffusion of reactants through a water channel, (5) fission of two micelles and the closing of layers of surfactant molecules, (6) chemical reactions between reactants, and (7) nucleation and growth of nanoparticles within a core of water droplets [18].

Mechanical milling, or ball milling, uses various types of mills to reduce the size, blend, and change the shape of particles. In this way, mechanically activate powder precursors form new metastable or stable compositions. The energy transfer from the mill to the powders can be controlled by the type of mill, milling speed, use of different solvents in the milling process, varying powder precursors, and by choosing type, size, and distribution of the milling balls. Temperature control can be applied during some of the milling processes, or temperature treatments can be applied to the final milled compositions [19].

The *laser ablation method* exploits strong laser radiations for the ablation of solid targets immersed in a liquid environment. The products of ablation are particles in the form of precipitated nanoparticles or colloidal solutions. Properties of the desired products can be tailored by changing different parameters of the laser radiation, such as laser energy and the pulse duration, as well as by using specific chemistry in terms of the target's and liquid's reactivity [20].

6.4.3 Preparation of Thin Films and Coatings

A thin film is a quasi two-dimensional material created by the deposition of individual atoms on a substrate, usually a single crystal substrate. Coatings (or thick films) can be regarded as a low-dimensional material developed by thinning a three-dimensional material or by assembly from large pieces (such as clusters or aggregates).

Advances in thin-film technology and coating technology crucially rely on the synthetic methods used for film fabrication. These methods differ substantially from techniques utilized for the preparation of the same materials in other forms (powders, crystals, etc.), and can be generally divided to physical and chemical methods, based on the nature of the employed deposition process. Phosphor coatings are readily obtained by the following processes: settling process, slurry process, dusting process, phototacky process, electrophoresis, and the doctor blade method [21]. The physical deposition processes include vacuum evaporation, laser ablation, molecular beam epitaxy, and sputtering. The chemical deposition processes comprise: chemical vapor deposition, atomic layer epitaxy, spray pyrolysis, sol–gel, spin-coating, and dip-coating. The detailed description of all preparation methods is beyond the scope of this book. Readers interested in these methods may find more information in [21,22]. Here, as an example, the pulsed laser deposition technique (PLD), schematically illustrated on Fig. 6.9, is highlighted.

The PLD technique is relatively simple. It is in fact a physical vapor deposition where a high-power pulsed laser beam (1 J per shot) is focused inside a vacuum chamber to strike a target of the material that is to be deposited. The

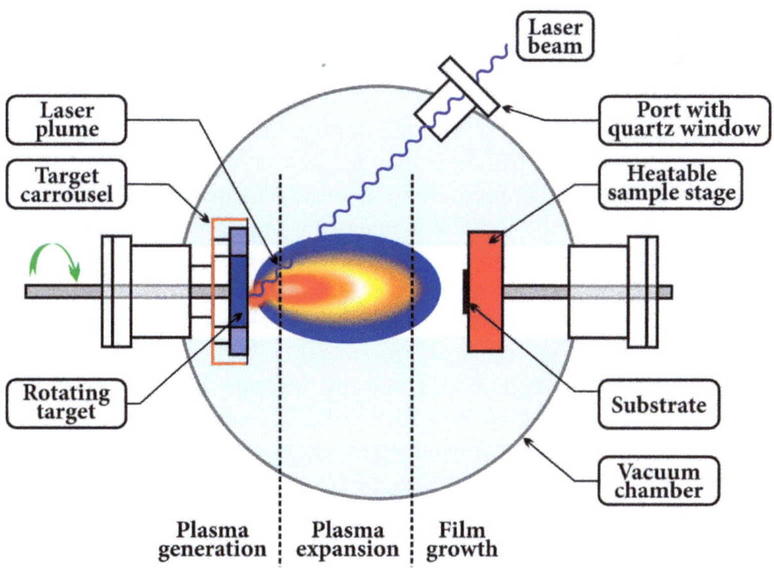

FIGURE 6.9 Schematic illustration of a typical pulsed laser deposition system used for the growth of thin films. *(Source: Reproduced from Liu WY,Li X,Chi D, Zhang H, Liu X. Lanthanide-doped upconversion materials: emerging applications for photovoltaics and photocatalysis. Nanotechnology 2014;25(48):16, Article ID: 482001, with permission from IOP Publishing).*

deposition process then occurs in four stages: laser ablation of the target material and creation of a plasma, plasma expansion dynamics, deposition of the ablation material on a substrate, and nucleation and growth of the film on the substrate [23]. Laser sources are usually ultraviolet-emitting ArF, KrF, and Nd-YAG (the third and fourth harmonic) lasers. This method provides several benefits to thin film preparations: relatively high deposition rates; easy, real-time control of film thickness; and the ability to perform the stoichiometric transfer of material from the target to the substrate. In addition, this process is exceptionally clean since it uses an external energy source, and facilitates the use of several important in situ characterization techniques.

6.4.4 Glasses and Glass-Ceramics

Rare earth and transition metal ion based phosphors in the form of glasses and glass-ceramics find important applications. For example, an Nd doped glass laser gives efficiencies of 2%, which is four times that of a ruby laser. Glasses and glass-ceramics doped with sensitizer and activator ions for upconversion may be introduced in photovoltaic cells to improve their efficiencies. In recent times, they have been used in WLEDs instead of coatings made from phosphor powder and binders. They are also often used in the form of fibers.

Even though silicate glasses are what we habitually consider as "glass," many other compounds can form glasses. For luminescence applications, as well as for other applications, their optical transparency is of great value. When lanthanide and transition metal ions are added to glass they basically determine its optical properties. The oldest established method (and still predominantly used) of producing a glass (amorphous solid) is to cool the molten form of the material quickly. This method is known as the melt-quenching process and its dominant feature is that the amorphous solid is formed by continuous hardening (i.e., an increase in viscosity) of the melt. Amorphous thin films may be prepared by thermal evaporation (a gas-to-solid transformation) and crystals can be turned to an amorphous phase by irradiation or pressure induced amorphization.

Glass-ceramics are composites of the glass phase (more precisely the amorphous phase) and crystalline phase (produced deliberately in the glass by so-called controlled crystallization). There are several routes for preparation of phosphors in the form of glass-ceramics. Among them, the glass crystallization technique, the low-temperature cosintering technique, and the sol–gel technique (the basics of which were explained in the previous section, Section 6.4.2.2) are the most commonly utilized. In recent times some novel methods have emerged, such as the screen-printing process [24]. In the glass crystallization process, the first step is to prepare precursor glass by melt quenching, and then the glass-ceramic is produced through controllable crystallization after heating the precursor glass to a suitable temperature. Low-temperature cosintering is a three-stage process. First, low-melting temperature precursor glass is prepared and then mixed with phosphors. Finally, the mixture is sintered at as low a temperature as possible. One should note that there is a wide range of commercial phosphor powders that could be mixed with glasses.

6.5 LANTHANIDE ION DOPED MATERIALS FOR TEMPERATURE SENSING

Both downshifting and upconversion emissions of lanthanide ion luminescence centers are vastly explored in terms of luminescence temperature sensing. Most of the research and applications have dealt with emissions from one activator ion, utilizing the luminescence intensity ratio (LIR), emission decay, and emission risetime schemes for temperature read-outs (for details see Chapter 4).

The LIR of emissions from two thermally coupled excited states of lanthanide ions, the so-called Boltzmann-type LIR, is the most popular temperature read-out scheme using lanthanide ions. This is because it can be realized with a simple experimental set-up and because of the straightforward data analysis and assessment of its figures of merit. In this technique one needs to observe two distinct emissions and follows their ratio as the temperature of the material changes. The list of lanthanide ions used for Boltzmann-type LIR, along with the summary of their excited levels and emissions, is given in Table 6.5. The LIR of emissions from higher energy (I_H) and lower energy (I_L) excited states, that is, the LIR of shorter wavelength and higher wavelength emissions, can be approximated by the following equation (see Chapter 4):

$$\mathrm{LIR}(T)=\frac{I_H(T)}{I_L(T)}=B\cdot\exp(-\Delta E/k_BT), \tag{6.1}$$

where ΔE is the energy difference between the involved excited states, k_B is Boltzmann's constant, T is absolute temperature, and B is a constant. High-energy emission increases in intensity with the rise of temperature (because the higher energy state becomes more populated), while the intensity of the lower energy emission remains unchanged. In this way, the LIR value increases with an increase in temperature. However, after a certain temperature, both emissions lessen in intensity due to temperature quenching (see Chapter 3) until one of them, or both of them, become undetectable.

The most important figures of merit of measurement, the absolute (S) and the relative sensitivity (S_R) (see Chapter 2), are easily derived from Eq. (6.1):

$$\begin{aligned} S(T)&=\left|\frac{d(\mathrm{LIR})}{dT}\right|=\frac{\Delta E}{k_BT^2}\cdot\mathrm{LIR}(T),\\ S_R(T)&=\frac{1}{\mathrm{LIR}}\left|\frac{d(\mathrm{LIR})}{dT}\right|\cdot 100\%=\frac{\Delta E}{k_BT^2}\cdot 100\%\cdot \end{aligned} \tag{6.2}$$

TABLE 6.5 Summary of Thermally Coupled Energy Levels of Different Trivalent Lanthanide Ions That Are Commonly Exploited by the Luminescence Intensity Ratio Technique, Emission Colors, Usual Energy Difference Between Coupled Levels (ΔE), and Theoretical Relative Sensitivity at 300 K

Ion	Coupled excited levels		Emission color	ΔE (cm^{-1})	Relative Sensitivity at 300 K ($\%K^{-1}$)
	High	Low			
Eu^{3+}	$^5D_1$	$^5D_0$	Green/red	1750	2.80
Dy^{3+}	$^4I_{15/2}$	$^4F_{9/2}$	Blue	1000	1.60
Sm^{3+}	$^4F_{3/2}$	$^4G_{5/2}$	Green/orange/red	1000	1.60
Nd^{3+}	$^4F_{5/2}$	$^4F_{3/2}$	NIR	1000	1.60
Nd^{3+}	$^4F_{3/2(b)}$	$^4F_{3/2(a)}$	NIR	100	0.16
Pr^{3+}	$^3P_1$	$^3P_0$	Green or red	580	0.93
Er^{3+}	$^2H_{11/2}$	$^4S_{3/2}$	Green	780	1.25
Ho^{3+}	$^5F_4$	$^5S_2$	Green	120	0.19
Tm^{3+}	$^3F_{2,3}$	$^3H_4$	Red/NIR	1700	2.72
Tm^{3+}	$^1G_{4(b)}$	$^1G_{4(a)}$	Blue	100	0.16
Yb^{3+}	$^2F_{5/2(b)}$	$^2F_{5/2(a)}$	NIR	680	1.09

NIR, Near-infrared. (a) low-energy Stark component, (b) high-energy Stark component.

It should be noted that both sensitivities are functions of temperature; when reporting their values, it is important to state at which temperature they were calculated. For example, the value of relative sensitivity at 300 K reduces by 20% at 335 K and by 50% at 424 K. What is more, in this temperature read-out scheme, the relative sensitivity depends only on the energy difference between thermally coupled excited states and temperature; the larger the ΔE and the lower the temperature the higher value of relative sensitivity (Fig. 6.10).

In other words, the relative sensitivity, in fact, is a consequence of an inherent property of the lanthanide ion used. The value of ΔE of lanthanide ions just slightly varies between different hosts (by a maximum of a few hundreds inverse centimeters), therefore, relative sensitivities of LIR using certain lanthanide ions will be pretty much the same irrespective of the host material, see Table 6.5 and inset in Fig. 6.10 for average values of the relative sensitivity at 300 K for different lanthanide ions. There are many literature reports on relative sensitivities that span wide range of values, and, quite frequently, significantly differ from the real values (which can be evaluated from emission spectrum as a difference between respective emission barycenters, from excitation spectrum, or from

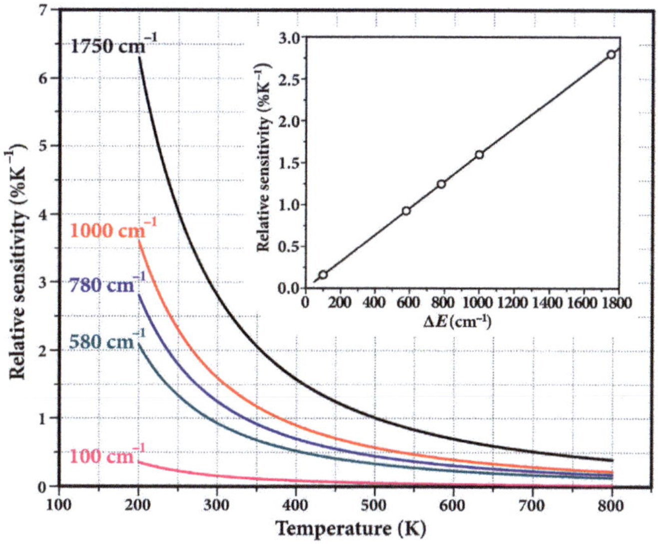

FIGURE 6.10 Temperature dependence of the relative sensitivity calculated for several ΔE values. Inset shows the dependence of the relative sensitivity ΔE calculated for a temperature of 300 K.

theoretical calculations). The error (difference) may be a consequence of a measurement error or poor fitting. It can be also an effect of ETs between thermally coupled levels, other levels, or charge-transfer bands, which occur along with the Boltzmann's re-distribution of the population between energy states. This difference can be quantified with a phenomenological error parameter δ [8,25]:

$$\delta(\%) = \frac{|\Delta E - \Delta E_T|}{\Delta E_T} \cdot 100\%, \tag{6.3}$$

where ΔE_T stands for the "true value" of the energy difference.

The temperature range of measurements is governed by properties of both the host material and lanthanide ion. It is also affected by the sensitivity of the measurement instrumentation, mostly by the ability of the instrumentation to measure emissions of low intensity with sufficient accuracy. At low temperatures, the higher energy state will not be populated because electrons do not have enough thermal energy (k_BT) to bridge the energy gap, and because the nonradiative relaxation rate from the higher to lower energy state is very large for the closely spaced states. Therefore, the LIR read-out scheme has a low-temperature bound of measurement (T_{min}), which can be assessed when Eq. (6.1) is transformed into the following form:

$$T = \frac{\Delta E}{k_B \cdot [\log(B) - \log(\mathrm{LIR_{min}})]}. \tag{6.4}$$

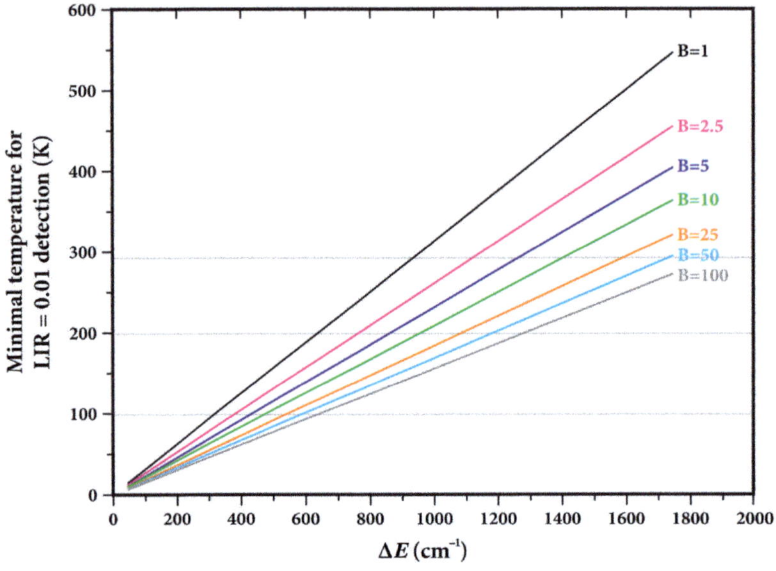

FIGURE 6.11 Dependence of the low-temperature bound on ΔE for different values of B for the hypothetical luminescence intensity ratio detection limit of 0.01.

Temperature is directly proportional to ΔE and inversely proportional to B (for the physical meaning of B please refer to Chapter 4), so that the smaller the ΔE and the higher the B, the lower the temperature from which LIR_{min} can be measured.

For example, if a measurement system enables accurate measurement of $LIR_{min} = 0.01$, that is, measurement of emission from a higher energy state which is 100 times smaller (1%) than an emission from a lower energy state, then with the Dy^{3+} ion ($\Delta E = 1000\ cm^{-1}$) and $B = 2.5$, the low-temperature bound is 260.6 K. However, under the same conditions and for the same value of B, using the Eu^{3+} ion ($\Delta E = 1750\ cm^{-1}$) the lower temperature bound has a much higher value of 456 K. Fig. 6.11 shows the dependence of the low-temperature bound on ΔE for different values of B calculated from Eq. (6.4) for the hypothetical LIR detection limit of 0.01. Horizontal lines are drawn for 100, 200, and 293 K temperatures. From their intersections with the ΔE–T family of lines one can estimate the conditions (B and ΔE) for which these temperatures can be measured with LIR_{min}.

The upper temperature bound of LIR applicability thus depends mainly on the phonon spectrum of the host material and the type of lanthanide ion involved. Generally, materials with low values of phonon energies and with a large energy gap between the ground and excited states of the activator ion provide LIR measurements with a high upper temperature limit. One should note that measurement uncertainty considerably increases near temperature bounds, that is, at both ends of the measurement range.

With a rising temperature during measurements, one should also be aware of an increase in blackbody radiation that may interfere with phosphor emissions.

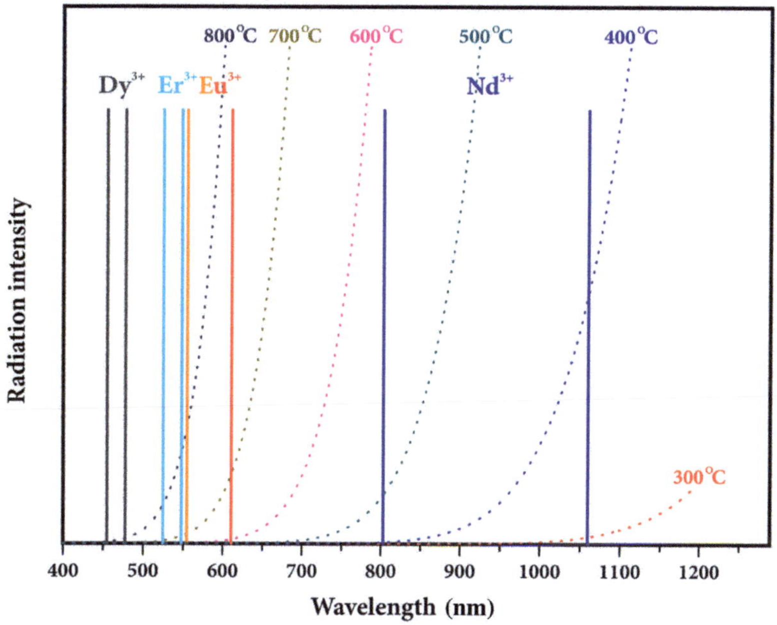

FIGURE 6.12 **Relation between Dy^{3+}, Er^{3+}, Eu^{3+}, and Nd^{3+} emissions used for luminescence intensity ratio and spectral radiance emittance (blackbody radiation) calculated for different temperatures.**

The measured signals are sums of both the blackbody radiation and the phosphor emission, and the ratio of signals does not correctly represent LIR. The problem can be overcome by measuring spectra both with and without excitation. In the absence of excitation, the signal will be composed only of the contribution of blackbody radiation, which can be later subtracted from the signal obtained with excitation.

For any given temperature, the background from blackbody radiation is higher at longer wavelengths, and, therefore, it is more effective to use phosphors whose emission wavelengths are as short as possible. The effect of blackbody radiation can be predicted using Planck's radiation law, where spectral radiant emittance (SRE) at different wavelengths and temperatures can be calculated as:

$$\mathrm{SRE}(\lambda,T)=\frac{2\pi hc^2}{\lambda^5}\cdot\frac{1}{\exp\left(\frac{hc}{\lambda k_B T}\right)-1}, \tag{6.5}$$

where λ is the wavelength of radiation, $h = 6.626\cdot10^{-34}$ J·s is Planck's constant, and $c = 2.998\cdot10^8$ m/s is the speed of light in a vacuum. Fig. 6.12 illustrates the increasing amount of blackbody radiation and its relationship with the characteristic emission pairs of Dy^{3+} (in the blue spectral region), Er^{3+} (in the green spectral region), Eu^{3+} (in the green–red spectral region), and Nd^{3+} (in the near-infrared spectral region). One can observe with ease that the LIR of short wavelength-emitting Dy^{3+} is affected much less than the LIR of near infrared-emitting Nd^{3+}.

6.5.1 Temperature Sensing Via Downshifting Emission

Practically all lanthanide ion luminescence centers may be used for Boltzmann-type LIR temperature sensing (Table 6.5). When high-energy excitation is feasible for application, downshifting emission is preferably used since signals are of much higher intensity than with upconversion. Up to now, most literature reports on Boltzmann-type downshifting LIR involve the use of Eu^{3+}, Dy^{3+}, and Nd^{3+} in a variety hosts of different chemical compositions and forms. Sm^{3+} and Pr^{3+} are used to a lesser extent.

The *Eu^{3+} ion* ($4f^6$ electron configuration) is well known for its strong luminescence and is an important ion for red-emitting phosphors. Starting levels of its transitions in both the absorption and the emission spectrum are nondegenerate ($J = 0$), which is a unique feature amid lanthanide ions with an even number of 4f electrons. Eu^{3+} luminescence is almost entirely due to the $^5D_0 \rightarrow {}^7F_J$ transitions ($J = 0$–6), among which the strongest emissions comes from the $^5D_0 \rightarrow {}^7F_1$ magnetic dipole transition near 600 nm and the $^5D_0 \rightarrow {}^7F_2$ forced electric dipole transition around 610–630 nm. Transitions from higher excited states ($^5D_1$, $^5D_2$, and $^5D_3$) are of considerably smaller intensities. The $^5D_1$ level is split by the crystal field in an identical manner to the $^7F_1$ level but with a magnitude of about five times smaller. Those interested in the detailed analyses of Eu^{3+} spectroscopic properties may read the excellent and comprehensive review of Binnemans [26].

For LIR, emissions from thermally coupled $^5D_1$ and $^5D_0$ states are of interest. The energy gap between these states is roughly 1750 cm^{-1}, and is the largest energy gap among all trivalent lanthanide ions used for LIR thermometry. For this reason, Boltzmann-type LIR with this ion reaches the highest relative sensitivity and highest temperature measurement bounds. LIR may be realized, in principle, by any combination that takes one of the $^5D_1 \rightarrow {}^7F_J$ and one of the $^5D_0 \rightarrow {}^7F_J$ emissions. Most frequently, for emissions of low-energy states, $^5D_0 \rightarrow {}^7F_2$ or $^5D_0 \rightarrow {}^7F_1$ emissions are used (since they are usually the strongest). But transitions commencing to other ^7F_J levels may be used, including $^7F_5$ and $^7F_6$ levels which sometimes are not observed, because they are outside of the detection range of most commercial instruments. The $^5D_1 \rightarrow {}^7F_{0,1,2}$ transitions provide emissions in the range 520–560 nm and any of them (if they are sufficiently intense) can be used in LIR as emissions from high-energy states. Other $^5D_1$ emissions have longer wavelengths and overlap $^5D_0$ emissions. One can conclude that Eu^{3+} offers versatile emissions suitable for LIR. On the other hand, there are more than a few serious drawbacks to its use. First, it cannot be used at low temperatures, since sufficiently high temperatures are needed to populate the $^5D_1$ level from $^5D_0$; emissions from the $^5D_1$ state are quite weak even at room temperature. Measurements of weak emissions are linked with reduced precision, and thus with lower temperature resolution. Second, in many compounds the $^5D_1$ emissions are too weak to be of practical use for LIR; in molecular compounds, for example, only transitions from the $^5D_0$ excited state are regularly observed at room temperatures. Also, high Eu^{3+} concentrations which provide strong emissions cannot be used because high concentrations favor quenching of emissions of the $^5D_1$ state (stronger than $^5D_0$). Last, one should note that emissions around 530 nm are, in fact, an overlap of the $^5D_1 \rightarrow {}^7F_1$ and $^5D_2 \rightarrow {}^7F_4$ emissions.

So far, Eu^{3+} emissions have been employed for Boltzmann-type LIR in $LiNbO_3$ crystals [27] and powders of $GdVO_4$ [28], Gd_2O_3 [29], $NaEuF_4$ [30], $GdAlO_3$ [31], $YNbO_4$ [32], and $CaEu_2(WO_4)_4$ [33]. It has also been used in nanocrystals of $Gd_2Ti_2O_7$ [34], $(Y_{0.75}Gd_{0.25})_2O_3$ [10], and Lu_2O_3 [35], as well as in calibo and tellurite glass [36].

The *Dy^{3+} ion* ($4f^9$ electron configuration) with its $^6H_{15/2}$ ground state displays efficient emissions due to transitions from the I and F levels. Due to two dominant blue (470–500 nm; $^4F_{9/2} \rightarrow {}^6H_{15/2}$) and yellow (570–600 nm; $^4F_{9/2} \rightarrow {}^6H_{13/2}$) emissions it offers potential use in WLEDs, mercury-free lamps, and long-lasting phosphors. It also presents strong near-infrared emissions from $^6H_{9/2},{}^6F_{11/2} \rightarrow {}^6H_{15/2}$, $^6H_{11/2} \rightarrow {}^6H_{15/2}$, and $^6H_{13/2} \rightarrow {}^6H_{15/2}$ transitions, the last of which may be utilized for lasing. Direct ultraviolet excitation of this ion is not effective since charge transfer and $4f^85d^1$ states have large energies, however, it can be efficiently excited by host complex ions, such as in the case of YVO_4.

Transitions from the $^4I_{15/2}$ and $^4F_{9/2}$ excited levels provide emissions that can be used for LIR. The energy gap between levels is relatively high, about 1000 cm^{-1}, and provides a 1.6%K^{-1} relative sensitivity at room temperature. It also enables use of Dy^{3+} at lower temperatures than is possible with Eu^{3+}. The $^4I_{15/2} \rightarrow {}^6H_{15/2}$ transition provides emission around 455 nm which is used as a high-energy emission in LIR. Blue emission around 488 nm (from the $^4F_{9/2} \rightarrow {}^6H_{15/2}$ transition) is used as a low-energy emission in most cases, since it is of strong intensity and because blackbody radiation at high temperatures causes less interference. Yellow emission around 570 nm (from the $^4F_{9/2} \rightarrow {}^6H_{13/2}$ transition) is usually also intense and thus may be utilized. Other emissions from the $^4F_{9/2}$ level are of smaller intensities, but may be of practical interest, and are in the red (around 660 nm; $^4F_{9/2} \rightarrow {}^6H_{11/2}$ transition) and near-infrared spectral region (around 735 nm; $^4F_{9/2} \rightarrow {}^6H_{9/2},{}^6F_{11/2}$ transition).

So far, Boltzmann-type LIR is demonstrated with Dy^{3+} in $Y_4Al_2O_9$ crystals [37], in $GdVO_4$ thin film prepared by the PLD technique [38], in powders of Al_2O_3 [39], $Y_3Al_5O_{12}$ [40,41], and $GdVO_4$ [28], and in nanocrystals of LaF_3 [42], $BaYF_5$ [42], $Gd_2Ti_2O_7$ [43], and $Y_3Al_5O_{12}$ [44].

The *Nd^{3+} ion* ($4f^3$ electronic configuration) with its $^4I_{9/2}$ ground state is used as a dopant in active media of many solid-state lasers; the most famous ones being Nd doped glass and YAG ($Y_3Al_5O_{12}$) lasers. This ion presents emission bands in the near-infrared spectral region, from 870 to 950 nm ($^4F_{3/2} \rightarrow {}^4I_{9/2}$ transition), from 1050 to 1120 nm ($^4F_{3/2} \rightarrow {}^4I_{11/2}$ transition), from 1230 to 1420 nm ($^4F_{3/2} \rightarrow {}^4I_{13/2}$ transition), and from 1700 to 1800 nm ($^4F_{3/2} \rightarrow {}^4I_{15/2}$ transition). Emissions from $^4F_{7/2}$ to $^4F_{5/2}$ are of lesser intensity and some of them overlap $^4F_{3/2}$ emissions. Regarding luminescence thermometry, Nd^{3+} emissions in near-infrared are valuable for biomedical applications since they fall in the spectral region where cells and tissues have the lowest absorption (so-called biological windows). Consequently, Nd^{3+} can be optically excited safely inside biological objects, for example with 808 nm laser radiation, which is a nonheating, nondamaging wavelength that can be provided using a relatively cheap laser diode. Also, Nd^{3+} emission will not be significantly absorbed and interfered by the autofluorescence of cells and tissues. Knowing that bioimaging with a range of Nd^{3+} doped particles has been successfully demonstrated [45], its use may facilitate the development of dual-thermal and luminescent bioimaging.

For LIR, the energy levels of interest for Nd^{3+} are $^4F_{7/2}$, $^4F_{5/2}$, and $^4F_{3/2}$. The energy difference of 1000 cm^{-1} provides a relative sensitivity of 1.6%K^{-1} at 300 K, the same as for the Dy^{3+}. In addition, the ratio of emissions from the $^4F_{3/2}$ Stark sublevels can be also exploited, but with a low relative sensitivity of about 0.16%K^{-1}, because of the small energy gap between the Stark sublevels (of the order of 100 cm^{-1}). To date, the LIR of Nd^{3+} emissions in several host materials has been tested in: La_2O_2S [46], β-$NaYF_4$ [47], silica fiber [48], and nanocrystals of LaF_3 [49,50], α-$NaYF_4$ [51], Gd_2O_3 [52], $LiNdP_4O_{12}$ [53], $Y_3Al_5O_{12}$ [54], and YVO_4 [55].

Pr^{3+} and Sm^{3+} ions. Pr^{3+} ($4f^2$ electronic configuration) with its $^3H_4$ ground state has interesting and advanced applications. It can provide photon cascade emission with an internal quantum efficiency larger than 100%, it is an activator in scintillator crystals and ceramics, it is an activator in conventional phosphors ($4f^2$ emissions), provides ultraviolet emission from $4f^15d^1 \rightarrow 4f^2$ transitions, is used in laser gain media, and to store information in quantum computing. The LIR method uses $4f^2$ emission of Pr^{3+} that originates from the $^3P_1 + {}^1I_6$ and $^3P_0$ levels, which are separated by approximately 580 cm^{-1}. This energy gap allows temperature measurements with a relative sensitivity of 0.93%K^{-1} on average. Literature reports exist on the use of Pr^{3+} doped powders of $(K_{0.5}Na_{0.5})NbO_3$ [56] and β-$NaYF_4$ [57], as well lithium tellurite glass [58,59].

Emissions of Sm^{3+} ($4f^5$ electronic configuration), with the $^6H_{5/2}$ ground state, have not found many applications, but may be valuable for luminescence temperature sensing. They are observed at about 560 nm ($^4G_{5/2} \rightarrow {}^6H_{5/2}$ transition), 610 nm ($^4G_{5/2} \rightarrow {}^6H_{7/2}$ transition), 650 nm ($^4G_{5/2} \rightarrow {}^6H_{9/2}$ transition), and 710 nm ($^4G_{5/2} \rightarrow {}^6H_{11/2}$ transition). Any of these emissions may be used in LIR as a low-energy emission, but usually the red emission from the $^4G_{5/2} \rightarrow {}^6H_{7/2}$ transition is the strongest and thus the preferred option. The $^4F_{3/2} \rightarrow {}^6H_{5/2}$ transition around 530 nm affords high-energy emission for LIR. The energy gap between the $^4F_{3/2}$ and $^4G_{5/2}$ levels is roughly 1000 cm^{-1}, as in the case of Nd^{3+} and Dy^{3+}, so it facilitates temperature measurements with 1.6%K^{-1} at 300 K. So far, powders of Sm^{3+} doped Lu_2O_3 [35], $YNbO_4$ [32], $GdVO_4$ [60], and Y_2O_2S [61] have been used for LIR.

Relative sensitivities of Boltzmann-type LIR temperature measurements are limited by the size of the energy gap between thermally coupled excited levels. The largest possible relative sensitivity is 2.8%K^{-1} at 300 K, and it becomes smaller as the temperature increases, which may be insufficient for many applications since low sensitivity may result in a low temperature resolution (see Chapter 2, for details on the figures of merit of temperature measurements). The strategy to increase measurement sensitivity of LIR involves the use of two emissions from different luminescence centers. Ideally, one emission should be strongly quenched by temperature (sensor temperature *I*), while the other one should be temperature independent or just slightly affected over the measurement range (reference emission I_R). This situation is easily theoretically explained. The temperature dependence of LIR follows changes of sensor emission intensity with temperature and can be formulated as (see Chapter 3):

$$\mathrm{LIR}(T) = C \cdot k_R \cdot \tau(T) = C \cdot \frac{k_R}{k_R + k_{NR}(T)}, \tag{6.6}$$

where k_R and k_{NR} are the radiative and nonradiative transition rates, τ is the lifetime of the excited state from which the sensed emission originates, and C is a constant. It turns out that temperature dependence of LIR is the same (only multiplied by a constant value) as the temperature dependence of the excited state's lifetime.

Such a situation exists when lanthanide ions are integrated into the wide band-gap nanocrystalline materials that show luminescence form its defect states. Temperature can be measured from the LIR of lanthanide emission (the sensing emission) and trap emission (the reference emission). Examples of this approach are Eu^{3+} and Sm^{3+} doped anatase TiO_2 nanocrystals [62,63], Dy^{3+} and Eu^{3+} doped $Gd_2Ti_2O_7$ nanocrystals [34,43], and Eu^{3+} doped $SrZrO_3$ nanocrystals [64]. The maximum relative sensitivity of 10.54%K^{-1} (at 65°C) is found for the LIR of the Sm^{3+} emission at 612 nm and the TiO_2 trap emission at 438 nm [63] (Fig. 6.13).

The LIR from two different centers may also be realized using two different lanthanide ions incorporated into one host material, such as in the case of β-$NaYF_4$:Ce^{3+},Tb^{3+},Eu^{3+} with the LIR of Tb^{3+} ($^5D_4 \rightarrow {}^7F_5$ transition) and Eu^{3+} ($^5D_0 \rightarrow {}^7F_2$ transition) emissions [65]. Similarly, in the Tb^{3+},Eu^{3+} doped $LiLaP_4O_{12}$ nanocrystals, a ratiometric temperature reading was achieved from the LIR of Tb^{3+} and Eu^{3+} emissions [66]. The LIR of near-infrared emissions from two lanthanide ions, the 1060 nm emission of Nd^{3+} and the 980 nm emission of Yb^{3+}, incorporated into the β-$NaYF_4$ core-shell nanoparticle, has been successfully demonstrated [67]. In this case the nanoparticle consisted of a Yb-Er codoped β-$NaYF_4$ core and a Yb-Nd codoped β-$NaYF_4$ shell. The approach based on the use of one lanthanide emission and one emission from a transition metal ion will be discussed later in the text.

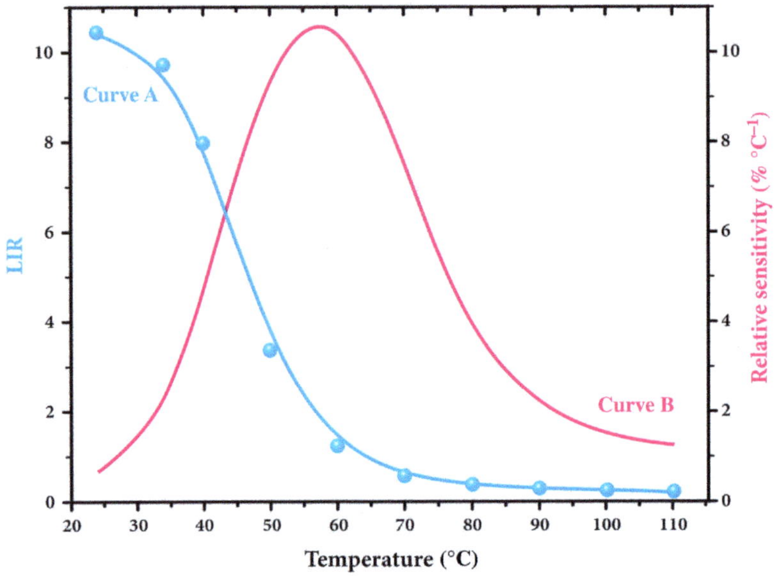

FIGURE 6.13 **Luminescence intensity ratio of Sm^{3+} doped anatase TiO_2 powder (*curve A*) obtained as the ratio between Sm^{3+} emission at 612 nm and TiO_2 trap emission at 438 nm, and corresponding relative sensitivity (*curve B*).** *(Source: Reproduced from Dramićanin MD, Antić Ž, Ćulubrk S, Ahrenkiel SP, Nedeljković JM. Self-referenced luminescence thermometry with Sm^{3+}doped TiO_2 nanoparticles. Nanotechnology 2014;25(48):6, Article ID: 485501, with permission from IOP Publishing).*

Temperature sensing from the temporal behavior of lanthanide luminescence is mostly used for high-temperature applications. Lifetimes of lanthanide excited states only slightly change at low temperatures, but significantly decrease at high temperatures (see Chapters 3 and 4 for more details). In fact, the lifetime based read-outs of high temperatures from the luminescence of lanthanides are superior to those from LIR in terms of accuracy, sensitivity, cross-sensitivity, and imaging ability (shot-to-shot and pixel-to-pixel precision) [68]. Depending of the nominal (room temperature) duration of the first excited state lifetime ($4f^n$ emissions, 5d emission for Ce^{3+}), lanthanide ions can be classified to: short (Ce^{3+}, Pr^{3+}) of microsecond order or less; moderate (Nd^{3+}, Ho^{3+}, Er^{3+}, and Tm^{3+}) up to a few 100 μs; and long (Eu^{3+}, Sm^{3+}, Tb^{3+}, Dy^{3+}, and Yb^{3+}) of the order of milliseconds or longer. Please note that the lifetime values may significantly differ between different host materials.

Since they are mainly used for surface temperature measurements, phosphors are prepared in the form of coatings, films, and paints. Coatings are adhered to the surface either by direct synthesis of the phosphor layer on the surface of interest, or by mixing of phosphor powder with a binder and then by subsequent application on the surface. A range of lanthanide doped materials is used for the lifetime read-out of temperature including the well-known

Dy^{3+} doped $Y_3Al_5O_{12}$, which enabled the highest ever temperature measured by luminescence thermometry of 2000 K [69]. Among these materials, the most widely used are aluminates ($Y_3Al_5O_{12}$, $YAlO_3$, and $GdAlO_3$), sesquioxides (Y_2O_3, Gd_2O_3, Lu_2O_3, and Sc_2O_3), phosphates ($LaPO_4$, $LuPO_4$, and $ScPO_4$), silicates (Y_2SiO_5, $Y_2Si_2O_7$), zirconates ($Y_2Zr_2O_7$, $Gd_2Zr_2O_7$), and vanadates (YVO_4, $GdVO_4$). Comprehensive surveys of materials used for lifetime-based luminescence can be found in [8,22,70]. The use of risetime temperature read-out is scarce even though it offers an appealing performance. So far this has been demonstrated with Eu^{3+} doped Y_2O_3, SrY_2O_4, and BaY_2ZnO_5 [71–73].

The use of time-resolved methods with lanthanide luminescence suffers some disadvantages. The signal may lack strength since excitation in pulsed form is only available for fractions of the time. The more serious problem occurs due to the high sensitivity of the excited state lifetimes of phosphor materials on their structural and morphological properties, and especially on the concentration of dopant ions. It is quite difficult to make phosphor reproducible regarding its excited state lifetimes, which complicates calibration and the routine practice of the method. Also, luminescence lifetimes of lanthanide ions show cross-sensitivities against their chemical environment and/or physical conditions (such as pressure). The changes in the physicochemical environment of phosphor materials may induce significant problems during temperature measurements.

6.5.2 Temperature Sensing Via Upconversion Emission

The temperature read-out schemes that utilize upconversion emission are the same as those using downshifting emission. The main difference is in the excitation of luminescence. For upconversion, strong near-infrared radiation sources are used, most commonly ones with 980 nm radiation where the Yb^{3+} ion has an absorption band. Low-energy, near-infrared excitations are not harmful to biological objects, therefore with upconversion materials, it is possible to measure temperature in different biological applications. For these uses, materials should be in the form of nanoparticles. However, the quantum efficiency of upconversion processes is much smaller than the efficiency of downshifting so thermometry may suffer from low signal strength. This drawback is amplified when using nanoparticles. They exhibit lower efficiencies of upconversion emission than their bulk counterparts. Considering the low efficiency of the ESA processes compared with the ET from a sensitizer to an activator ion, different hosts, usually those with low phonon energies, codoped with Yb^{3+},Re^{3+} (Re = Er, Ho, and Tm) are the materials of choice. Upconversion emission decay times are not a good option for thermometry. Most often they are not of single exponential shape and are shorter than those in downshifting process with the same lanthanide ions. Accordingly, LIR is the favorable option and almost exclusively employed.

The *Yb^{3+} ion* ($4f^{13}$ electronic configuration) has just two states, $^2F_{5/2}$ (the excited state) and $^2F_{7/2}$ (the ground state), with energy differences between them of approximately 10,000 cm^{-1}, which is the result of a large spin-orbit coupling. This energy difference is similar to the difference between some combinations of energy levels in Er^{3+}, Ho^{3+}, and Tm^{3+}, easing the ETs between Yb^{3+} and these ions when they are located close to each other. The absorption near 980 nm is relatively strong (compared with the absorption of other lanthanide ions) and is used for sensitizing upconversion emission. One should note that Yb^{3+} ions can be exploited for downshifting thermometry with the LIR of near-infrared emissions from the $^2F_{5/2}$ Stark sublevels, since they have an energy difference of roughly 680 cm^{-1}.

The *Er^{3+} ion* ($4f^{11}$ electronic configuration) with its $^4I_{15/2}$ ground state, has a dense structure of energy levels. Energy transitions between these levels emit ultraviolet, visible, and infrared radiation. It is widely used for upconversion, but also in optical fibers for the optical amplification of 1550 nm semiconductor laser light. It is the most used ion for luminescence thermometry, with or without sensitizing by Yb^{3+}. The LIR method with Er^{3+} ions exploit the green emissions from $^2H_{11/2}$ and $^4S_{3/2}$ levels (which are separated by about 780 cm^{-1}) to the ground state centered near 525 and 548 nm, respectively. Literature reports on the use of the Er^{3+} ion for upconversion LIR include the following hosts: α-$NaYF_4$ [74], ZnO [75], Al_2O_3 [76], Y_2O_3 [77–79], $CaWO_4$ [80], $SrWO_4$ [81], $BaMoO_4$ [82], and YVO_4 [83].

The *Ho^{3+} ion* ($4f^{10}$ electronic configuration) with its $^5I_8$ ground level also has an abundant number of closely spaced energy levels which can facilitate upconversion emission. In the majority of host materials, it intensively emits green luminescence around 540 nm due to the $^5F_4,^5S_2 \rightarrow {^5I_8}$ transition. The spectra also present red emission centered at 660 nm (the $^5F_5 \rightarrow {^5I_8}$ transition), which in some rare cases may be stronger than the green emission, and relatively weak near-infrared emission near 755 nm (the $^5F_4,^5S_2 \rightarrow {^5I_7}$ transition). Sometimes it is possible to observe emissions in the ultraviolet-blue spectral region due to transitions from higher energy excited states. These include $^5G_4 \rightarrow {^5I_8}$ (390 nm), $^3D_3 \rightarrow {^5I_6}$ (420 nm), $^5G_5 \rightarrow {^5I_8}$ (428 nm), and $^5F_3 \rightarrow {^5I_8}$ (490 nm), and are the consequence of an upconversion processes which involves more than two excitation photons. Also, one should note that for several crystals, laser action has been achieved using the $^5F_5 \rightarrow {^5I_8}$ transition.

Regarding LIR, the Ho^{3+} ion offers three pairs of thermally coupled excited levels: $^5F_4$ and $^5S_2$, $^5F_{2,3}$ and $^3K_8$, and $^5G_6$ and $^5F_1$. The energy difference between them is relatively small, of the order of 120 cm^{-1}; consequently, relative sensitivities of measurements are also small, about 0.19%K^{-1} at room temperature. In the case of the Ho^{3+} ion, it is also possible to analyze the LIR of emissions from thermally coupled levels to other excited states and the LIR of emission from thermally coupled levels to the ground state using the four-level scheme [84,85]. So far, LIRs of Ho^{3+} upconversion emissions have been demonstrated in: TeO_2 [86] and $LiTeO_2$ [87] glass, and powders of Y_2O_3 [85] and $Ba_{0.77}Ca_{0.23}TiO_3$ [88].

Tm^{3+} ion ($4f^{12}$ electron configuration) with its $^3H_6$ ground state, dominantly emits in the blue spectral region near 475 nm ($^1G_4 \rightarrow {^3H_6}$ transition) and the near-infrared region around 800 nm (the $^3H_4 \rightarrow {}^3H_6$ transition), with small deep-red emissions near 650 nm (the $^1G_4 \rightarrow {}^3F_4$ transition) and 700 nm (the $^3F_{2,3} \rightarrow {}^3H_6$ transition). The efficiency of the ultraviolet and blue emission of Tm^{3+} is relatively low due to efficient competitive infrared emission. In some materials with low-phonon energy, ultraviolet upconversion emission centered at 360 nm ($^1D_2 \rightarrow {}^3H_6$ transition) may be evidenced. The $^3H_4 \rightarrow {}^3H_6$ transition in some crystals has been utilized as a near-infrared laser transition.

Two couples of excited levels are of interest for temperature reading from the LIR of their emissions. Lower lying $^3F_{2,3}$ and $^3H_4$ excited levels are populated by the two-photon upconversion process and have a large energy difference between them of about 1700 cm^{-1}, which provides a 2.72%K^{-1} relative sensitivity at 300 K. A higher positioned couple are the two Stark components of the $^1G_4$ level which are populated by the three-photon upconversion process, with a very small energy difference between them of about 100 cm^{-1}, and, therefore providing low temperature sensitivity of 0.16%K^{-1} at 300 K. Examples of Tm^{3+} use in upconversion LIR thermometry include: Y_2O_3 [85,89], $NaNbO_3$ [90], and $NaLuF_4$ [91].

6.6 TRANSITION ION DOPED MATERIALS FOR TEMPERATURE SENSING

The luminescence of transition metal ion centers is more strongly quenched by temperature than the luminescence of lanthanide ions. Several options are available to take advantage of this lavish temperature dependence of luminescence for self-referencing thermometry. One is associated with the measurements of the spectral positions of *R* lines, their emission ratios, or ratios of the emission sidebands of the *R* lines at different temperatures.

Cr^{3+} and Mn^{4+} ion emissions ($3d^3$ electronic configuration) in many host materials have been investigated for this purpose so far. Ruby (Cr^{3+} doped Al_2O_3) was among the first. The *R* lines of ruby are two strong emission lines at 692.7 and 694.3 nm. It was found that the ratio of intensities of these lines changes with temperature and can be used for temperature sensing [92,93]. *R* lines in Cr^{3+} doped YAG ($Y_3Al_5O_{12}$) are redshifted with increasing temperature and the intensities of their sidebands vary with temperature [94]. The temperature dependence of the Cr^{3+} emission in a strong crystal field is primarily governed by the thermalization between its $^4T_2$ and ^2E states, and in some instances by the phonon induced $^2E \rightarrow {}^4A_2$ transitions [95], see the simplified Tanabe–Sugano diagram in Fig. 6.14A. A similar situation occurs in Mn^{4+} doped Mg_4FGeO_6. Fig. 6.14B displays emission spectra measured at three temperatures, where the intensity of the 631-nm line increases with temperature relative to the intensity of the 657-nm line [96]. The intensities of these lines increase linearly with temperature in the 303–373 K temperature range [96] (Fig. 6.14C). However, the relative sensitivity of the LIR is relatively poor, only 0.4%K^{-1} at 303 K.

Borisov et al. [97] have demonstrated that both the LIR and lifetime reading of temperature can be accomplished with Cr^{3+} doped $YAl_3(BO_3)_4$ (YAB).

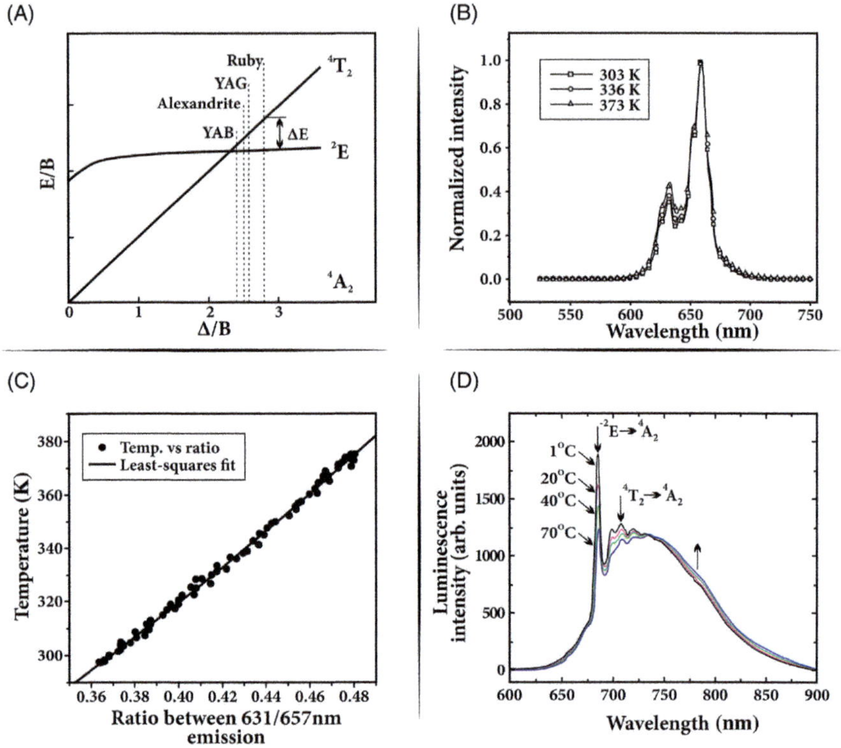

FIGURE 6.14 (A) Simplified Tanabe–Sugano diagram for the $3d^3$ electronic configuration *(Adapted with permission from Zhang ZY, Grattan KT, Palmer AW, Fernicola V, Crovini L. Temperature dependence of the YAG:Cr^{3+}fluorescence lifetime over the range 77 to 900 K. Phys Rev B 1995;51(5):2656–60, copyright (1995) American Physical Society).* (B) The emission of Mn^{4+} doped Mg_4FGeO_6 at 303, 336, and 373 K. (C) The linear dependence on temperature of the intensity ratio of the 631-nm and 657-nm lines in Mn^{4+} doped Mg_4FGeO_6 *(Reproduced with permission from Omrane A, Juhlin G, Ossler F, Aldén M. Temperature measurements of single droplets by use of laser-induced phosphorescence. Appl Opt 2004;43(17):3523–9, copyright (2004) Optical Society of America).* (D) Temperature dependence of the luminescence intensity for Cr^{3+} doped YAB. *(Adapted from Borisov SM, Gatterer K, Bitschnau B, Klimant I. Preparation and characterization of Chromium(III)-Activated Yttrium Aluminum Borate: a new thermographic phosphor for optical sensing and imaging at ambient temperatures. J Phys Chem C 2010;114(19):9118–24, copyright (2010) American Chemical Society).*

Sharp emission from the $^2E \rightarrow {}^4A_2$ transition (*R* line), centered at 685 nm, lessens in intensity when temperature rises from 273 K (Fig. 6.14D). On the contrary, the broadband emission of lower energy, which occurs due to the vibronic $^4T_2 \rightarrow {}^4A_2$ transition, was almost temperature insensitive. The ratio of these emissions showed a sensitivity to temperature. However, owing to its high brightness (compared to that of the well-known Mn^{4+} doped magnesium fluorogermanate), existence of an excitation band in the red spectral region, and the absence of cross-sensitivity to oxygen, this phosphor is a promising candidate for luminescence thermometry applications. A recent report [98], showed that Cr^{3+} doped $Bi_2Ga_4O_9$ mullite powders can also be also for intensity ratio temperature sensing in the first biological window in the same manner as Cr^{3+} doped $YAl_3(BO_3)_4$. The material shows sharp emissions (*R* lines) at 702 and 710 nm from the spin-forbidden $^2E \rightarrow {}^4A_2$ transition, and at broadband emission of 800 nm from the spin-allowed $^4T_2 \rightarrow {}^4A_2$ transition. The ratio of these emissions changes with temperature over the range 77–450 K, and has a 0.7%K^{-1} relative sensitivity at room temperature.

In addition to relatively poor sensitivity, the drawback of these approaches lies in the difficult determination of *R* lines in some host materials—those not presenting strong crystal fields and showing only a broad emission band.

However, the codoping of lanthanide ions in such crystals or codoping of transition metal ions to some other materials facilitates the LIR of dual-emission centers and may pave a new direction to self-referencing temperature sensing with high sensitivity. Three alternatives are foreseen for the use of transition metal ions in materials with two emission centers: (1) transition metal ion doped nanoparticles that display intrinsic defect state emission, (2) materials activated with both transition metal and lanthanide ions, and (3) transition metal doped quantum dots (QDs). In all three cases, the temperature can be derived from the LIR of emissions coming from different luminescence centers. The first two alternatives will be described further in this chapter. The use of transition metal doped QDs for temperature sensing is outlined in Chapter 7.

The concept of temperature sensing using the ratio of the transition metal ion emission and the trap emission of the host material has been originally demonstrated with Mn^{2+} doped Zn_2SiO_4 [99] (Fig. 6.15A). Due to tetrahedral coordination and a weak crystal field in this host, Mn^{2+} ions ($3d^5$ electron configuration) emit a green broadband emission at 527 nm (the $^4T_1(G) \rightarrow {}^6A_1(S)$ transition), which is strongly quenched by temperature over the 303–373 K range. Broad and low-intensity Zn_2SiO_4 trap emission (maximum at 437 nm), on the other hand, shows no temperature dependence and serves as an excellent reference. The emission intensity ratio proved to be highly sensitive to temperature. This material also provides the opportunity for multiple read-outs of temperature. In addition to the intensity ratio read-out, emission decay rapidly changes over the entire temperature range, and the Mn^{2+} emission band position shifts toward smaller wavelengths with a temperature increase of about 0.011 nm/K.

YAG powder codoped with transition metal and lanthanide ions is one interesting perspective to consider in terms of luminescence thermometry [100]. The ratio of highly temperature-sensitive emissions of Mn^{4+} or Cr^{3+} over low-sensitive Eu^{3+}, Tb^{3+}, or Dy^{3+} emissions shows an exceptional temperature-sensing performance with relative sensitivities as high as 2.32–4.81%K^{-1}. In all combinations, both transition metal and lanthanide ions can be excited with the same radiation. Fig. 6.15B shows a comparison of the Mn^{4+} and Eu^{3+} emissions in YAG; Mn^{4+} emission intensity decreases quickly and Eu^{3+} emission intensity only slightly when temperature is increased from room temperature to 400 K [100].

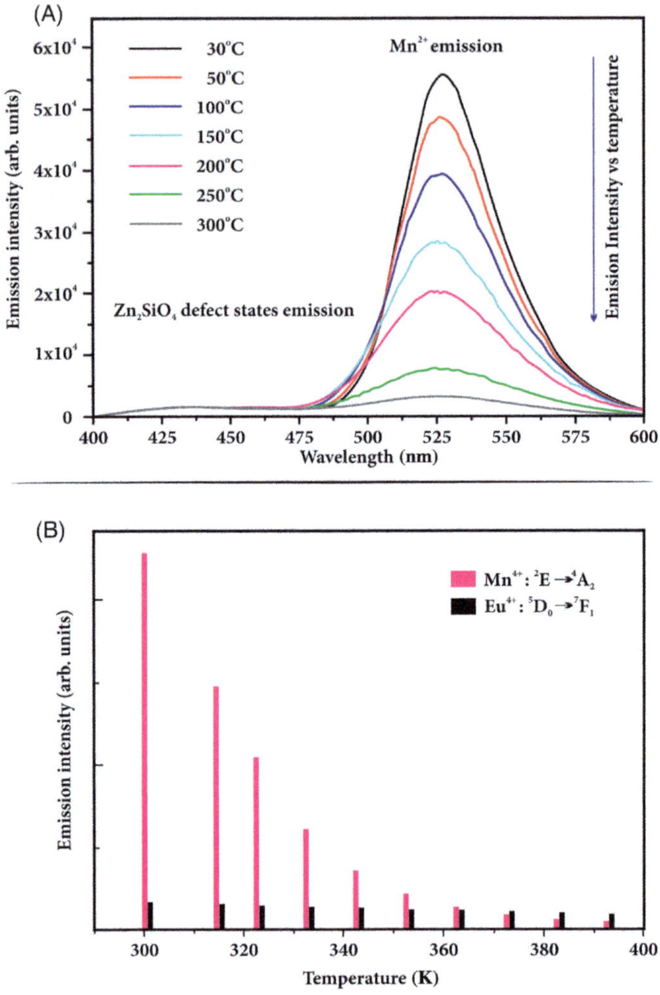

FIGURE 6.15 (A) Emission spectra of Zn_2SiO_4:Mn^{2+} powder at different temperatures taken under 370-nm excitation (Reprinted from Lojpur V, Nikolić MG, Jovanović D, Medić M, Antić Ž, Dramićanin MD. Luminescence thermometry with Zn_2SiO_4:Mn^{2+}powder. Appl Phys Lett 2013;103(14):3, Article ID: 141912, with permission from AIP Publishing). (B) Comparison of Mn^{4+} and Eu^{3+} emission intensities at different temperatures from Mn^{4+},Eu^{3+} codoped YAG ($Y_3Al_5O_{12}$) powder. *(Adapted from Chen D, Liu S, Zhou Y, Wan Z, Huang P, Ji Z. Dual-activator luminescence of RE/TM:$Y_3Al_5O_{12}$ (RE = Eu^{3+}, Tb^{3+}, Dy^{3+}; TM = Mn^{4+}, Cr^{3+}) phosphors for self-referencing optical thermometry. J Mater Chem C 2016;4(38):9044–51, with permission from The Royal Society of Chemistry).*

The LIR of near-infrared emissions of transition metal and lanthanide ions have been recently demonstrated in $LiLaP_4O_{12}$:Cr^{3+},Nd^{3+} nanocrystals [101]. The ratio is made from part (820–840 nm) of the Cr^{3+} emission (the $^4T_2 \rightarrow {}^4A_2$ transition) and Nd^{3+} emission centered at 1048 nm (the $^4F_{3/2} \rightarrow {}^4I_{11/2}$ transition). It shows strong temperature dependence, about 4.89%K^{-1} at 323 K. The intensity ratio of transition metals and lanthanide ions is also demonstrated with Cr^{3+} and Yb^{3+} in the same host nanomaterial [102]. A peculiarity was in the use of two excitations with different energies, one to excite the Cr^{3+} ions and other for Yb^{3+}.

Even though intensity ratio from dual centers is becoming more popular these days, self-referencing luminescence thermometry can also be also constructed from temperature-induced changes in emission decay times. It is the traditionally preferred technique with transition metal ions. So far, the performance of many materials has been tested in different applications and over wide temperature ranges—from cryogenic to very high temperatures. In most cases materials were coated on the surface of interest or attached to the tip of an optical fiber probe. As in the LIR method, luminescence of Cr^{3+} and Mn^{4+} ions were almost exclusively utilized. For example, lifetime temperature sensing was demonstrated with Cr^{3+} doped Al_2O_3 (ruby) [103,104], $Be_3Al_2(SiO_3)_6$ (emerald) [105], $BeAl_2O_4$ (alexandrite) [106], $Y_2Al_5O_{12}$ [107,108], $YAlO_3$ [109], $YAl_3(BO_3)_4$ and $GdAl_3(BO_3)_4$ [97], $Gd_3Ga_5O_{12}$ [110], $LiCaAlF_6$ and $LiSrAlF_6$ [111], $MgAl_2O_4$ [112], and Mg_2SiO_4 [113]; and Mn^{4+} doped Mg_4FGeO_6 [114], $TiMg_2O_4$ [115,116], $LiAl_5O_8$ [116], and $ZnGa_2O_4$ [116]. In some cases, lifetime readings of temperature may be achieved from doubly doped materials, like in the approach already considered with the intensity ratio method. Typical examples are alumina films codoped with Cr^{3+} and Dy^{3+}, and Cr^{3+} and Tm^{3+} [117]. This approach enables significant extension of measurement range.

REFERENCES

[1] Avram NM, Brik MG. Optical properties of 3d-ions in crystals: spectroscopy and crystal field analysis. Berlin Heidelberg: Springer-Verlag; 2013. p. VIII, 268.

[2] Maldiney T, Bessière A, Seguin J, Teston E, Sharma SK, Viana B, et al. The in vivo activation of persistent nanophosphors for optical imaging of vascularization, tumours and grafted cells. Nat Mater 2014;13(4):418–26.

[3] Medić MM, Brik MG, Dražić G, Antić ZM, Lojpur VM, Dramićanin MD. Deep-red emitting Mn^{4+} doped Mg_2TiO_4 nanoparticles. J Phys Chem C 2015;119(1):724–30.

[4] Henderson B, Imbusch GF. Spectroscopy of lanthanide (rare-earth) and actinide ions in solids. Optical spectroscopy of inorganic solids. Oxford: Clarendon Press; 2006. p. 387–407.

[5] Judd BR. Optical absorption intensities of rare-earth ions. Phys Rev 1962;127(3):750–61.

[6] Ofelt GS. Intensities of crystal spectra of rare-earth ions. J Chem Phys 1962;37(3):511–20.

[7] Dorenbos P. The 5d level positions of the trivalent lanthanides in inorganic compounds. J Lumin 2000;91(3–4):155–76.

[8] Dramićanin MD. Sensing temperature via downshifting emissions of lanthanide-doped metal oxides and salts. A review. Methods Appl Fluoresc 2016;4(4):23. 42001.

[9] Srivastava AM, Ronda CR. Quantum-splitting systems. Luminescence: from theory to applications. Wiley-VCH Verlag GmbH & Co. KGaA; 2007. p. 75–103.
[10] Nikolić MG, Lojpur V, Antić Ž, Dramićanin MD. Thermographic properties of a Eu^{3+} doped $(Y_{0.75}Gd_{0.25})_2O_3$ nanophosphor under UV and X-ray excitation. Phys Scr 2013;87(5). 55703.
[11] Wold A, Dwight K. Solid state chemistry: synthesis, structure, and properties of selected oxides and sulfides. Netherlands: Springer; 1993. p. XII, 246.
[12] Kimura T. Molten salt synthesis of ceramic powders. Advances in ceramics—synthesis and characterization, processing and specific applications. Croatia: INTECH; 2011. p. 75–100.
[13] Saengkerdsub S, Im H-J, Willis C, Dai S. Pechini-type in-situ polymerizable complex (IPC) method applied to the synthesis of Y_2O_3:Ln (Ln = Ce or Eu) nanocrystallites. J Mater Chem 2004;14(7):1207–11.
[14] Kakihana M, Yoshimura M. Synthesis and characteristics of complex multicomponent oxides prepared by polymer complex method. Bull Chem Soc Jpn 1999;72(7):1427–43.
[15] Cushing BL, Kolesnichenko VL, O'Connor CJ. Recent advances in the liquid-phase syntheses of inorganic nanoparticles. Chem Rev 2004;104(9):3893–946.
[16] Suchanek W, Riman R. Hydrothermal synthesis of advanced ceramic powders. Adv Sci Technol 2006;45:184–93.
[17] Kodas TT, Hampden-Smith MJ. Aerosol processing of materials. Wiley-VCH; 1999. p. 712.
[18] Dhand C, Dwivedi N, Loh XJ, Jie Ying AN, Verma NK, Beuerman RW, et al. Methods and strategies for the synthesis of diverse nanoparticles and their applications: a comprehensive overview. RSC Adv 2015;5(127):105003–37.
[19] Prasad Yadav T, Manohar Yadav R, Pratap Singh D. Mechanical milling: a top down approach for the synthesis of nanomaterials and nanocomposites. J Nanosci Nanotechnol 2012;2(3):22–48.
[20] Semaltianos NG. Nanoparticles by laser ablation. Crit Rev Solid State Mater Sci 2010;35(2):105–24.
[21] Yen WM, Shionoya S, Yamamoto H. Phosphor handbook. CRC Press/Taylor and Francis; 2007. p. 1051.
[22] Brübach J, Pflitsch C, Dreizler A, Atakan B. On surface temperature measurements with thermographic phosphors: a review. Prog Energy Combust Sci 2013;39(1):37–60.
[23] Liu WY, Li X, Chi D, Zhang H, Liu X. Lanthanide-doped upconversion materials: emerging applications for photovoltaics and photocatalysis. Nanotechnol 2014;25(48):16. 482001.
[24] Yang L, Chen M, Lv Z, Wang S, Liu X, Liu S. Preparation of a YAG:Ce phosphor glass by screen-printing technology and its application in LED packaging. Opt Lett 2013;38(13):2240–3.
[25] Wang X, Liu Q, Bu Y, Liu C-S, Liu T, Yan X. Optical temperature sensing of rare-earth ion doped phosphors. RSC Adv 2015;5(105):86219–36.
[26] Binnemans K. Interpretation of europium(III) spectra. Coord Chem Rev 2015;295:1–45.
[27] Liang Z, Qin F, Zheng Y, Zhang Z, Cao W. Noncontact thermometry based on downconversion luminescence from Eu^{3+} doped $LiNbO_3$ single crystal. Sens Actuators A Phys 2016;238:215–9.
[28] Nikolić MG, Jovanović DJ, Dramic´anin MD. Temperature dependence of emission and lifetime in Eu^{3+} and Dy^{3+}-doped $GdVO_4$. Appl Opt 2013;52(8):1716–24.
[29] Nikolić MG, Al-Juboori AZ, Đor ević V, Dramićanin MD. Temperature luminescence properties of Eu^{3+}-doped Gd_2O_3 phosphors. Phys Scr 2013;T157:5. 14056.
[30] Tian Y, Tian B, Cui C, Huang P, Wang L, Chen B. Excellent optical thermometry based on single-color fluorescence in spherical $NaEuF_4$ phosphor. Opt Lett 2014;39(14):4164–7.

[31] Lojpur V, Ćulubrk S, Medić M, Dramicanin M. Luminescence thermometry with Eu^{3+} doped $GdAlO_3$. J Lumin 2016;170:467–71.

[32] Đačanin LR, Dramićanin MD, Lukić-Petrović SR, Petrović DM, Nikolić MG. Eu^{3+} doped $YNbO_4$ phosphor properties for fluorescence thermometry. Radiat Meas 2013;56:143–6.

[33] Meert KW, Morozov VA, Abakumov AM, Hadermann J, Poelman D, Smet PF. Energy transfer in Eu^{3+} doped scheelites: use as thermographic phosphor. Opt Express 2014;22(S3):A961–72.

[34] Lojpur V, Ćulubrk S, Dramićanin MD. Ratiometric luminescence thermometry with different combinations of emissions from Eu^{3+} doped $Gd_2Ti_2O_7$ nanoparticles. J Lumin 2016;169:534–8.

[35] Lojpur V, Antić Z, Krsmanović R, Medić M, Nikolić MG, Dramićanin MD. Thermographic properties of Eu^{3+}- and Sm^{3+}-doped Lu_2O_3 nanophosphor. J Serbian Chem Soc 2012;77(12):1735–46.

[36] Rai KV, Rai A. Temperature sensing behavior of Eu^{3+} doped tellurite and calibo glasses. Appl Phys B 2007;86(2):333–5.

[37] Boruc Z, Kaczkan M, Fetlinski B, Turczynski S, Malinowski M. Blue emissions in Dy^{3+} doped $Y_4Al_2O_9$ crystals for temperature sensing. Opt Lett 2012;37(24):5214–6.

[38] Antić Ž, Dramićanin MD, Prashanthi K, Jovanović D, Kuzman S, Thundat T. Pulsed laser deposited dysprosium-doped gadolinium–vanadate thin films for noncontact, self-referencing luminescence thermometry. Adv Mater 2016;28(35):7745–52.

[39] Rakov N, Bispo LRA, Maciel GS. Temperature sensing performance of dysprosium doped aluminum oxide powders. Opt Commun 2012;285(7):1882–4.

[40] Hasegawa R, Sakata I, Yanagihara H, Johansson B, Omrane A, Aldén M. Two-dimensional gas-phase temperature measurements using phosphor thermometry. Appl Phys B 2007;88(2):291–6.

[41] Kwong WY, Steinberg A, Chin YH. Effect of B^{3+}-N^{3-} on YAG:Dy thermographic phosphor luminescence. Opt Lett 2014;39(21):6166–9.

[42] Bu YY, Cheng SJ, Wang XF, Yan XH. Optical thermometry based on luminescence behavior of Dy^{3+}-doped transparent LaF_3 glass ceramics. Appl Phys A 2015;121(3):1171–8.

[43] Ćulubrk S, Lojpur V, Ahrenkiel SP, Nedeljković JM, Dramićanin MD. Non-contact thermometry with Dy^{3+} doped $Gd_2Ti_2O_7$ nano-powders. J Lumin 2016;170:395–400.

[44] Chong J-Y, Zhang Y, Wagner BK, Kang Z. Co-precipitation synthesis of YAG:Dy nanophosphor and its thermometric properties. J Alloys Compd 2013;581:484–7.

[45] Del Rosal B, Pérez-Delgado A, Misiak M, Bednarkiewicz A, Vanetsev AS, Orlovskii Y, et al. Neodymium-doped nanoparticles for infrared fluorescence bioimaging: the role of the host. J Appl Phys 2015;118(14). 143104.

[46] Jiang G, Wei X, Zhou S, Chen Y, Duan C, Yin M. Neodymium doped lanthanum oxysulfide as optical temperature sensors. J Lumin 2014;152:156–9.

[47] Tian X, Wei X, Chen Y, Duan C, Yin M. Temperature sensor based on ladder-level assisted thermal coupling and thermal-enhanced luminescence in $NaYF_4$: Nd^{3+}. Opt Express 2014;22(24):30333–45.

[48] Wade SA, Muscat JC, Collins SF, Baxter GW. Nd^{3+}-doped optical fiber temperature sensor using the fluorescence intensity ratio technique. Rev Sci Instrum 1999;70(11):4279–82.

[49] Rocha U, Jacinto da Silva C, Ferreira Silva W, Guedes I, Benayas A, Martínez Maestro L, et al. Subtissue thermal sensing based on neodymium-doped LaF_3 nanoparticles. ACS Nano 2013;7(2):1188–99.

[50] Jaque D, Jacinto C. Luminescent nanoprobes for thermal bio-sensing: towards controlled photo-thermal therapies. J Lumin 2016;169:394–9.

[51] Wawrzynczyk D, Bednarkiewicz A, Nyk M, Strek W, Samoc M. Neodymium(III) doped fluoride nanoparticles as non-contact optical temperature sensors. Nanoscale 2012;4(22):6959–61.

[52] Balabhadra S, Debasu ML, Brites CDS, Nunes LAO, Malta OL, Rocha J, et al. Boosting the sensitivity of Nd^{3+}-based luminescent nanothermometers. Nanoscale 2015;7(41):17261–7.

[53] Marciniak L, Prorok K, Bednarkiewicz A, Kowalczyk A, Hreniak D, Strek W. Water dispersible $LiNdP_4O_{12}$ nanocrystals: new multifunctional NIR–NIR luminescent materials for bioapplications. J Lumin 2016;176:144–8.

[54] Benayas A, Del Rosal B, Pérez-Delgado A, Santacruz-Gómez K, Jaque D, Hirata GA, et al. Nd:YAG near-infrared luminescent nanothermometers. Adv Opt Mater 2015;3(5):687–94.

[55] Kolesnikov IE, Golyeva EV, Kurochkin MA, Lähderanta E, Mikhailov MD. Nd^{3+}-doped YVO_4 nanoparticles for luminescence nanothermometry in the first and second biological windows. Sens Actuators B Chem 2016;235:287–93.

[56] Tang W, Wang S, Li Z, Sun Y, Zheng L, Zhang R, et al. Ultrahigh-sensitive optical temperature sensing based on ferroelectric Pr^{3+}-doped $(K_{0.5}Na_{0.5})NbO_3$. Appl Phys Lett 2016;108(6):5. 61902.

[57] Zhou S, Jiang G, Wei X, Duan C, Chen Y, Yin M. Pr^{3+}-doped-$NaYF_4$ for temperature sensing with fluorescence intensity ratio technique. J Nanosci Nanotechnol 2014;14(5):3739–42.

[58] Rai KV, Rai SB. A comparative study of FIR and FL based temperature sensing schemes: an example of Pr^{3+}. Appl Phys B 2007;87(2):323–5.

[59] Rai VK, Rai DK, Rai SB. Pr^{3+} doped lithium tellurite glass as a temperature sensor. Sens Actuators A hys 2006;128(1):14–7.

[60] Nikolić MG, Jovanović DJ, Đor ević V, Antić Ž, Krsmanović RM, Dramićanin MD. Thermographic properties of Sm^{3+}-doped $GdVO_4$ phosphor. Phys Scr 2012;T149(8):4. 14063.

[61] Feist JP, Heyes AL. The characterization of Y_2O_2S:Sm powder as a thermographic phosphor for high temperature applications. Meas Sci Techno 2000;11(7):942–7.

[62] Nikolić MG, Antić Ž, Ćulubrk S, Nedeljković JM, Dramićanin MD. Temperature sensing with Eu^{3+} doped TiO_2 nanoparticles. Sens Actuators B Chem 2014;201:46–50.

[63] Dramićanin MD, Antić Ž, Ćulubrk S, Ahrenkiel SP, Nedeljković JM. Self-referenced luminescence thermometry with Sm^{3+}doped TiO_2 nanoparticles. Nanotechnology 2014;25(48). 485501.

[64] Das S, Som S, Yang C-Y, Chavhan S, Lu C-H. Structural evaluations and temperature dependent photoluminescence characterizations of Eu^{3+} activated $SrZrO_3$ hollow spheres for luminescence thermometry applications. Sci Rep 2016;6:13. 25787.

[65] Ding M, Zhang H, Chen D, Junhua Xi QH, Ji Z. Color-tunable luminescence, energy transfer and temperature sensing behavior of hexagonal $NaYF_4:Ce^{3+}/Tb^{3+}/Eu^{3+}$ microcrystals. J Alloys Compd 2016;672:117–24.

[66] Marciniak L, Bednarkiewicz A. The influence of dopant concentration on temperature dependent emission spectra in $LiLa_{1-x-y}Eu_xTb_yP_4O_{12}$ nanocrystals: toward rational design of highly-sensitive luminescent nanothermometers. Phys Chem Chem Phys 2016;18(23):15584–92.

[67] Marciniak L, Prorok K, Frances-Soriano L, Perez-Prieto J, Bednarkiewicz A. A broadening temperature sensitivity range with a core-shell YbEr@YbNd double ratiometric optical nanothermometer. Nanoscale 2016;8(9):5037–42.

[68] Fuhrmann N, Brübach J, Dreizler A. Phosphor thermometry: a comparison of the luminescence lifetime and the intensity ratio approach. Proc Combust Inst 2013;34(2):3611–8.

[69] Cates MR, Allison SW, Jaiswal SL, Beshears DL. YAG: Dy and YAG:Tm fluorescence to 1700 C. ISA 49th International Instrumentation Symposium; 2003Orlando.

[70] Chambers MD, Clarke DR. Doped oxides for high-temperature luminescence and lifetime thermometry. Annu Rev Mater Res 2009;39(1):325–59.

[71] Ranson RM, Evangelou E, Thomas CB. Modeling the fluorescent lifetime of Y_2O_3:Eu. Appl Phys Lett 1998;72(21):2663–4.
[72] Lojpur V, Antić Z, Dramićanin MD. Temperature sensing from the emission rise times of Eu^{3+} in SrY_2O_4. Phys Chem Chem Phys 2014;16(46):25636–41.
[73] Li X, Wei X, Qin Y, Chen Y, Duan C, Yin M. The emission rise time of BaY_2ZnO_5:Eu^{3+} for non-contact luminescence thermometry. J Alloys Compd 2016;657:353–7.
[74] Vetrone F, Naccache R, Zamarrón A, Juarranz de la Fuente A, Sanz-Rodríguez F, Martinez Maestro L, et al. Temperature sensing using fluorescent nanothermometers. ACS Nano 2010;4(6):3254–8.
[75] Wang X, Kong X, Yu Y, Sun Y, Zhang H. Effect of annealing on upconversion luminescence of ZnO:Er^{3+} nanocrystals and high thermal sensitivity. J Phys Chem C 2007;111(41):15119–24.
[76] Dong B, Liu DP, Wang XJ, Yang T, Miao SM, Li CR. Optical thermometry through infrared excited green upconversion emissions in Er^{3+}–Yb^{3+} codoped Al_2O_3. Appl Phys 2007;90(18). 181117.
[77] Du P, Luo L, Yue Q, Li W. The simultaneous realization of high- and low-temperature thermometry in Er^{3+}/Yb^{3+}-codoped Y_2O_3 nanoparticles. Mater Lett 2015;143:209–11.
[78] Lojpur V, Nikolić MG, Dramićanin MD. Luminescence thermometry below room temperature via up-conversion emission of Y_2O_3:Yb^{3+},Er^{3+} nanophosphors. J Appl Phys 2014;115(20). 203106.
[79] Dey R, Pandey A, Rai VK. Er^{3+}–Yb^{3+} and Eu^{3+}–Er^{3+}–Yb^{3+} codoped Y_2O_3 phosphors as optical heaters. Sens Actuators B Chem 2014;190:512–5.
[80] Xu W, Zhang Z, Cao W. Excellent optical thermometry based on short-wavelength upconversion emissions in Er^{3+}/Yb^{3+} codoped $CaWO_4$. Opt Lett 2012;37(23):4865–7.
[81] Pandey A, Rai VK, Kumar V, Kumar V, Swart HC. Upconversion based temperature sensing ability of Er^{3+}–Yb^{3+} codoped $SrWO_4$: an optical heating phosphor. Sens Actuators B Chem 2015;209:352–8.
[82] Soni AK, Kumari A, Rai VK. Optical investigation in shuttle like $BaMoO_4$:Er^{3+}–Yb^{3+} phosphor in display and temperature sensing. Sens Actuators B Chem 2015;216:64–71.
[83] Mahata MK, Kumar K, Rai VK. Er^{3+}–Yb^{3+} doped vanadate nanocrystals: a highly sensitive thermographic phosphor and its optical nanoheater behavior. Sens Actuators B Chem 2015;209:775–80.
[84] Haro-González P, León-Luis SF, González-Pérez S, Martín IR. Analysis of Er^{3+} and Ho^{3+} codoped fluoroindate glasses as wide range temperature sensor. Mater Res Bull 2011;46(7):1051–4.
[85] Lojpur V, Nikolic M, Mancic L, Milosevic O, Dramicanin MD. Y_2O_3:Yb,Tm and Y_2O_3:Yb,Ho powders for low-temperature thermometry based on up-conversion fluorescence. Ceram Int 2013;39(2):1129–34.
[86] Singh AK. Ho^{3+}:TeO_2 glass, a probe for temperature measurements. Sens Actuators A Phys 2007;136(1):173–7.
[87] Singh AK, Rai SB. Upconversion and optical thermometry in Ho^{3+}:TeO_2 glass, effect of addition of PbO_2 and $BaCO_3$. Appl Phys B 2007;86(4):661–6.
[88] Du P, Luo L, Yu JS. Low-temperature thermometry based on upconversion emission of Ho/Yb-codoped $Ba_{0,77}Ca_{0,23}TiO_3$ ceramics. J Alloys Compd 2015;632:73–7.
[89] Li D, Wang Y, Zhang X, Yang K, Liu L, Song Y. Optical temperature sensor through infrared excited blue upconversion emission in Tm^{3+}/Yb^{3+} codoped Y_2O_3. Opt Commun 2012;285(7):1925–8.
[90] Pereira AF, Kumar KU, Silva WF, Santos WQ, Jaque D, Jacinto C. Yb^{3+}/Tm^{3+} co-doped $NaNbO_3$ nanocrystals as three-photon-excited luminescent nanothermometers. Sens Actuators B Chem 2015;213:65–71.

[91] Zheng K, Liu Z, Lv C, Qin W. Temperature sensor based on the UV upconversion luminescence of Gd^{3+} in Yb^{3+}–Tm^{3+}–Gd^{3+} codoped $NaLuF_4$ microcrystals. J Mater Chem C 2013;1(35):5502–7.
[92] Weinstein BA. Ruby thermometer for cryobaric diamond-anvil cell. Rev Sci Instrum 1986;57(5):910.
[93] Grattan KT, Selli RK, Palmer AW. Ruby fluorescence wavelength division fiber-optic temperature sensor. Rev Sci Instrum 1987;58(7):4. 1231.
[94] Hehir JP, Henry MO, Larkin JP, Imbusch GF. Nature of the luminescence from YAG:Cr^{3+}. J Phys C Solid State Phys 1974;7(12):8. 2241.
[95] Zhang ZY, Grattan KT, Palmer AW, Fernicola V, Crovini L. Temperature dependence of the YAG:Cr^{3+} fluorescence lifetime over the range 77 to 900 K. Phys Rev B 1995;51(5):2656–60.
[96] Omrane A, Juhlin G, Ossler F, Aldén M. Temperature measurements of single droplets by use of laser-induced phosphorescence. Appl Opt 2004;43(17):3523–9.
[97] Borisov SM, Gatterer K, Bitschnau B, Klimant I. Preparation and characterization of Chromium(III)-Activated Yttrium Aluminum Borate: a new thermographic phosphor for optical sensing and imaging at ambient temperatures. J Phys Chem C 2010;114(19):9118–24.
[98] Back M, Trave E, Ueda J, Tanabe S. Ratiometric optical thermometer based on dual near-infrared emission in Cr^{3+}-doped bismuth-based gallate host. Chem Mater 2016;28(22):8347–56.
[99] Lojpur V, Nikolić MG, Jovanović D, Medić M, Antić Ž, Dramićanin MD. Luminescence thermometry with Zn_2SiO_4:Mn^{2+} powder. Appl Phys Lett 2013;103(14). 141912.
[100] Chen D, Liu S, Zhou Y, Wan Z, Huang P, Ji Z. Dual-activator luminescence of RE/TM:$Y_3Al_5O_{12}$ (RE = Eu^{3+}, Tb^{3+}, Dy^{3+}; TM = $Mn4^{+}$, Cr^{3+}) phosphors for self-referencing optical thermometry. J Mater Chem C 2016;4(38):9044–51.
[101] Marciniak L, Bednarkiewicz A, Kowalska D, Strek W. A new generation of highly sensitive luminescent thermometers operating in the optical window of biological tissues. J Mater Chem C 2016;4(24):5559–63.
[102] Marciniak L, Bednarkiewicz A. Nanocrystalline NIR-to-NIR luminescent thermometer based on Cr^{3+},Yb^{3+} emission. Sens Actuators B Chem 2017;243:388–93.
[103] Sholes RR, Small JG. Fluorescent decay thermometer with biological applications. Rev Sci Instrum 1980;51(7):882–4.
[104] Aizawa H, Ohishi N, Ogawa S, Katsumata T, Komuro S, Morikawa T, et al. Fabrication of ruby sensor probe for the fiber-optic thermometer using fluorescence decay. Rev Sci Instrum 2002;73(10):3656–8.
[105] Fonger WH, Struck CW. Temperature dependences of Cr^{3+} radiative and nonradiative transitions in ruby and emerald. Phys Rev B 1975;11(9):3251–60.
[106] Zhang Z, Grattan KT, Palmer AW. Fiber-optic high-temperature sensor based on the fluorescence lifetime of alexandrite. Rev Sci Instrum 1992;63(8):3869–73.
[107] Zhang ZY, Grattan KT. Temperature dependence of YAG:Cr^{3+} fluorescence lifetime up to 550 K. J Lumin 1994;62(6):263–9.
[108] Allison SW, Beshears DL, Cates MR, Paranthaman M, Gilles GT. LED-induced fluorescence diagnostics for turbine and combustion engine thermometry. In: Proceedings SPIE 4448, Optical Diagnostics for Fluids, Solids, and Combustion; 2001. p. 28–35.
[109] Uchiyama H, Aizawa H, Katsumata T, Komuro S, Morikawa T, Toba E. Fiber-optic thermometer using Cr-doped $YAlO_3$ sensor head. Rev Sci 2003;74(8):3883–5.
[110] Fuhrmann N, Kissel T, Dreizler A, Brübach J. $Gd_3Ga_5O_{12}$:Cr—a phosphor for two-dimensional thermometry in internal combustion engines. Meas Sci Technol 2011;22(4). 45301.
[111] Zhang Z, Grattan KT, Palmer AW. Temperature dependences of fluorescence lifetimes in Cr^{3+}-doped insulating crystals. Phys Rev B 1993;48(11):7772–8.

[112] Aizawa H, Ohishi N, Ogawa S, Watanabe E, Katsumata T, Komuro S, et al. Characteristics of chromium doped spinel crystals for a fiber-optic thermometer application. Rev Sci Instrum 2002;73(8):3089–92.
[113] Fernicola VC, Zhang ZY, Grattan KT. Fiber optic thermometry based on Cr-fluorescence in olivine crystals. Rev Sci Instrum 1997;68(6):2418–21.
[114] Cates MR, Beshears DL, Allison SW, Simmons CM. Phosphor thermometry at cryogenic temperatures. Rev Sci Instrum 1997;68(6):2412–7.
[115] Fuhrmann N, Baum E, Brübach J, Dreizler A. High-speed phosphor thermometry. Rev Sci Instrum 2011;82(10). 104903.
[116] Brübach J, Kissel T, Frotscher M, Euler M, Albert B, Dreizler A. A survey of phosphors novel for thermography. J Lumin 2011;131(4):559–64.
[117] Eckert C, Pflitsch C, Atakan B. Sol–gel deposition of multiply doped thermographic phosphor coatings Al_2O_3:(Cr^{3+}, M^{3+}) (M = Dy, Tm) for wide range surface temperature measurement application. Prog Org Coatings 2010;67(2):116–9.

Chapter 7

Luminescence Temperature Sensing Using Semiconductor Quantum Dots

7.1 INTRODUCTION

Research on synthesis, properties, and applications of semiconductor quantum dots (QDs) has developed over the past 3 decades into one of the key fields in science. QDs are composed of a few hundred to a few hundred thousand atoms, being as such between a single atom and solid-state materials, and displaying a mix of atomic and solid-state properties. One of the defining features of QDs is a quantization effect, responsible for their size-tunable properties and, consequently, for their exceptional potential in many modern technologies. Simultaneously, the quantization effect has boosted fundamental research on scaling the chemical and physical properties of materials, including those related to the temperature dependence of semiconductor luminescence. Today, the principal applications of QDs are in optoelectronic, catalytic devices, solar cells, thermoelectric materials, biolabels, and energy storage devices.

More than any other characteristics, the massive size-dependent change in the optical properties of QDs, the high luminescence efficiency, and the sensitivity of luminescence on temperature, has fostered applications of QDs for chemical, biological, and temperature sensing. QDs have been extensively researched in terms of luminescence thermometry over the past years, with a particular focus on their use in the nanoscale environment (in nanoelectronics and nanofluids) and in biomedicine. It has been revealed that QD luminescence shows substantial size-dependent temperature sensitivity. This feature may be considered as a valuable and beneficial add-on to the already well-known set of size-dependent properties of QDs.

At the beginning of this chapter, we will consider the basic description of semiconductor nanostructures with an explanation of quantum confinement. Luminescence of semiconductor QDs is elaborated with an emphasis on how temperature variations affect optical transitions. Most of this chapter is devoted to analyses of temperature-induced changes in emission peak spectral positions, bandwidths, intensity, and emission decays of QDs. It is shown that some of the Mn^{2+}-doped QDs show two emissions, one from exciton recombination and the

Luminescence Thermometry. http://dx.doi.org/10.1016/B978-0-08-102029-6.00007-5

other from a transition metal ion, something which can be exploited for sensitive intensity ratiometric thermometry. Examples include both core QDs and heterostructural QDs, those emitting in the visible spectral range and those applicable for near-infrared sensing, as well as relatively novel, low-toxicity QDs.

7.2 SEMICONDUCTOR NANOSTRUCTURES AND QUANTUM CONFINEMENT

Low-dimensional materials are fascinating for both fundamental research and applications. They present altered structural, optical, chemical, and electronic properties with respect to the corresponding bulk materials due to quantum confinement phenomena which result from their unique dimensionality. Since confinement occurs on the nanometer scale, low-dimensional materials are frequently termed nanostructures. Quantum confinement appears in three ways, with potential barriers in one, two, or three spatial dimensions; the assemblies which exhibit such confinement are respectively said to have two-dimensional, one-dimensional, and zero-dimensional geometric dimensionality (Fig. 7.1). Structures of different dimensionalities show markedly distinctive optical properties because of the drastically different dependence of electron density of states (DOS) on energy (E). In bulk materials DOS is proportional to $E^{1/2}$, in two-dimensional structures DOS is described by a step function (E^0 dependence for the single quantized state), in one-dimensional structures DOS is proportional to $E^{-1/2}$ for each quantized state. In zero-dimensional structures (quantum dots) DOS is a δ-function, since the three-dimensional confinement provides states only at discrete energies [1].

Luminescence of semiconductors is detailed in Chapter 3. Of interest here are optical properties of semiconductor QDs (zero-dimensional structures), which show dramatic quantization effects from the geometric confinement of electrons and holes as independently acting “wave particles” or as bound pairs known as excitons [2]. Upon decreasing the size of QDs, their effective bandgap becomes larger and discrete energy levels develop at the band edges of both the conduction band and valence band (Fig. 7.2) [3]. Consequently, the color of luminescence changes with size decrease, as can be visualized from photographs of CdSe QDs in Fig. 7.2.

The most probable distance between the electron and hole in an exciton is called an exciton Bohr radius, by analogy to the hydrogen atom. Its value in different semiconductors (Table 7.1) can be used to assess the size of QD for which confinement occurs, but also to estimate the strength of the confinement regime. To this end, it is useful to account for three different Bohr radii (a_B): one for the electron (a_e), one for the hole (a_h), and one for the exciton (a_{exc}), which can be calculated as

$$a_B = \frac{m_0 \varepsilon_r}{\mu} \cdot a_0, \tag{7.1}$$

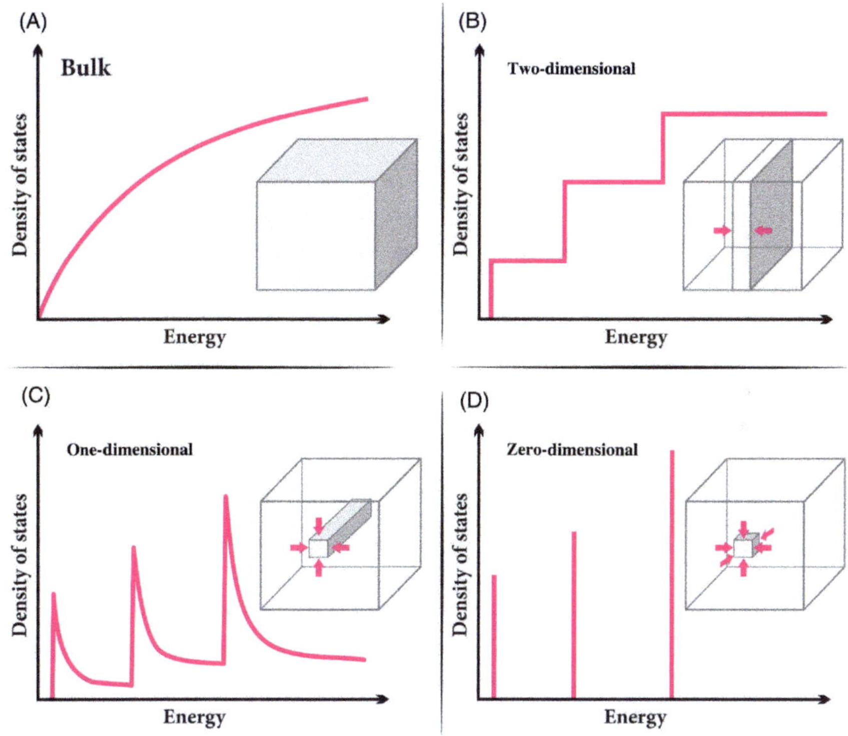

FIGURE 7.1 Energy dependence of the electronic density of states for materials of different geometrical dimensionalities: bulk (no confinement), two-dimensional (one-dimensional spatial confinement), one-dimensional (two-dimensional spatial confinement), and zero-dimensional (three-dimensional spatial confinement). *(Reprinted from Mino L, Agostini G, Borfecchia E, Gianolio D, Piovano A, Gallo E, et al. Low-dimensional systems investigated by x-ray absorption spectroscopy: a selection of 2D, 1D and 0D cases. J Phys D Appl Phys 2013;46(42):423001 [1], with permission from IOP Publishing).*

where m_0 is the rest mass of the electron, ε_r is the relative permittivity (relative dielectric constant) of the semiconductor, μ is the reduced mass of the particle, and $a_0 = 0.0529$ nm is the Bohr radius of a hydrogen atom. Values of exciton Bohr radius along with values of the energy gap at room temperature and temperature change of the energy gap per unit change of temperature for selected bulk semiconductors are given in Table 7.1.

The onset of quantum confinement effects occurs at different nanocrystal dimensions in different materials. Semiconductors which have a wide energy bandgap possesses a shorter exciton Bohr radius than narrow bandgap semiconductors, and insulators, which have large bandgaps and $a_{exc} < 1$ nm, show quantum confinement only when they are of cluster sizes. Depending on QD size (R) and values of Bohr radii, three different confinement regimes can be identified. First is the *weak-confinement* regime where R is larger than the exciton Bohr

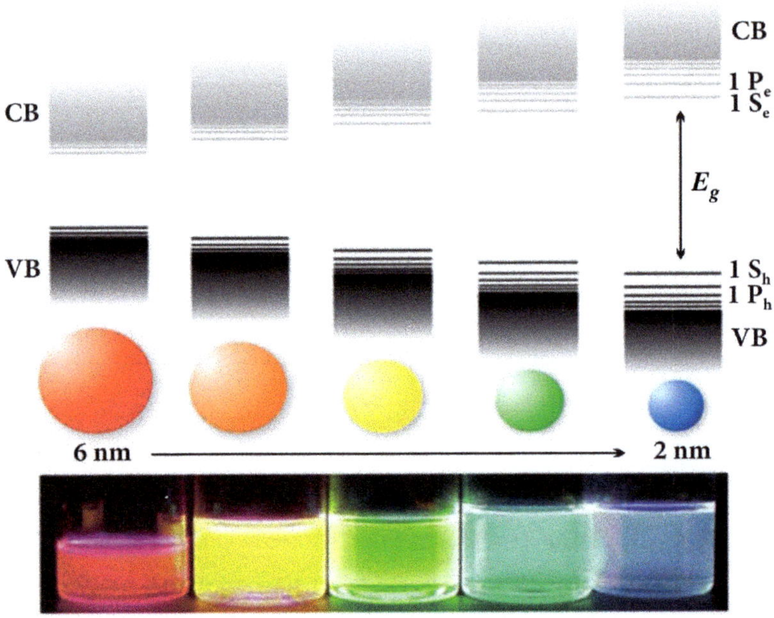

FIGURE 7.2 **Illustration of the quantum confinement effects: with a size decrease of quantum dots, the effective bandgap increases and discrete energy levels develop at the band edges.** The luminescence of the dispersions of CdSe quantum dots of different sizes is shown in the photographs at the bottom. *(Reproduced from de Mello Donegá C. Synthesis and properties of colloidal heteronanocrystals. Chem Soc Rev 2011;40(3):1512–46 [4], with permission from The Royal Society of Chemistry).*

TABLE 7.1 Energy Gap at Room Temperature, Type of Energy Gap, Exciton Bohr Radius, and Change of Energy Gap per Unit Change of Temperature for a Selection of Bulk Semiconductors

Semiconductor	Type of bandgap	Bandgap at RT (eV)	Exciton Bohr radius (nm)	dE_g/dT (meV/K)
CuCl	Direct	3.35	0.7	
CdS	Direct	2.42	2.5	−50
CdSe	Direct	1.74	5.5	−46
CdTe	Direct	1.49	7.5	−41
PbS	Direct	0.37	20	+40
InAs	Direct	0.36	34	−28
PbSe	Direct	0.27	46	+40
PbTe	Direct	0.25	50	+40
ZnS	Direct	3.7	2.2	−53

TABLE 7.1 Energy Gap at Room Temperature, Type of Energy Gap, Exciton Bohr Radius, and Change of Energy Gap per Unit Change of Temperature for a Selection of Bulk Semiconductors (*cont.*)

Semiconductor	Type of bandgap	Bandgap at RT (eV)	Exciton Bohr radius (nm)	dE_g/dT (meV/K)
ZnSe	Direct	2.72	4.5	−72
ZnTe	Direct	2.25	6.7	
ZnO	Direct	3.2	2.3	−95
InSb	Direct	0.18	54	−28
InP	Direct	1.42	11	−46
InN	Direct	0.65	11.4	
GaSb	Direct	0.7	60	−35
GaP	Indirect	2.27	29	−54
GaN	Direct	3.39	3.1	
AlSb	Indirect	1.63	64	−35
AlAs	Indirect	2.15	41	
AlP	Indirect	2.45	37	
AlN	Direct	6.2	1.6	
$CuInS_2$	Direct	1.5	4.1	
$CuInSe_2$	Direct	1.04	10.6	
$AgInS_2$	Direct	1.8	5.5	
$AgInSe_2$	Direct	1.24	5–6	−14
Ag_2S	Direct	0.9–1.1	2.2	
Ag_2Se	Direct	0.15	2.9	
Si	Indirect	1.17	4.3	−23
Ge	Indirect	0.67	11.5	−37
GaAs	Direct	1.43	12.4	−50

Source: Values compiled from data in Yen WM, Shionoya S, Yamamoto H. Phosphor handbook. Boca Raton, FL: CRC Press, Taylor and Francis Group; 2007. p. 1055; Haynes WM. CRC handbook of chemistry and physics. 97th ed; 2016 p. 2652. Adachi S. Handbook on physical properties of semiconductors: volume 3: II–VI compound semiconductors. Springer US; 2004. p. 1472. Wise FW. Lead salt quantum dots: the limit of strong quantum confinement. Acc Chem Res 2000;33(11):773–80. Jaque Garcia D, Garcia Sole J. Quantum dot fluorescence thermometry. In: Thermometry at the nanoscale: techniques and selected applications. London: Royal Society of Chemistry; 2016. p. 83–123. Vossmeyer T, Katsikas L, Giersig M, Popovic IG, Diesner K, Chemseddine A, et al. CdS nanoclusters: synthesis, characterization, size dependent oscillator strength, temperature shift of the excitonic transition energy, and reversible absorbance shift. J Phys Chem 1994;98(31):7665–73. Chang Y-M, Shieh J, Chu P-Y, Lee H-Y, Lin C-M, Juang J-Y. Enhanced free exciton and direct band-edge emissions at room temperature in ultrathin ZnO films grown on Si nanopillars by atomic layer deposition. ACS Appl Mater Interfaces 2011;3(11):4415–9. Gurinovich LI, Gurin VS, Ivanov VA, Bodnar IV, Molochko AP, Solovej NP. Crystal structure and optical properties of CuInS2 nanocrystals in a glass matrix. Phys Status Solidi 1998;208(2):533–40. Xu G, Zeng S, Zhang B, Swihart MT, Yong K-T, Prasad PN. New generation cadmium-free quantum dots for biophotonics and nanomedicine. Chem Rev 2016;116(19):12234–327 [5–13].

radius a_{exc}. The *intermediate-confinement* regime occurs when R is larger than the Bohr radii of the exciton and one particle, but smaller than the Bohr radius of another particle; the one particle is then strongly confined while the other is not. Finally, the *strong-confinement* regime is characterized with QD size being smaller than all three Bohr radii ($R < a_{exc}$, a_e, and a_h). In this case, QDs demonstrate a ladder of discrete energy levels, as in molecular systems, rather than energy bands. The electrons and holes are then treated as independent particles, and for excited states we refer to electron–hole pairs (EHP) rather than excitons [2]. One should note that besides the size of semiconductor nanocrystals, their shape also influences quantum confinement [14], but further elaboration of this phenomenon is beyond the scope of this chapter.

The prerequisite for confinement of charge carriers in QDs is a higher potential energy surrounding the QD than the potential energy inside the QD. The manner by which the potential barrier is raised at the QD surface can be used to classify them into three basic types: (1) colloidal QDs in dispersion, (2) epitaxially grown QDs of one semiconductor in the matrix of another semiconductor, and (3) laterally gated QDs (for details, see Section 2.4.1.1 in Ref. [3]). QDs can be engineered from one or several semiconductors that have different bandgaps. The later ones are called heterostructured QDs and can be further classified based on the type of charge carrier localization after photoexcitation. The localization is governed by the band alignment between the semiconductors that are combined at junctions in the heterostructure. QDs which trap simultaneously electrons and holes are called type I. Contrary to these structures, QDs of type II are characterized by the spatial separation of electrons and holes on different sides of the heterojunction. Finally, in the quasi-type II QDs, one carrier is confined in one of the semiconductors, while the other is delocalized over the whole heterostructure [3].

Vast work has been performed on the development of the growth/synthesis techniques of semiconductor nanostructures over the years. Now, we can tailor the structure and the composition of QDs; we are able to control their size and make them monodispersed, while all the time ensuring their complete reproducibility. Also, the surface chemistry of QDs is well developed; it is possible to coat dots with a variety of ligands to functionalize them for targeted applications. As in the case of phosphor materials, preparation of QD can be accomplished by assorted "physical" and "chemical" methods. Some of these methods are already illustrated in Chapter 6; therefore, syntheses methods will not be elaborated here. Those interested in the subject of semiconductor QD synthesis should refer to some of many available excellent books and review articles, such as those in Refs. [4,13,15–18].

7.3 LUMINESCENCE OF QUANTUM DOTS

Luminescence in semiconductor QDs comes mainly from the direct radiative recombination of excitons in their ground states (the $1S_e1S_h$ exciton). This near-bandgap luminescence is recognized in emission spectra of QDs as a distinct peak of energy adjacent to the absorption transition of lowest energy. Broadband emission of energy lower than exciton emission, and of quite low efficiency, may additionally arise in QDs from the radiative recombination of excitons trapped in a defect or surface state. The temperature dependence of QD luminescence, which is a phenomenon of particular relevance to luminescence thermometry, is not only a consequence of the temperature dependence of radiative exciton recombination, but also of other optical transitions involved in a complex process which follows photon absorption.

When a photon of sufficiently large energy is absorbed, the electrons and holes (usually in the form of excitons) are created with an excess energy above the lowest exciton energy; these energetic carriers are termed hot carriers or hot excitons. The hot carriers quickly (<1 ps) relax to their ground states by means of a cascade of intraband nonradiative relaxation steps and their excess energy is dissipated as heat through electron–phonon interactions or Auger processes. However, the excess energy (if it is at least twice the bandgap of a QD) may also produce a second electron–hole pair, or the electrons and holes may separate and the excess energy be converted into enlarged electrical free energy by a photovoltaic effect or be stored as extra chemical free energy by facilitating more endoergic electrochemical surface reactions [19]; these processes are valuable for some applications of QDs. The radiative recombination of excitons competes with a nonradiative recombination and exciton energy transfer, which are also temperature dependent processes. Nonradiative recombination occurs mainly via defects, impurities, and surface states, and, to a much less extent, via exciton coupling to phonons. Exciton energy transfer occurs between two QDs which are in close proximity through dipole–dipole interaction [3].

Having in mind that near-bandgap luminescence dominates the overall luminescence of QDs, it is reasonable to expect that temperature-induced changes in the effective bandgap of QDs will have a major impact on luminescence variations. The temperature dependence of the energy gap of bulk semiconductors has been the subject of numerous theoretical and experimental studies. For bulk materials, the bandgap change per unit increase in temperature $\mathrm{d}E_g/\mathrm{d}T$ is usually negative (a decrease in bandgap with temperature; Table 7.1). The effect arises from the thermal expansion of lattice and electron–phonon interactions. However, much less work has been done on the variations caused by temperature of the electronic energies of QDs. Studies generally report that the gap depends on temperature similarly to the energy gap of the bulk material, with $\mathrm{d}E_g/\mathrm{d}T$ not differing by more than 30% of the value of the respective bulk material [20]. In many cases of semiconductor nanocrystals, the changes in energy gap with temperature have been described by the Varshni relation [21], which was developed for bulk semiconductors:

$$E_g(T) = E_g(0) - \frac{\alpha T^2}{\beta + T}, \tag{7.2}$$

where α is the temperature coefficient, $E_g(0)$ is the bandgap at 0 K, and β approximates the Debye temperature of the material. However, as will be seen in the next Section, the temperature dependence of QD luminescence is a more complex phenomenon, which includes also the effect of temperature-induced changes in QD size, the effect which is not present for bulk semiconductors.

7.4 TEMPERATURE SENSING USING VARIOUS LUMINESCENCE FEATURES OF SEMICONDUCTOR QUANTUM DOTS

Temperature notably influences all aspects of QD luminescence: peak energy shifts, broadening of emission bands, quenching of emission intensity, and a shortening of lifetimes with temperature evolution. Emission intensity undergoes the largest variations with temperature, peak energies are less affected, but still enough to facilitate temperature readings, and bandwidths suffer small increases. The striking manifestation of these changes is their linearity with temperature, as will be evidenced by several examples later in the text. Temperature measurements, thus, benefit from an invariable sensitivity over the complete measurement range, a characteristic which materializes from linearity. The extent of temperature-induced changes in luminescence is different for different QDs, but, importantly, it strongly depends on their size and surface chemistry.

Emission bands of QDs have a Gaussian shape with peak energies determined by their composition, size, and local environment. Absorption and emission peaks of the first exciton are separated by a small Stokes shift, and all three peak features are influenced by temperature. Changes in an energy of a first exciton absorption peak with temperature is a consequence of five main processes: (1) expansion/contraction of the QD lattice, (2) changes in electron–phonon coupling (interband and intraband), (3) size increase/decrease of dots, (4) changes in mechanical strain, and (5) the thermally induced variation of a solvent's refractive index (for colloidal QDs). The first two also occur with bulk semiconductors, while the others are either specific or more pronounced for QDs. The detailed elaboration of these processes can be found in Refs. [9,20]. Their temperature sensitivities dE_g/dT (if we term E_g as the transition energy of the lowest electron–hole pair) are governed by the size of QDs, as can be evidenced from the Fig. 7.3, where the temperature sensitivity of the first exciton peak of PbS QDs is plotted against QD diameter [9,20]. For ease of following spectral changes, temperature sensitivity E_g may be converted to temperature sensitivity of spectral peak maximum λ_{max} using the following equation [22]:

$$\frac{d\lambda_{max}}{dT}(\text{nm/K}) = -\frac{\lambda_{max}^2}{1239.8} \times \frac{dE_g}{dT}(\text{eV/K}). \tag{7.3}$$

Temperature sensitivity increases with the size of the QD (full line in Fig. 7.3) coming near to the bulk value of 500 μeV/K in the largest dots (15.5 nm). In PbS dots of this size, electron–hole pairs are only weakly confined, since they are just smaller than the exciton Bohr radius (20 nm). For small QDs (diameter

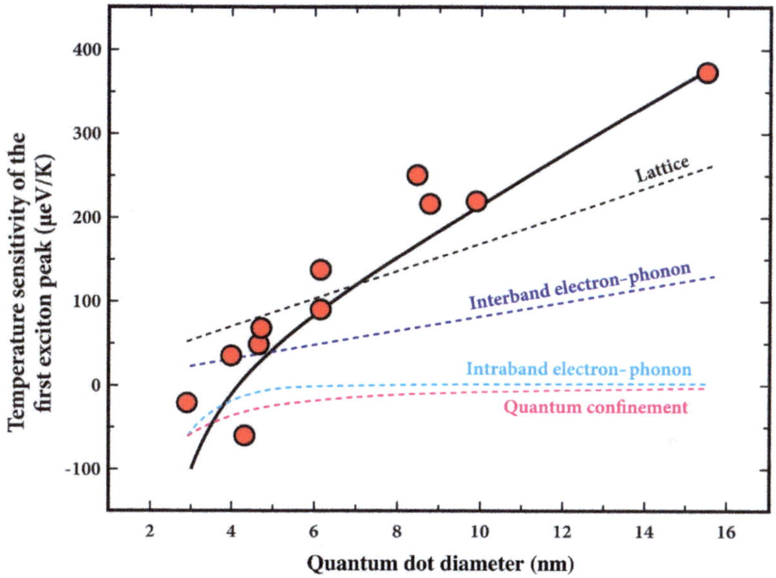

FIGURE 7.3 Temperature sensitivity of the first exciton peak in PbS quantum dots of different sizes [20]. *Broken lines* represent sensitivities of individual processes (lattice expansion/contraction, electron–phonon coupling, and size increase/decrease of particles), while the *solid line* represents their sum. *(Reproduced from Jaque Garcia D, Garcia Sole J. Quantum dot fluorescence thermometry. In: Thermometry at the nanoscale: techniques and selected applications. London: Royal Society of Chemistry; 2016. p. 83–123 [9], with permission from The Royal Society of Chemistry).*

<5 nm) sensitivity values are negative, meaning that E_g decreases as temperature increases—peaks are redshifted. Blueshift occurs with larger dots (diameter >5 nm) and becomes more pronounced with an increase in dot size. In fact, negative temperature sensitivity values and peak redshifts are characteristics of QDs with strong confinement of electron and holes, while positive sensitivities and peak blueshifts are characteristic of QDs in which electron and hole confinement is weak. This small analysis exposes the amazing feature of QD luminescence—both character and magnitude of temperature-induced changes in band spectral positions depend on QDs size!

Fig. 7.3 also shows that in large QDs temperature-induced changes to the crystal lattice dimension dominates over the contribution of other processes, similar to the case of bulk semiconductors [22]. When the size of dots is decreased, that is, with a stronger confinement, this effect becomes weaker. The situation is in accordance with the fact that particle size crucially influences the spacing between the confined energy levels in small-sized QDs.

The temperature sensitivity of the spectral band positions have also been analyzed in PbSe QDs [23]. In small 3.9 nm PbSe nanocrystals strong confinement occurs since the exciton Bohr radius is 46 nm. For this reason, temperature

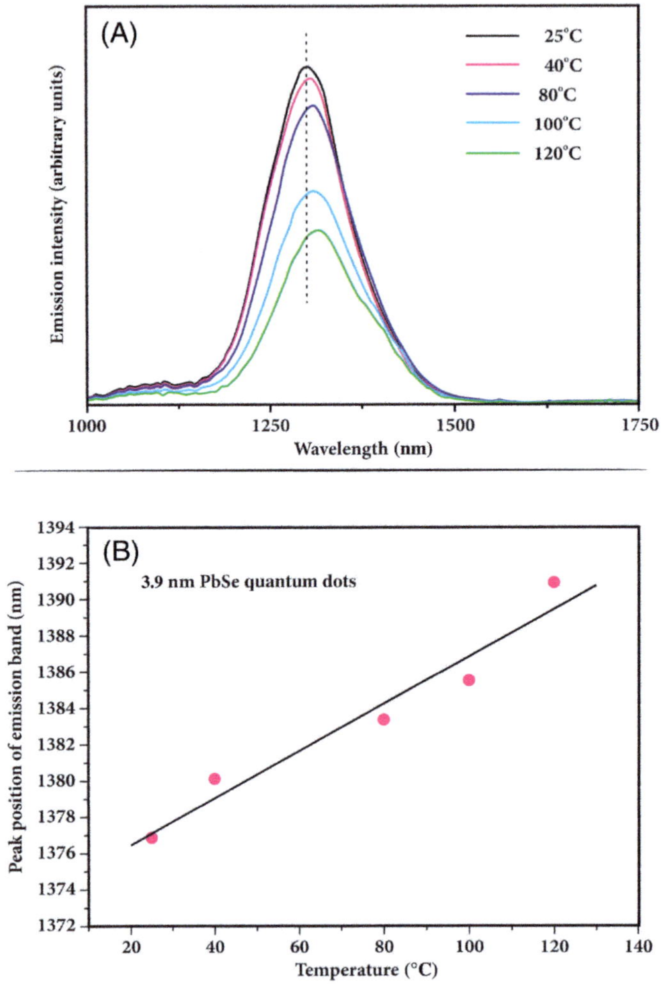

FIGURE 7.4 (A) Near-infrared emission spectra of 3.9 nm PbSe quantum dots at different temperatures. (B) Temperature-induced shift of emission peak position. *(Part A: Adapted from the Supporting Information in Dai Q, Zhang Y, Wang Y, Hu MZ, Zou B, Wang Y, et al. Size-dependent temperature effects on PbSe nanocrystals. Langmuir 2010;26(13):11435–40 [23], copyright (2010) American Chemical Society); part B: data extracted from Dai Q, Zhang Y, Wang Y, Hu MZ, Zou B, Wang Y, et al. Size-dependent temperature effects on PbSe nanocrystals. Langmuir 2010;26(13):11435–40 [23]).*

sensitivity of the emission peak position is negative. Fig. 7.4A presents near-infrared emission spectra of 3.9 nm PbSe QDs measured at different temperatures from 25°C to 120°C. Emission intensity markedly lessens with rise in temperature; this effect will be discussed later in the text. An increase in the QD temperature redshifts the near-infrared emission (1376 nm), and the peak position linearly evolves with temperature for about 0.13 nm/°C (Fig. 7.4B).

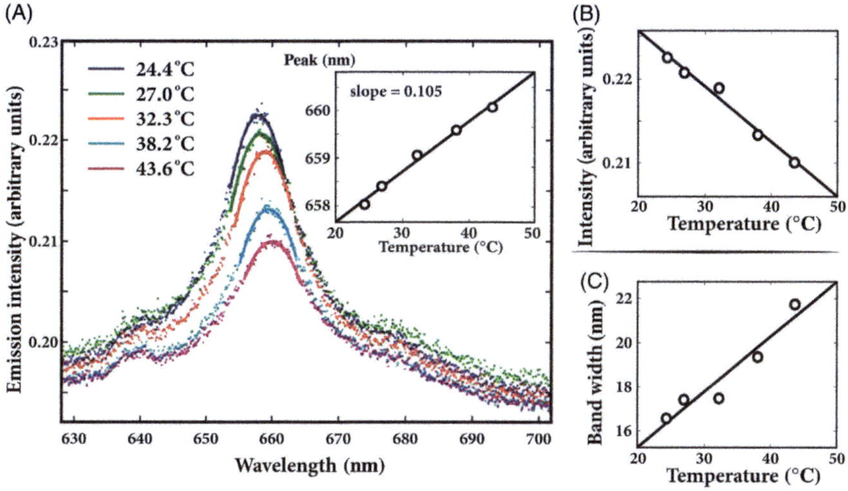

FIGURE 7.5 Emission of a single CdSe/ZnS core–shell quantum dot determined in the 24.4°C–43.6°C temperature range. (A) Emission spectra where experimental data are given with dots and Gaussian fits of the emission band peak are presented with *solid lines*; the linear dependence of the emission peak wavelength on temperature is given in the inset. (B) Linear dependence of the emission intensity on temperature. (C) Linear dependence of the emission bandwidth on temperature. *(Adapted from Li S, Zhang K, Yang J-M, Lin L, Yang H. Single quantum dots as local temperature markers. Nano Lett 2007;7(10):3102–5 [24], copyright (2007) American Chemical Society).*

Emissions from PbSe QDs of larger size is not feasible for use in temperature sensing because of the extremely weak intensity.

Exciton Bohr radii of CdS, CdSe, and CdTe (II–VI semiconductors) are about 2.5, 5.5, and 7.5 nm (Table 7.1), much smaller than those of PbS and PbSe (IV–VI semiconductors). Negative sensitivity of E_g has also been observed in CdS QDs with diameters of 1.3–4 nm [10]. For temperatures above 150 K, sensitivity of E_g on temperature is −0.55 meV/K for the smallest QDs and −0.40 meV/K for the largest. CdSe QDs ranging from 2.5 to 4 nm [22] produce a redshifting of spectral peaks of 0.1 nm/K on average.

A thorough investigation of the influence of temperature on luminescence features of a single QD has been performed with commercially available water-soluble CdSe/ZnS core–shell QDs (Invitrogen; nominal diameter of ~7–12 nm for core + shell; emission maximum at ~655 nm) coated with a layer of organic polymer, and conjugated with streptavidin [24]. Emission spectra for single dots have been acquired in a biologically compatible temperature range from 24.4°C to 43.6°C (Fig. 7.5A). With a temperature increase, the emission peak linearly shifts to higher wavelengths at a rate of 0.1 nm/K (see inset in Fig. 7.5A). In addition, emission lessens with intensity (Fig. 7.5A) and the emission bandwidth increases (Fig. 7.5C).

This investigation [24] presented additional and very important findings. Values of emission peak position at room temperature considerably varied between individual dots. For 71 QDs, the mean value of the emission peak was 659 nm with a quite large standard deviation of 2.4 nm. The variation in emission peak positions between dots was larger than the peak shift over the complete temperature range, and is a consequence of the distribution of the size and shape of wet chemistry-synthesized QDs. The huge uncertainty in the measurement practically restricts the use of a single QD emission for temperature sensing. However, temperature measurements are feasible with larger ensembles of QDs; for a temperature read-out based on emission peak shifts, an uncertainty in temperature of 1°C can be achieved using luminescence from 1200 QDs [24]. Indubitably, the number of required QDs depends on the distribution of their size and shape. For uniform QDs, smaller ensembles may provide a similar or even better uncertainty.

An interesting comparative analysis of temperature sensitivity of emissions from CdSe QDs (4 nm in diameter; commercial Invitrogen Inc.—Qtracker 655) under one-photon and two-photon excitation has been performed over a temperature range of 30°C–70°C [25]. For both types of excitations, emission peak wavelength linearly increased with temperature at a rate of 0.16 nm/°C (Fig. 7.6A). On the other hand, reduction of emission intensity with temperature increase was much stronger under a two-photon than a one-photon excitation, the dependences being linear in both cases.

Quite a large temperature-induced redshift of the 510 nm emission (0.8 nm/°C) was noted in ultrasmall, 1.2 nm CdTe QDs [26]. In principle, such a high rate of peak position change facilitates measurements with a 0.2°C temperature resolution. Larger CdTe QDs exhibit a smaller rate of change, 0.4 nm/°C for 3.6 nm dots and 0.2 nm/°C for 7.8 nm ones. Temperature sensitivity of spectral shift CdTe QDs is governed by surface conditions in addition to dot size. Water-dispersible CdTe QDs that were decorated with per-6-thio-β-cyclodextrin (β-CD) showed a more than double improvement in temperature sensitivity than nondecorated, or monothiolligand-decorated, ones [27]. Glucopyranose rings rotate around C—O bonds in β-CD, which when attached to a QD surface produces a torsional force against the lattice of the QDs. In this way, CD contributes to a temperature-driven lattice dilatation and enhances the change of E_g and spectral shift with temperature.

Regarding the temperature-induced spectral shift of QDs, one should note that exposure of QDs to elevated temperatures may cause their growth. Then, the observed spectral shifts cannot be categorically related to temperature change. It is, therefore, necessary to conduct repeated luminescence measurements in several heating/cooling cycles to ascertain whether the emission changes are only due to temperature variations.

Temperature affects the homogeneous emission and absorption bandwidths of QDs. The central contribution to homogeneous bandwidths comes from the exciton–phonon coupling, which is temperature dependent. When temperature increases, the number of phonons increases, discrete phonon spectra turn

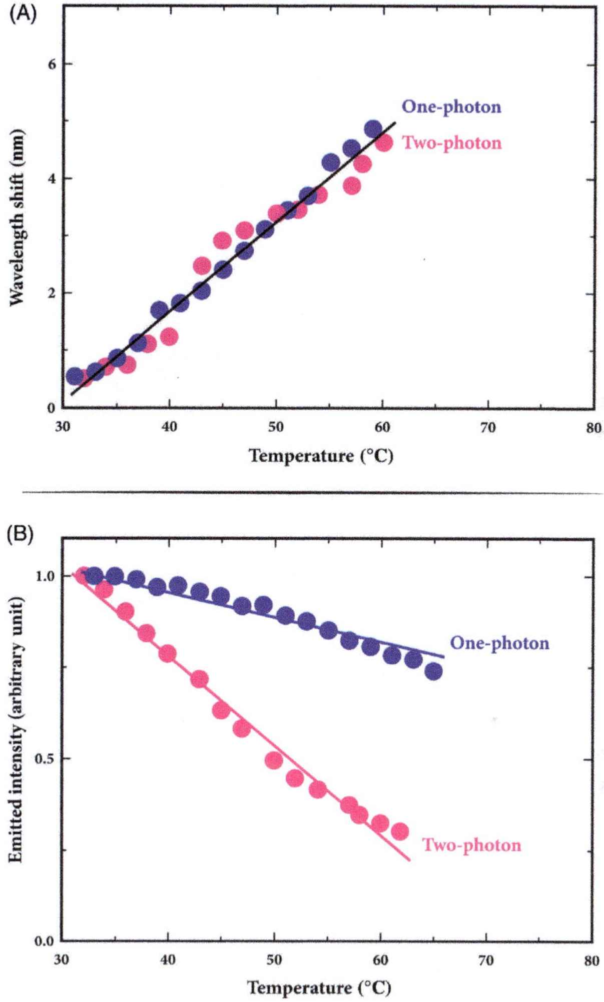

FIGURE 7.6 Changes with temperature of one-photon and two-photon excited emission from 4 nm CdSe quantum dots. (A) Temperature dependence of the emission peak wavelength, and (B) emission intensity decrease with temperature. *(Adapted from Maestro LM, Rodríguez EM, Rodríguez FS, la Cruz MCI, Juarranz A, Naccache R, et al. CdSe quantum dots for two-photon fluorescence thermal imaging. Nano Lett 2010;10(12):5109–15 [25], copyright (2010) American Chemical Society).*

into continuous, and the nonradiative relaxation of excitons through interactions with phonons becomes more efficient. However, widths of QD emission and absorption bands are dominated by inhomogeneous components which are much larger than homogenous ones. Variations in size, shape, and symmetry of individual dots within an ensemble cause a small modification of energy levels which results in an inhomogeneous band broadening. If we assume that

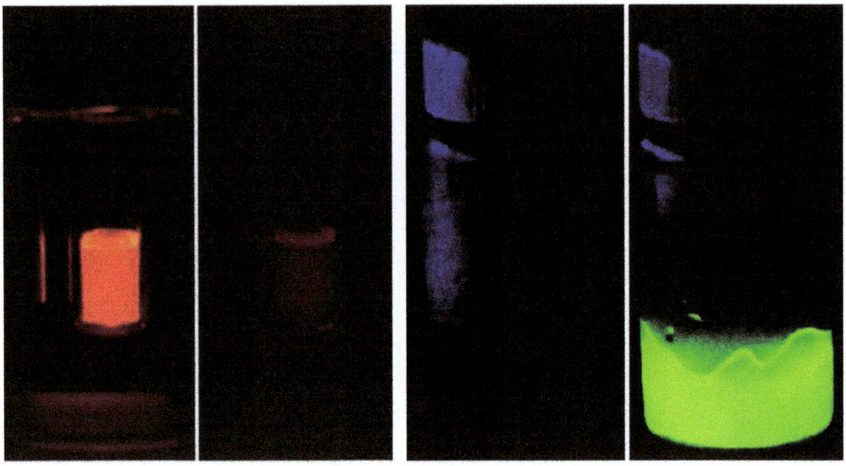

50°C | 250°C | 77 K | 265 K

FIGURE 7.7 Temperature quenching and antiquenching of semiconductor quantum dot luminescence. (A) Photographs of emission quenching taken under the ultraviolet illumination of a vial of CdSe/CdS/ZnS core–shell–shell quantum dots incorporated in cross-linked poly(lauryl methacrylate) taken at 50°C and 250°C. (B) Photographs of emission antiquenching in a vial of aqueous solution of green-emitting aminoethanethiol-capped CdTe quantum dots illuminated with 365 nm light at temperatures of 77 and 265 K. *(Part A: Adapted from Zhao Y, Riemersma C, Pietra F, Koole R, de Mello Donegá C, Meijerink A. High-temperature luminescence quenching of colloidal quantum dots. ACS Nano 2012;6(10):9058–67 [37], copyright (2012) American Chemical Society; part B: adapted from Wuister SF, de Mello Donegá C, Meijerink A. Luminescence temperature antiquenching of water-soluble cdte quantum dots: role of the solvent. J Am Chem Soc 2004;126(33):10397–402 [35], copyright (2004) American Chemical Society).*

temperature does not influence the size distribution of dots, it is possible to conclude that the sensitivity of QDs bandwidths on temperature is small. It is believed that colloidal QDs present broader bands than epitaxially grown dots. Still, for both types of QDs, the temperature reading from emission bandwidths is not utilized due to poor temperature sensitivity and more complex data processing than in other temperature read-outs (Chapter 4). Temperature sensitivity of a single QD emission bandwidth may be appreciable for the application (Fig. 7.5C), however, as already mentioned, variations in size and shape between dots cause large measurement uncertainties.

Apart from the change in emission energies of QDs, temperature variations also alter their emission intensity (quantum yield) (Fig. 7.7). Mechanisms responsible for emission quenching in QDs are not the same as in lanthanide- or transition metal ion-doped materials, which are discussed in Chapter 6. In QDs, the main nonradiative deexcitation channels are: (1) transfer of electrons and holes to trap and surface nonradiative states, (2) interaction between excitons and longitudinal optical phonons (so-called thermal escape) [28], (3) Auger nonradiative scattering [29], (4) Förster resonant energy transfer between different sized dots, and (5) exciton energy transfer [30]. The first two have the largest impact under excitation densities relevant for luminescence thermometry, and grow in magnitude at elevated temperatures. Considering only these two processes, temperature dependence of emission *I*(*T*) can be described by following equation [31,32]:

$$I(T)=\frac{I_0}{1+a_1\cdot\exp(-E_a/k_BT)+a_2\cdot[\exp(E_{LO}/k_BT)-1]^{-m}},\qquad(7.4)$$

where I_0 is the intensity of the emission at 0 K, E_a is the activation energy of the nonradiative process at the surface, m is the number of longitudinal optical phonons involved in the interaction with the carriers, E_{LO} is the energy of the phonons, and $a_{1,2}$ are factors standing as weights for the probabilities of these processes. One should note that previous equation (Eq. 7.4) takes into account a single thermally activated process at the surface while sometimes more complex process may be involved, such as trapping in a distribution of defects among which each defect has a slightly different energy.

The surface of a QD acts like a large defect because of a significant rearrangement of atoms, which consequently cause the appearance of energy levels within the energy gap [33]. Electrons and holes are trapped by these surface states. Therefore, one can consider that the surface of QDs plays a dominant role in the temperature quenching of emission. The way to eliminate surface states is to bond surface atoms to other material which has a larger bandgap. These core–shell QDs have been the subject of extensive investigation over the past 2 decades. It is now established that they present significantly greater quantum efficiencies of emission over the classical dots, which is a benefit for luminescence thermometry since it eases detection and improves the signal-to-noise ratio. On the other hand, the sensitivity of emission intensity on temperature is greatly reduced by the removal of surface traps. One should note that QDs in some cases may exhibit an antiquenching effect [34]. Emission of CdTe QDs is completely quenched when cooled below 230 K [35] (Fig. 7.7). This effect can be assigned to the inter-QD dipole–dipole interactions [36] and the surface reconstruction of QDs by capping layers [34].

That surface plays a strong role in temperature quenching of luminescence of semiconductor QDs has been demonstrated with CdSe/CdS/ZnS core–shell–shell dots dispersed in solution (octadecene) and solid matrices (cross-linked poly(methyl methacrylate) and poly(lauryl methacrylate)) [37]. Quenching of emission starts at lower temperatures (70°C) for QDs in solution than for those in solid matrices (110°C), and is much stronger. This suggests that the surrounding medium influences the quenching of QD emissions. On the other hand, quenching temperatures are the same for the two different solid matrices, so it is reasonable to conclude that physical properties of the polymer did not play any role in a temperature quenching of emission. Upon cooling from high temperatures, the emission of QDs in solution was completely lost, while it was partially recovered for QDs in solid matrices. The irreversible quenching in solution is triggered by surface states generated by the detachment of capping ligands, while in a polymer, the passivating ligands cannot detach and leave the surface [37]. Temperature quenching of emission from colloidal CdSe/ZnS QDs embedded in polystyrene has shown that no additional nonradiative channel occurs when compared to emission quenching of bulk CdSe [31]. These results show that incorporation of colloidal QDs in polymers may be a valuable route for the development of hybrid organic–inorganic luminescent thermometry probes.

Temperature quenching of emission has been investigated in a significant number of QDs which diverged in composition, size, and architecture, with emissions spanning over the visible and near-infrared spectral range. So far, these studies have covered a wide temperature range, from temperatures as low as 0.3 K up to temperatures of 500 K or higher. Even though low-temperature behavior of QD emissions can reveal some interesting features, such as bright–dark state splitting and the already mentioned antiquenching effect, the luminescence response above room temperature is of much greater interest for luminescence thermometry.

With increased awareness of the harmful effects of toxic heavy metals, their use in many applications is becoming undesirable. As a result, cadmium-, lead-, and mercury-free QDs have gained popularity, especially for biomedical applications. Among them, low-toxicity ternary I−III−VI QDs, such as $AgInS_2$ and $CuInS_2$, and alloy $ZnS-AgInS_2$ and $ZnS-CuInS_2$ QDs [13] show appealing properties due to their broad and strong absorption bands and tunable emissions across visible and near-infrared wavelengths. Moreover, these QDs are able to sense temperature, as demonstrated recently with $ZnS–AgInS_2$ (ZAIS) QDs [38]. In alloy $ZnS-AgInS_2$ QDs, the energy gap changes with varying ZnS content providing emissions with peak wavelengths from 530 to 750 nm. Emissions linearly quench with temperature with a sensitivity of about 1 %/K (Fig. 7.8). These QDs also exhibit a negligible cross-sensitivity to oxygen.

Using semiconductor QDs, temperature mapping of different bodies with high spatial resolution may be performed along with luminescence imaging. Considering that QDs are the most utilized nanoprobes for luminescence imaging, biomedicine could benefit greatly from this added measurement capability, since temperature fundamentally regulates the dynamics and properties of any biosystem. It is now well established that two-photon luminescence microscopy provides better spatial resolution than conventional one-photon microscopy. Also, two-photon near-infrared excitations cause less damage to cells and tissues. We have already seen that emissions from CdSe QDs shift with temperature in a similar manner under both one-photon and two-photon excitation (Fig. 7.6A). However, emission intensity decreases much more apparently under two-photon excitation than under one-photon excitation (Fig. 7.6B) [25]. The reason for this discrepancy is the decrease in the two-photon absorption cross section of QDs with temperature rise, which is an additional effect to those effects occurring under one-photon excitation. One should note that in this case, as for most QDs, both spectral shift and emission intensity change with

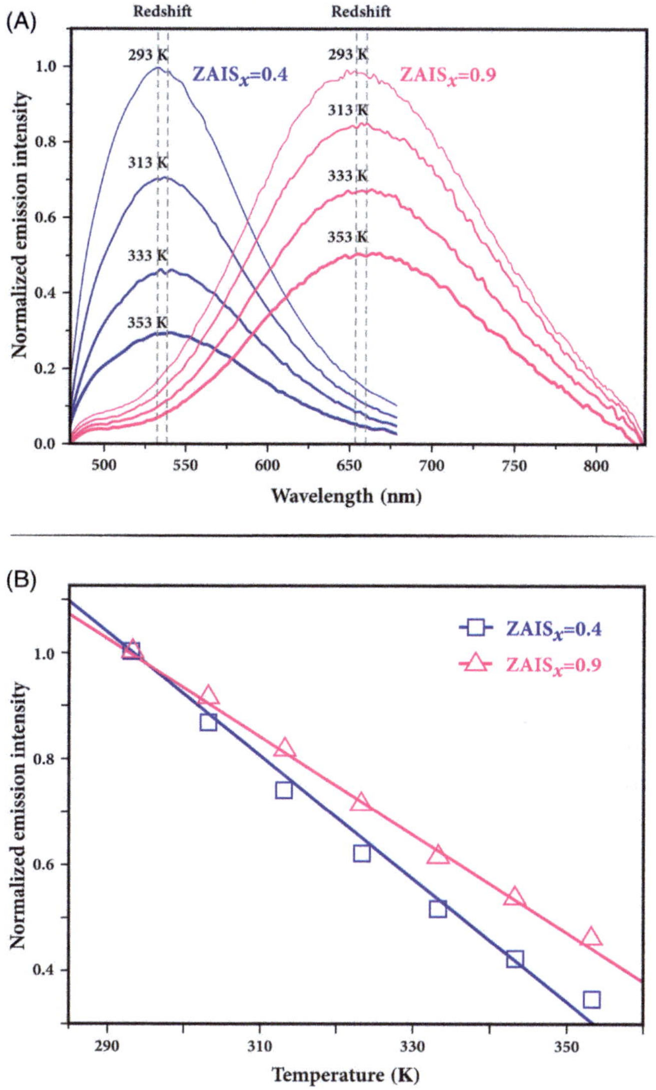

FIGURE 7.8 Luminescence of ZnS–$AgInS_2$ alloy quantum dots of two ZnS contents. (A) Temperature variation of emission spectra. The *broken lines* show the redshift of the emission peak. (B) The linear dependence of the emission intensity on temperature. Intensities are normalized to the value of the emission intensity measured at 293 K. *(Adapted from Matsuda Y, Torimoto T, Kameya T, Kameyama T, Kuwabata S, Yamaguchi H, et al. ZnS–AgInS$_2$ nanoparticles as a temperature sensor. Sensors Actuators B Chem 2013;176:505–8 [38], copyright (2013), with permission from Elsevier).*

temperature, making these probes "dual thermal sensors" capable of reading temperature using two complementary methods.

For application in luminescence thermometry, QDs can be used in a solution, but also in an immobilized form. Intensity of emission of CdSeS/ZnS QDs encapsulated in polymer (poly(methyl methacrylate-co-methacrylic acid) has been found to be sensitive to temperature changes. Over the range 25°C–45°C, emission intensity linearly decreases with temperature at a rate of 1.55%/°C. However, more importantly, emission intensity exhibits strong resistance against physiological pH and ionic strength variations of the environment [39], making these QDs promising candidates for biologically applicable nanothermometry. Other examples include CdSe/ZnS QDs immobilized in a highly temperature-resistant SiO_2 matrix [40], in a polymerized mixture of 3-glycidoxypropyl trimethoxysilane and 3-aminopropyl trimethoxysilane [41], in a polystyrene matrix [42], in a poly(lauryl methacrylate) matrix [43], and in anodized aluminum [44].

Combined imaging and high-resolution *temperature sensing* over the physiological temperature range (25°C–60°C) is feasible with PbS/CdS/ZnS QDs emissions (1270 nm) that fall *in the second biological window* (1000–1350 nm), frequently abbreviated as II-BW or NIR-2 [45]. This spectral range allows large penetration depths, since the radiation is only weakly absorbed and scattered by biological tissues, and the impact of tissue autofluorescence and other background noise is negligible. Fig. 7.9 shows the linear decrease with temperature of 1270 nm emission intensity (normalized to a value measured at 60°C) for

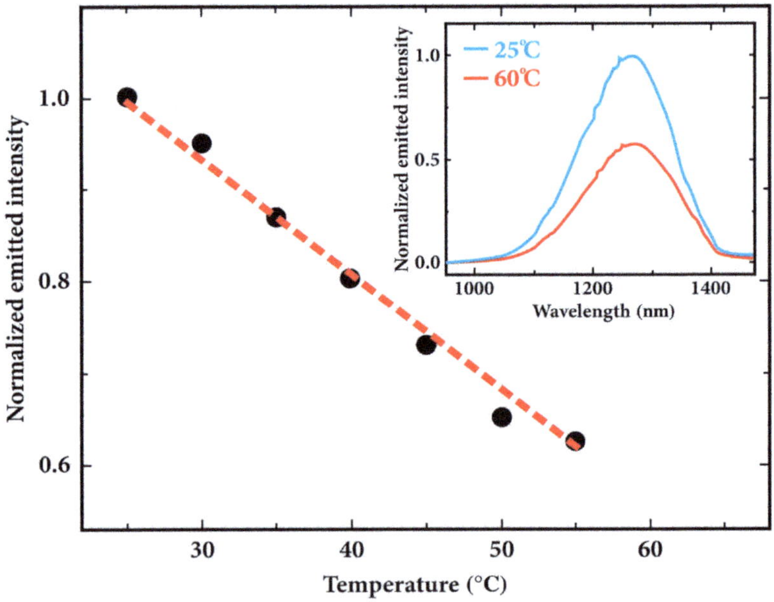

FIGURE 7.9 Emission intensity (normalized to the values measured at room temperature) collected from PbS/CdS/ZnS quantum dots emitting at 1270 nm as a function of temperature. The inset shows two spectra recorded at the two extreme values (25°C and 60°C) of the temperature span studied. *(Reprinted from Benayas A, Ren F, Carrasco E, Marzal V, del Rosal B, Gonfa BA, et al. PbS/CdS/ZnS quantum dots: a multifunctional platform for in vivo near-infrared low-dose fluorescence imaging. Adv Funct Mater 2015;25(42):6650–9 [45], with permission from Wiley).*

PbS/CdS/ZnS QDs (with a temperature sensitivity 1 %/K). The inset in Fig. 7.9 displays QD emission spectra measured at 25°C and 60°C. This QD platform has proven to be capable (ex vivo and in vitro) of high-resolution thermal sensing in the physiological temperature range. The emission intensity variations of these QDs are completely reversible, proving that no changes in the structure and the morphology of the dots has occurred during heating.

Ag_2S and Ag_2Se I−VI semiconductors possess narrow bandgaps of 0.9 and 0.15 eV, respectively (Table 7.1), and their QDs emit in the II-BW. The peak position of Ag_2S QD emission shows little dependence on size due to the heavy carrier effective mass and relatively small exciton Bohr radius (2.2 nm; Table 7.1) [46]. Therefore, small dots (<4 nm) are needed to achieve strong quantum confinement. This was realized with QDs composed of a thin Ag_2S shell on a Ag core [34]. Thin films, prepared by spin coating of Ag/Ag_2S QDs on a polyethylene terephthalate substrate, showed a strong emission in the 1000–1400 nm range under 800 nm laser excitation, with an intensity strongly decreasing in the 15°C–50°C range. The relative temperature sensitivity of 3 %/K is almost 3 times higher than that of PbS/CdS/ZnS QDs which are discussed in the previous paragraph. Moreover, the emission peak of Ag/Ag_2S shows a remarkable temperature-induced redshift of about 2 nm/K (which corresponds to a very large change in energy gap of −170 meV/K) which also enables sensitive temperature readings to be taken.

The key problem for temperature measurements from changes in QD exciton emission intensities, as is the case for other materials, is that observed intensities are substantially influenced by fluctuations in measurement conditions. This problem can be solved by ratiometric intensity temperature sensing as described in Chapter 4, but for ratiometric sensing two emissions with different temperature behaviors are needed. Additional emission in semiconductor QDs may arise from low-energy, trap-related states located within the bandgap. However, these emissions have sufficient intensity for practical measurements only at very low temperatures and can be observed in very small QDs, as has been demonstrated in the case of CdTe QDs embedded in NaCl [47]. To date, ratiometric temperature sensing with QDs has been addressed by doping them with transition metal or lanthanide ions, or by the use of hybrid organic/inorganic nanostructures containing both phosphor nanoparticles and QDs.

Complex emission spectra arise when semiconductors are doped with impurities. In such cases, intensities of emissions and shapes of emission spectra are governed by the values of exciton energy of QDs and energies of impurity excited states, as well as by relative position between them. To date, temperature read-out from the ratio of emissions from excitons and impurities in QDs has been almost exclusively tested with Mn^{2+}-doped QDs, except for the case of Eu^{3+}-doped ZnS nanoparticles. Emission of Mn^{2+} is dominantly due to the $^4T_1(^4G) \rightarrow ^6A_1(^6S)$ d–d electron transition and is strongly quenched by temperature. The energy of this transition depends on the Mn^{2+} coordination, the crystal field strength that acts upon Mn^{2+}, and also by the nephelauxetic effect (see Chapter 6).

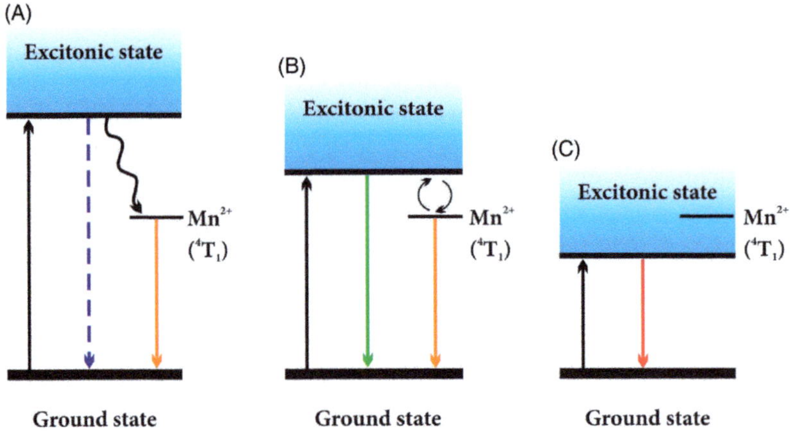

FIGURE 7.10 Energy level configurations and optical transitions of Mn^{2+}-doped semiconductor quantum dots: the Mn^{2+} $^4T_1(^4G)$ level (A) inside the energy gap, (B) close to the exciton energy, and (C) outside the energy gap.

In semiconductor QDs, the Mn^{2+} $^4T_1(^4G)$ level can be located (1) within the semiconductor energy gap, (2) within the gap, but close to exciton energy states, and (3) outside of gap (Fig. 7.10).

The first case occurs when Mn^{2+} ions are doped into wide-gap QDs such as ZnS. Excitation over the bandgap of QDs is followed with an efficient energy transfer to Mn^{2+} which radiatively deexcites showing its characteristic emission (Fig. 7.10A). This energy configuration provides strong sensitization of luminescence due to the highly efficient absorption cross section of QDs and an efficient energy transfer to Mn^{2+}.

In the second case, excited populations are shared between excitonic states and the $^4T_1$ level of Mn^{2+}, that is, they are in thermal equilibrium (Fig. 7.10B). The condition of thermal equilibrium is fulfilled at a certain minimal temperature which depends on the difference between the energies of the $^4T_1$ level and the excitonic states; the smaller the difference, the lower the equilibrium temperature. Then, the emission spectrum of an Mn^{2+}-doped QD consists of one exciton recombination emission band and one lower energy emission band (at a higher wavelength) from the Mn^{2+} luminescent center; the ratio of these emissions is exceptionally sensitive to temperature. The temperature range in which QDs provide dual emission can be tuned by altering the exciton–Mn^{2+} energy difference, for example, by adjusting the size of the QDs during synthesis. Large QDs result in a reduced energy difference between these states, that is, in a lower equilibrium temperature, so that luminescence sensing can be employed from lower temperatures. Both configurations can be utilized for luminescence temperature sensing, however, only the later one involves exciton emission from QDs. The third case, when excited energy levels of Mn^{2+} lie outside the gap (Fig. 7.10C), is not of interest for ratiometric luminescence thermometry because only exciton emission can be observed.

In 10 nm large uncapped Mn^{2+}-doped ZnS nanoparticles only Mn^{2+} emission (peaking around at 589 nm) was used for temperature sensing [48]. Emission intensity showed a linear and reversible response to temperature, and decreased by factor of 2 over the 30°C–150°C range (Fig. 7.11A). Ratiometric temperature sensing has been realized with the double-doped ZnS:Mn^{2+},Eu^{3+} nanoparticles [48] where the ratio of Mn^{2+} emission at 595 nm and Eu^{3+} emission at 612 nm has been used as an indication of temperature (Fig. 7.11B). Both cases may be classified to the first energy level configuration (Fig. 7.10A).

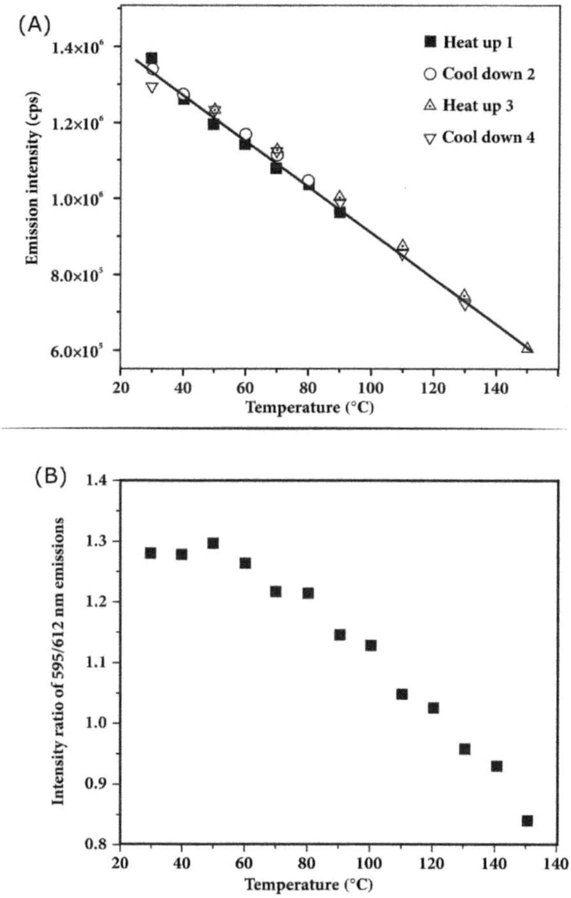

FIGURE 7.11 (A) Emission peak intensity at 589 nm as a function of temperature for 10 nm uncapped ZnS:Mn^{2+} nanoparticles, (B) the intensity ratio of the Mn^{2+} emission at 595 nm and the Eu^{3+} emission at 612 nm at different temperatures of ZnS:Mn^{2+}, Eu^{3+} nanoparticles. *(Adapted from Wang S, Westcott S, Chen W. Nanoparticle luminescence thermometry. J Phys Chem B 2002;106(43):11203–9 [48], copyright (2002) American Chemical Society).*

Mn^{2+} can be doped either at the interface between the CdS core and ZnS shell or in the ZnS shell in CdS/ZnS core/shell QDs (core 3.6 nm in diameter, shell 2.1 nm in thickness) [49]. With dopant located at the interface, relatively high sensitivity (0.5 %/K at 293 K) is found for the ratio of Mn^{2+} emission intensities measured at 600 and 650 nm. The temperature measurement range with these dots spans 77–380 K.

In colloidal Mn^{2+}:CdSe QDs, quantum confinement allows continuous tuning of the first excitonic excited state from below all Mn^{2+} excited states (large particles in which emission comes from solely exciton recombination) to above the lowest $^4T_1$ Mn^{2+} excited state (small particles which show sensitized Mn^{2+} emission), simply by reducing the QD diameter [50].

Temperature-dependent dual emission suitable for temperature sensing via the intensity ratio has been observed in ZnSe:Mn^{2+}/ZnCdSe core/shell QDs (core diameter 3.7 nm, shell thickness ~0.4 nm, Mn content 0.01% with respect to Zn ions) [51]. For these QDs, the energy level configuration corresponds to the one depicted in Fig. 7.10B, and emission spectra consists of an Mn^{2+} emission band around 560 nm and an exciton emission band around 500 nm. With a temperature increase in the 223–403 K range, Mn^{2+} emission lessens in intensity while exciton emission increases (Fig. 7.12). Thus, the respective intensity ratio provides a temperature measurement with a resolution of 0.2 K. One should note that for cases

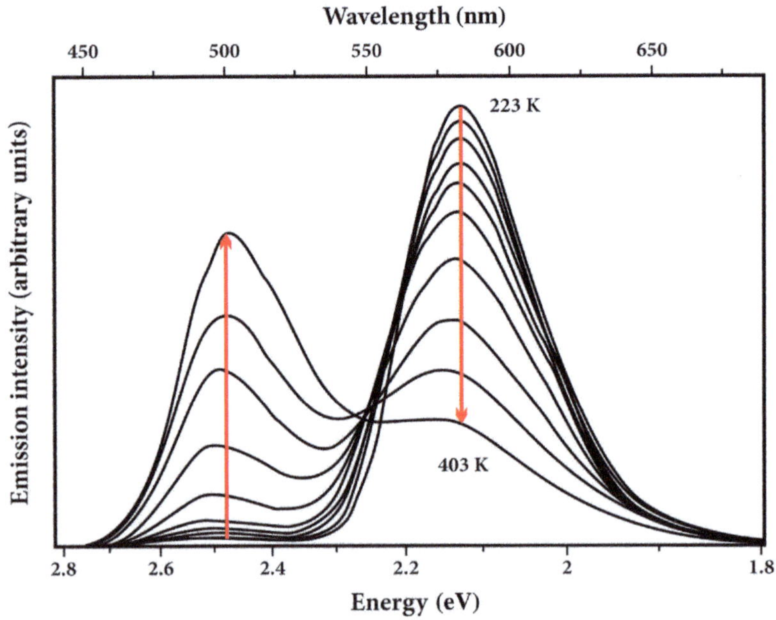

FIGURE 7.12 Emission spectra of ZnSe:Mn^{2+}/ZnCdSe core/shell quantum dots showing the temperature-dependent Mn^{2+} emission band around 560 nm and the exciton emission band around 500 nm. *(Reproduced from Vlaskin VA, Janssen N, van Rijssel J, Beaulac R, Gamelin DR. Tunable dual emission in doped semiconductor nanocrystals. Nano Lett 2010;10(9):3670–4 [51], copyright (2010) American Chemical Society).*

in which the Mn^{2+} $^4T_1$ level exists within the gap, but has excitonic states within thermal reach, the general rule that the excitonic emission is quenched does not hold [50]—doping may have precisely the opposite effect. This rule is only valid when the energy gap greatly exceeds kT, so that the thermal back energy transfer cannot occur within the lifetime of the dopant excited state [50].

CdSSe/ZnS:Mn^{2+} core/shell QDs [52] can sense temperature by the same mechanism as the previously discussed ZnSe:Mn^{2+}/ZnCdSe core/shell QDs, even though they possess a different chemical structure and Mn^{2+} ions are located in the shell. By changing the thickness of the ZnS shell, it is possible to change the relative intensity of exciton emission, and also to redshift the Mn^{2+} emission band and blueshift the exciton band. The concentration of Mn^{2+} and the shell thickness is found to influence the sensitivity of temperature measurements. Examples of ratiometric intensity temperature sensing also includes water-soluble, citrate-capped ZnSe:Mn^{2+}/ZnS/CdS/ZnS QDs ($d \approx 5$ nm) [53]. These dots possess good photo and thermal stability, and the ratio of 520 nm exciton emission and 600 nm Mn^{2+} emission shows strong temperature sensitivity over the $-140°C$ to $90°C$ range.

That Mn^{2+} doping can significantly increase quantum efficiency of exciton emission has been recently demonstrated with $CsPbCl_3$ and $CsPb(Cl/Br)_3$ perovskite QDs [54]. Fig. 7.13 displays the emission spectra of undoped and Mn^{2+}-doped $CsPbCl_3$ QDs dispersed in hexanes along with photographs of QDs in vial taken under ultraviolet illumination. While undoped QDs exhibit only the narrow exciton emission around 400 nm (lower line in Fig. 7.13), Mn-doped ones exhibit an additional broad emission around 600 nm from the Mn^{2+} $^4T_1 \rightarrow ^6A_1$ transition

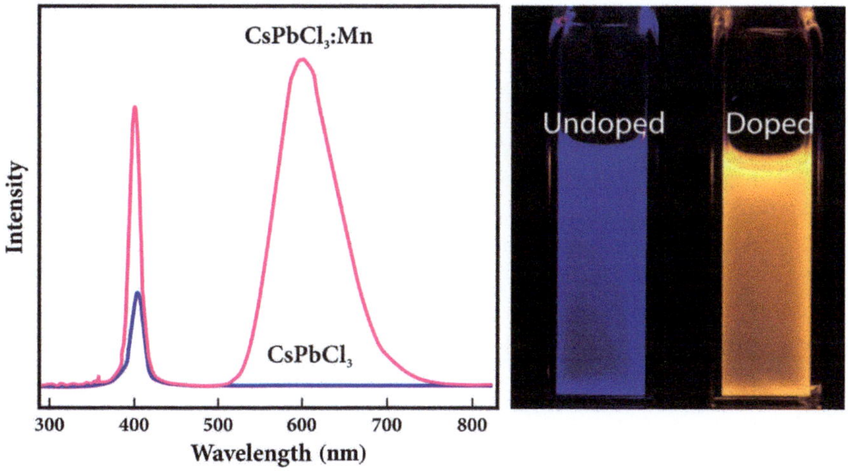

FIGURE 7.13 (A) Emission spectra of Mn-doped *(upper line)* and undoped *(lower line)* $CsPbCl_3$ perovskite quantum dots dispersed in hexane. (B) Photographs of samples under ultraviolet excitation. *(Adapted from Parobek D, Roman BJ, Dong Y, Jin H, Lee E, Sheldon M, et al. Exciton-to-dopant energy transfer in Mn-doped cesium lead halide perovskite nanocrystals. Nano Lett 2016;16(12):7376–80 [54], copyright (2016) American Chemical Society).*

(upper line in Fig. 7.13). The dual emission of these QDs may be promising for ratiometric temperature sensing, although this has not been tested to date.

Emissions from different types of particles can be used for ratiometric intensity temperature sensing if these particles are located close to each other so they experience the same temperature changes, and if their luminescence presents markedly different thermal responses over the temperature measurement range. These *hybrid luminescence temperature sensors* can be fabricated by the mild sintering of different particles, by incorporation of particles in glasses or transparent polymers, by the growth of multilayer films and coatings, and so on.

A hybrid luminescence temperature nanosensor (with an average size of 150 ± 70 nm) has been recently fabricated from Nd^{3+}-doped $NaGdF_4$ 15 nm large phosphor particles and 4 nm large PbS/CdS/ZnS QDs by encapsulating them in poly(lactic-co-glycolic acid) (PLGA) [55]. The luminescence excitation and emission spectra of Nd^{3+}:$NaGdF_4$ and PbS/CdS/ZnS QDs (Fig. 7.14A) show that both particles can be excited in the first biological window (I-BW, 650–950 nm) and that both emit in the II-BW. These luminescence features facilitate the biological application of hybrid nanostructures, especially since they are formed from a biocompatible and biodegradable polymer (PLGA). Emission intensity temperature variations (Fig. 7.14B) follow the general behavior of lanthanide-doped phosphors (see Chapter 6) and QDs (detailed earlier in this chapter). During the temperature increase over the 20°C–70°C range, intensity of emission of Nd^{3+} was only slightly affected (decreased), while emission intensity of QDs suffered a strong decline. When combined in the hybrid nanostructure, particles showed a complex emission spectrum in the II-BW under 808 nm excitation, but retained their intrinsic luminescence behavior (Fig. 7.14C). The ratio of intensities of their emissions, peaking at 1060 nm (Nd^{3+}) and 1250 nm (QDs), linearly decreases with temperature (Fig. 7.14D). An excellent relative sensitivity of about 2.5 %/K is obtained at 30°C, which, theoretically, provides a good temperature resolution of 0.2 K. The importance of this hybrid nanostructure is that it provides a bioimaging ability in the II-BW and simultaneously acts as a temperature sensor with a subdegree temperature resolution.

A hybrid luminescence nanothermometer has been fabricated from a red-emitting CdSe−CdS QD/QR (quantum rod) heterostructure passivated with an amphiphilic polymer shell, to which is appended a far-red-emitting cyanine dye (Alexa Fluor 647) [56]. Thermoresponsive dual emission is conferred by a Förster resonant energy transfer (FRET) process between CdSe−CdS QDs−QRs as donors and cyanine dyes as acceptors, which are conjugated to QDs−QRs using an amphiphilic polymer coating. The ratio of QD−QR emission over the 630–640 nm range and dye emission over the 664–674 nm range is strongly temperature dependent from 20°C to 40°C, and presents a 2.4 %/K relative sensitivity and a resolution better than 0.2°C.

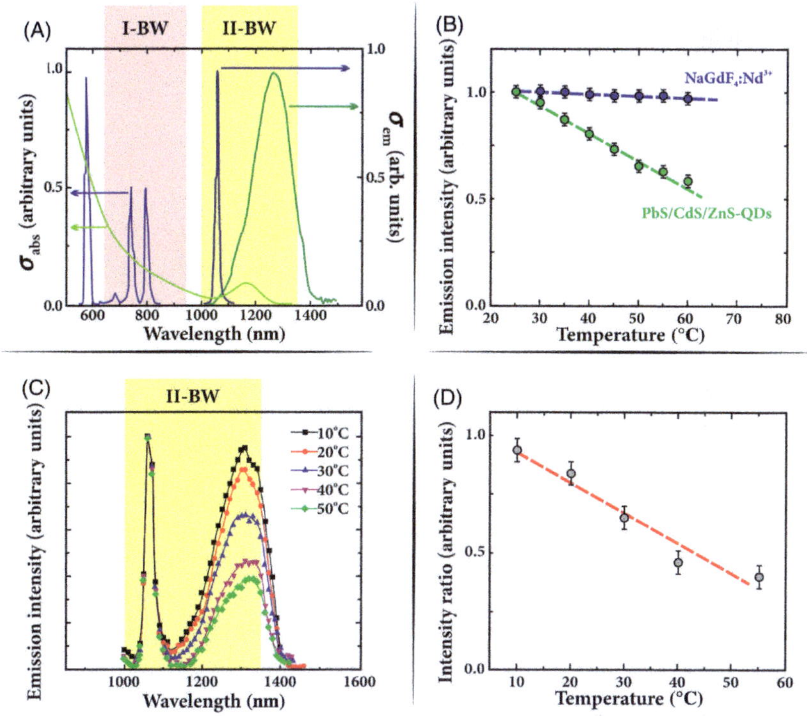

FIGURE 7.14 (A) Emission and absorption spectra of Nd^{3+}-doped $NaGdF_4$ nanoparticles (blue lines) and PbS/CdS/ZnS quantum dots (green lines). (B) Temperature dependence of emission intensities of Nd^{3+}:$NaGdF_4$ nanoparticles and PbS/CdS/ZnS quantum dots. Symbols represent experimental data and dashed lines indicate linear temperature dependence. (C) Emission spectra measured under 808 nm excitation at different temperatures of the hybrid nanostructure which consists of Nd^{3+}:$NaGdF_4$ and PbS/CdS/ZnS nanoparticles encapsulated into poly(lactic-co-glycolic acid). (D) Temperature variation in the ratio between 1060 and 1250 nm emitted intensities. Symbols represent experimental data and dashed line the best linear fit. *(Adapted from Cerón EN, Ortgies DH, del Rosal B, Ren F, Benayas A, Vetrone F, et al. Hybrid nanostructures for high-sensitivity luminescence nanothermometry in the second biological window. Adv Mater 2015;27(32):4781–7 [55], with permission from Wiley).*

As described in Chapter 4, temperature sensing can be achieved by monitoring temporal changes of emission intensity. QDs normally have emission decays on a nanosecond timescale, with a nonexponential shape and with average lifetime values that are strongly size dependent. For example, Fig. 7.15A shows emission decays of different sizes CdTe QDs [57] and illustrates the nonexponential decay shape and its shortening with a reduction of QD size. It has been found [57] that lifetime values decrease with temperature for all QD sizes, and that the temperature-induced lifetime reduction is much larger for smaller dots (Fig. 7.15B); a 7.8 nm CdTe QD has a temperature sensitivity of 0.08 %/K while a 1.2 nm dot has a temperature sensitivity of 1.7 %/K.

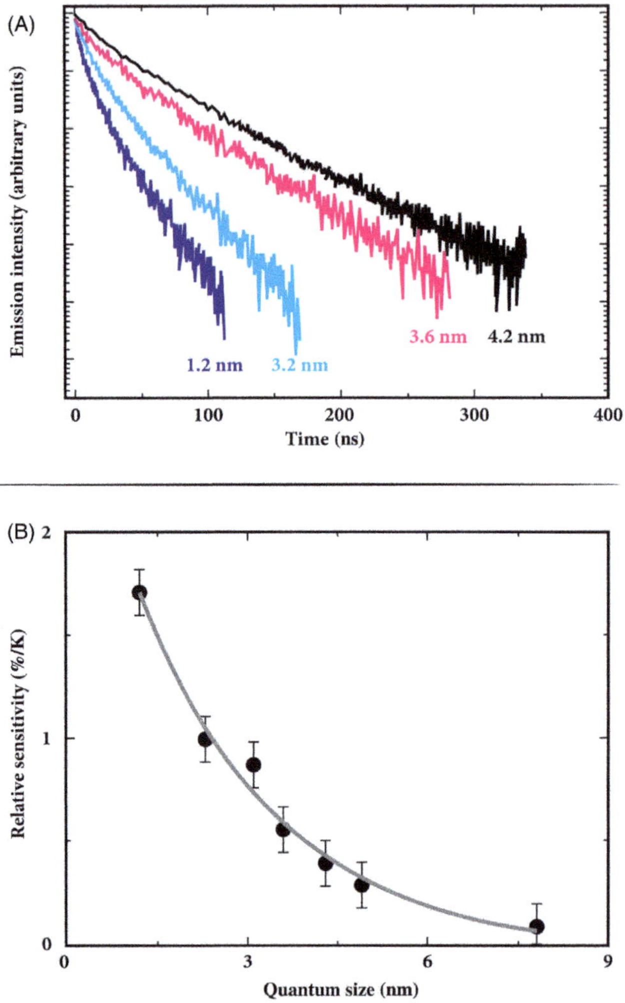

FIGURE 7.15 (A) Emission decays of CdTe quantum dots of different diameters (3.2, 3.8, and 4.2 nm) (B) Relative temperature sensitivity of lifetime as a function of the CdTe quantum dot diameter. *(Adapted from Haro-González P, Martínez-Maestro L, Martín IR, García-Solé J, Jaque D. High-sensitivity fluorescence lifetime thermal sensing based on CdTe quantum dots. Small 2012;8(17):2652–8 [57], with permission from Wiley).*

A strong size dependence of exciton emission lifetimes of CdTe QDs has been confirmed by photoluminescence analysis of powders prepared by embedding CdTe dots in NaCl [47]. However, this analysis also showed that, besides exciton emission, the small size CdTe dots present broad trap-related emissions at longer wavelengths with emission decays on the order of several microseconds (Fig. 7.16A) and with a high sensitivity of lifetime on temperature (1.9 nm/K) over the 80–360 K range (Fig. 7.16B).

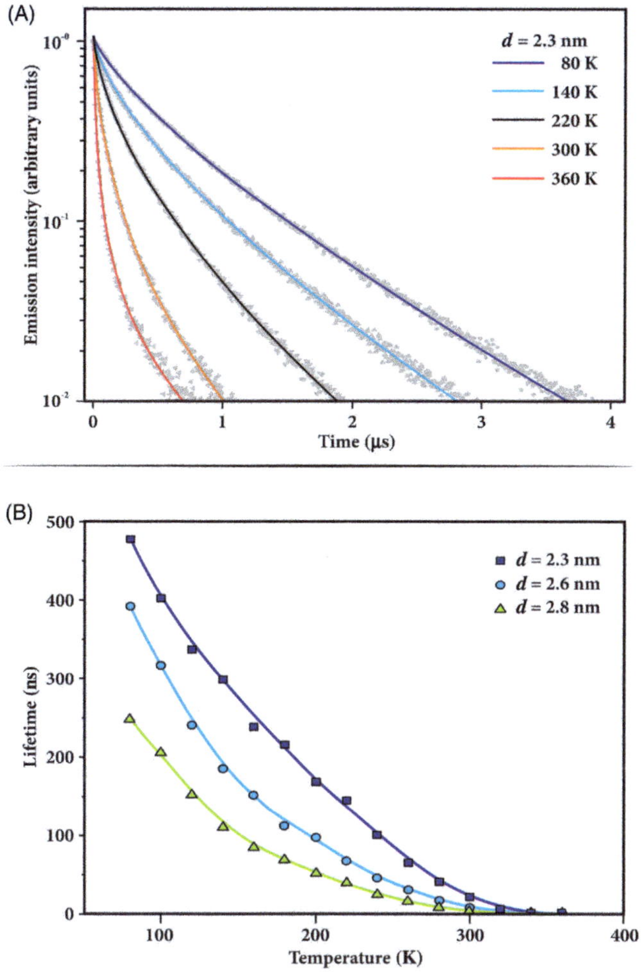

FIGURE 7.16 (A) Temporal dependence of trap-related emission of 2.3 nm CdTe quantum dots embedded in NaCl. (B) Temperature dependence of average lifetime values of trap-related emission of CdTe quantum dots embedded in NaCl for 2.3, 2.6, and 2.8 nm dots. *(Adapted from Kalytchuk S, Zhovtiuk O, Kershaw S V, Zbořil R, Rogach AL. Temperature-dependent exciton and trap-related photoluminescence of CdTe quantum dots embedded in a NaCl matrix: implication in thermometry. Small 2016;12(4):466–76 [47], with permission from Wiley).*

REFERENCES

[1] Mino L, Agostini G, Borfecchia E, Gianolio D, Piovano A, Gallo E, et al. Low-dimensional systems investigated by x-ray absorption spectroscopy: a selection of 2D, 1D and 0D cases. J Phys D Appl Phys 2013;46(42):423001.

[2] Yoffe AD. Low-dimensional systems: quantum size effects and electronic properties of semiconductor microcrystallites (zero-dimensional systems) and some quasi-two-dimensional systems. Adv Phys 2002;51(2):799–890.

[3] Koole R, Groeneveld E, Vanmaekelbergh D, Meijerink A, de Mello Donegá C. Size effects on semiconductor nanoparticles. Nanoparticles: workhorses of nanoscience. Berlin, Heidelberg: Springer; 2014. 13–51.

[4] de Mello Donegá C. Synthesis and properties of colloidal heteronanocrystals. Chem Soc Rev 2011;40(3):1512–46.

[5] Yen WM, Shionoya S, Yamamoto H. Phosphor handbook. Boca Raton, FL: CRC Press, Taylor & Francis Group; 2007. 1055.

[6] Haynes WM. CRC handbook of chemistry and physics. 97th ed. Boca Raton, FL: CRC Press, Taylor & Francis Group; 2016. 2652.

[7] Adachi S. Handbook on physical properties of semiconductors: volume 3: II–VI compound semiconductors. The Netherlands: Springer US, Kluwer Academic Publishers Group, Dordrecht; 2004. 1472.

[8] Wise FW. Lead salt quantum dots: the limit of strong quantum confinement. Acc Chem Res 2000;33(11):773–80.

[9] Jaque Garcia D, Garcia Sole J. Quantum dot fluorescence thermometry. Thermometry at the nanoscale: techniques and selected applications. London: Royal Society of Chemistry; 2016. 83–123.

[10] Vossmeyer T, Katsikas L, Giersig M, Popovic IG, Diesner K, Chemseddine A, et al. CdS nanoclusters: synthesis, characterization, size dependent oscillator strength, temperature shift of the excitonic transition energy, and reversible absorbance shift. J Phys Chem 1994;98(31):7665–73.

[11] Chang Y-M, Shieh J, Chu P-Y, Lee H-Y, Lin C-M, Juang J-Y. Enhanced free exciton and direct band-edge emissions at room temperature in ultrathin ZnO films grown on Si nanopillars by atomic layer deposition. ACS Appl Mater Interfaces 2011;3(11):4415–9.

[12] Gurinovich LI, Gurin VS, Ivanov VA, Bodnar IV, Molochko AP, Solovej NP. Crystal structure and optical properties of $CuInS_2$ nanocrystals in a glass matrix. Phys Status Solidi 1998;208(2):533–40.

[13] Xu G, Zeng S, Zhang B, Swihart MT, Yong K-T, Prasad PN. New generation cadmium-free quantum dots for biophotonics and nanomedicine. Chem Rev 2016;116(19):12234–327.

[14] Kan S, Mokari T, Rothenberg E, Banin U. Synthesis and size-dependent properties of zincblende semiconductor quantum rods. Nat Mater 2003;2(3):155–8.

[15] Green M, O'Brien P. Semiconductor quantum dots: organometallic and inorganic synthesis. London: Royal Society of Chemistry; 2014. 295.

[16] Rogach A. Semiconductor nanocrystal quantum dots: synthesis, assembly, spectroscopy and applications. Springer-Verlag, Wien; 2008. XI, p. 372.

[17] Klimov VI. . Semiconductor and metal nanocrystals: synthesis and electronic and optical properties. Boca Raton, FL: CRC Press, Taylor & Francis Group; 2003. 500.

[18] Jing L, Kershaw SV, Li Y, Huang X, Li Y, Rogach AL, et al. Aqueous based semiconductor nanocrystals. Chem Rev 2016;116(18):10623–730.

[19] Nozik J, Micic O. Quantum dots and quantum dot arrays: synthesis, optical properties, photogenerated carrier dynamics, multiple exciton generation and applications to solar photon conversion. Nanocrystal quantum dots. Boca Raton, FL: CRC Press, Taylor & Francis Group; 2010. 311-68.

[20] Olkhovets A, Hsu R-C, Lipovskii A, Wise FW. Size-dependent temperature variation of the energy gap in lead-salt quantum dots. Phys Rev Lett 1998;81(16):3539–42.

[21] Varshni YP. Temperature dependence of the energy gap in semiconductors. Physica 1967;34(1):149–54.

[22] Dai Q, Song Y, Li D, Chen H, Kan S, Zou B, et al. Temperature dependence of band gap in CdSe nanocrystals. Chem Phys Lett 2007;439(1–3):65–8.

[23] Dai Q, Zhang Y, Wang Y, Hu MZ, Zou B, Wang Y, et al. Size-dependent temperature effects on PbSe nanocrystals. Langmuir 2010;26(13):11435–40.

[24] Li S, Zhang K, Yang J-M, Lin L, Yang H. Single quantum dots as local temperature markers. Nano Lett 2007;7(10):3102–5.

[25] Maestro LM, Rodríguez EM, Rodríguez FS, la Cruz MCI, Juarranz A, Naccache R, et al. CdSe quantum dots for two-photon fluorescence thermal imaging. Nano Lett 2010;10(12):5109–15.

[26] Maestro LM, Jacinto C, Silva UR, Vetrone F, Capobianco JA, Jaque D, et al. CdTe quantum dots as nanothermometers: towards highly sensitive thermal imaging. Small 2011;7(13): 1774–8.

[27] Zhou D, Lin M, Liu X, Li J, Chen Z, Yao D, et al. Conducting the temperature-dependent conformational change of macrocyclic compounds to the lattice dilation of quantum dots for achieving an ultrasensitive nanothermometer. ACS Nano 2013;7(3):2273–83.

[28] Yang W, Lowe-Webb RR, Lee H, Sercel PC. Effect of carrier emission and retrapping on luminescence time decays in InAs/GaAs quantum dots. Phys Rev B 1997;56(20):13314–20.

[29] Klimov VI, McBranch DW. Auger-process-induced charge separation in semiconductor nanocrystals. Phys Rev B 1997;55(19):13173–9.

[30] Kagan CR, Murray CB, Bawendi MG. Long-range resonance transfer of electronic excitations in close-packed CdSe quantum-dot solids. Phys Rev B 1996;54(12):8633–43.

[31] Valerini D, Cretí A, Lomascolo M, Manna L, Cingolani R, Anni M. Temperature dependence of the photoluminescence properties of colloidal CdSe/ZnS core/shell quantum dots embedded in a polystyrene matrix. Phys Rev B 2005;71(23):235409.

[32] Morello G, De Giorgi M, Kudera S, Manna L, Cingolani R, Anni M. Temperature and size dependence of nonradiative relaxation and exciton−phonon coupling in colloidal CdTe quantum dots. J Phys Chem C 2007;111(16):5846–9.

[33] Alivisatos AP. Semiconductor clusters, nanocrystals, and quantum dots. Science 1996;271(5251):933–7.

[34] Wuister SF, van Houselt A, de Mello Donegá C, Vanmaekelbergh D, Meijerink A. Temperature antiquenching of the luminescence from capped CdSe quantum dots. Angew Chemie 2004;43(23):3029–33.

[35] Wuister SF, de Mello Donegá C, Meijerink A. Luminescence temperature antiquenching of water-soluble CdTe quantum dots: role of the solvent. J Am Chem Soc 2004;126(33): 10397–402.

[36] Wang X, Wolfbeis OS, Meier RJ. Luminescent probes and sensors for temperature. Chem Soc Rev 2013;42(19):7834–69.

[37] Zhao Y, Riemersma C, Pietra F, Koole R, de Mello Donegá C, Meijerink A. High-temperature luminescence quenching of colloidal quantum dots. ACS Nano 2012;6(10):9058–67.

[38] Matsuda Y, Torimoto T, Kameya T, Kameyama T, Kuwabata S, Yamaguchi H, et al. ZnS–$AgInS_2$ nanoparticles as a temperature sensor. Sens Actuators B Chem 2013;176:505–8.

[39] Liu H, Fan Y, Wang J, Song Z, Shi H, Han R, et al. Intracellular temperature sensing: an ultrabright luminescent nanothermometer with non-sensitivity to pH and ionic strength. Sci Rep 2015;5:14879.

[40] Gupta DP-T, BMW, MC. CdSe(ZnS) nanocomposite luminescent high temperature sensor. Nanotechnology 2011;22(18):185503.

[41] Duong HD, Rhee JIl. Enhancement of the sensitivity of a quantum dot-based fiber optic temperature sensor using the sol–gel technique. Sens Actuators B Chem 2008;134(2):423–6.

[42] Liwang L, Zhong K, Meng L, Van Hemelrijck D, Wang L, Glorieux C. Temperature-sensitive photoluminescent CdSe–ZnS polymer composite film for lock-in photothermal characterization. J Appl Phys 2016;119(22):224902.

[43] Walker GW, Sundar VC, Rudzinski CM, Wun AW, Bawendi MG, Nocera DG. Quantum-dot optical temperature probes. Appl Phys Lett 2003;83(17):3555–7.

[44] Sakaue H, Aikawa A, Iijima Y. Anodized-aluminum as quantum dot support for global temperature sensing from 100 to 500 K. Sens Actuators B Chem 2010;150(2):569–73.

[45] Benayas A, Ren F, Carrasco E, Marzal V, del Rosal B, Gonfa BA, et al. PbS/CdS/ZnS quantum dots: a multifunctional platform for in vivo near-infrared low-dose fluorescence imaging. Adv Funct Mater 2015;25(42):6650–9.

[46] Lin S, Feng Y, Wen X, Zhang P, Woo S, Shrestha S, et al. Theoretical and experimental investigation of the electronic structure and quantum confinement of wet-chemistry synthesized Ag_2S nanocrystals. J Phys Chem C 2015;119(1):867–72.

[47] Kalytchuk S, Zhovtiuk O, Kershaw SV, Zbořil R, Rogach AL. Temperature-dependent exciton and trap-related photoluminescence of CdTe quantum dots embedded in a NaCl matrix: implication in thermometry. Small 2016;12(4):466–76.

[48] Wang S, Westcott S, Chen W. Nanoparticle luminescence thermometry. J Phys Chem B 2002;106(43):11203–9.

[49] Park Y, Koo C, Chen H-Y, Han A, Son DH. Ratiometric temperature imaging using environment-insensitive luminescence of Mn-doped core-shell nanocrystals. Nanoscale 2013;5(11):4944–50.

[50] Beaulac R, Archer PI, van Rijssel J, Meijerink A, Gamelin DR. Exciton storage by Mn^{2+} in colloidal Mn^{2+}-doped CdSe quantum dots. Nano Lett 2008;8(9):2949–53.

[51] Vlaskin VA, Janssen N, van Rijssel J, Beaulac R, Gamelin DR. Tunable dual emission in doped semiconductor nanocrystals. Nano Lett 2010;10(9):3670–4.

[52] Hsia C-H, Wuttig A, Yang H. An accessible approach to preparing water-soluble Mn^{2+}-doped (CdSSe)ZnS (core)shell nanocrystals for ratiometric temperature sensing. ACS Nano 2011;5(12):9511–22.

[53] McLaurin EJ, Vlaskin VA, Gamelin DR. Water-soluble dual-emitting nanocrystals for ratiometric optical thermometry. J Am Chem Soc 2011;133(38):14978–80.

[54] Parobek D, Roman BJ, Dong Y, Jin H, Lee E, Sheldon M, et al. Exciton-to-dopant energy transfer in Mn-doped cesium lead halide perovskite nanocrystals. Nano Lett 2016;16(12):7376–80.

[55] Cerón EN, Ortgies DH, del Rosal B, Ren F, Benayas A, Vetrone F, et al. Hybrid nanostructures for high-sensitivity luminescence nanothermometry in the second biological window. Adv Mater 2015;27(32):4781–7.

[56] Albers AE, Chan EM, McBride PM, Ajo-Franklin CM, Cohen BE, Helms BA. Dual-emitting quantum dot/quantum rod-based nanothermometers with enhanced response and sensitivity in live cells. J Am Chem Soc 2012;134(23):9565–8.

[57] Haro-González P, Martínez-Maestro L, Martín IR, García-Solé J, Jaque D. High-sensitivity fluorescence lifetime thermal sensing based on CdTe quantum dots. Small 2012;8(17):2652–8.

Chapter 8

Luminescence Temperature Sensing Using Organic Materials

8.1 INTRODUCTION

Ever since the discovery of organic fluorescent compounds in the late 19th century, much work has been made to adopt them in everyday use. Today we know thousands of fluorescent organic compounds, both synthetic and natural, and use many of them in important applications, such as bioimaging and biosensing, chemical and physical sensing, lighting, fluid tracing, laser devices, etc. For example, organic light-emitting diodes (OLEDs) are used today in mobile phones, digital cameras, virtual-reality headsets, tablets, laptops, and TVs as dominant display materials.

Fluorescent organic materials can be efficiently excited since they have strong and wide absorption bands, which appear in many cases in the visible spectral range. Depending on their structure, they may provide intense emissions with quantum yields approaching one, and large Stokes shifts. Electronic excited states of organic molecules are severely influenced by their environment. Therefore, temperature changes in the environment (as well as other changes, for example, pH) strongly affect fluorescence features of organic materials (band positions and shapes, lifetimes, and polarization) and make them excellent materials for fluorescence temperature probes. Consequently, many organic compounds and organic–inorganic hybrids have been methodically tested for luminescence thermometric performance, already having found applications mainly in the fields of biomedicine, optoelectronics and nanoelectronics, and microfluidics and nanofluidics.

In this chapter, we discuss basics of fluorescence of organic compounds and explain differences in fluorescence response to temperature changes due to differences in their structure. The chapter gives several examples of luminescence thermometry with different types of organic and organic–inorganic probes. These include organic dyes and pigments, dyes incorporated in functional polymers (including semiconducting polymers), discrete metal–organic complexes, and metal–organic frameworks (MOFs).

Luminescence Thermometry. http://dx.doi.org/10.1016/B978-0-08-102029-6.00008-7

8.2 DYES AND ORGANIC PIGMENTS

Dyes and organic pigments are organic molecules with extended conjugation containing a framework of σ-bonds and an associated π-system, which have specific colors due to the selective absorption of visible light. The difference between the two arises from their solubility. Dyes are soluble, while pigments are not. Today, dyes and pigments are an important class of materials extensively used in many industries, but mostly they are used for the coloration of textiles, paints, printing inks, plastics, and cosmetics. The majority of dyes do not demonstrate notable fluorescence since their fluorescence is easily quenched by environmental factors. However, a great number of fluorescent dyes are known today, and new ones are constantly being developed, mostly for use in biomedical labeling and chemical and physical sensing. The extensive list of fluorescent dyes along with their chemical and spectroscopic data is freely available on the fluorphores.org database (http://www.fluorophores.tugraz.at). Interestingly, the first identified fluorescent organic molecule was quinine sulfate, which was described by Sir John Herschel in 1845 [1], several years before Sir George Stokes introduced the term fluorescence. Sometime later, in 1871, a German chemist Adolf von Bayer synthesized the famous fluorescent dye 3′,6′-dihydroxyspiro[isobenzofuran-1(3H),9′-[9H]xanthen]-3-one and named it fluorescein (see Fig. 8.1A and B). This dye has a quantum yield of 0.79 and is still much used today along with its derivatives fluorescein isothiocyanate (FITC) and 6-FAM phosphoramidite. In water, fluorescein has an absorption maximum at 494 nm and an emission maximum at 512 nm (see Fig. 8.1C).

Like most organic compounds that exhibit efficient fluorescence, fluorescein has a rigid structure (it has an ether link between two aromatic rings) which blocks the loss of excited state energy through torsional vibrations. In contrast, phenolphthalein, for example, displays no emission because its excited state loses energy by vibrations of the benzene rings [2]. The properties of some well-known fluorescent dyes are listed in Table 8.1. In addition to visible-emitting dyes, there are many organic molecules that fluoresce in the ultraviolet and near-infrared spectral range. Near-infrared emitting compounds are of huge importance for imaging in the first and second biological windows due to the small tissue absorption in these spectral regions.

Electronic transitions between molecular orbitals are responsible for the absorption and emission of light from organic molecules. Molecular orbitals arise from two types of atomic orbital overlaps: direct and lateral. Direct (i.e., collinear with the symmetry axis) overlapof two s atomic orbitals, or one s and one p atomic orbital, or two p atomic orbitals, produces σ molecular orbital. The bond formed in this way is called a σ bond. In the ground state of a molecule the low-energy σ orbital (bonding orbital) is occupied by two electrons, while the high-energy σ* orbital (antibonding orbital) is unoccupied. Lateral overlap of p atomic orbitals (e.g., from two singly occupied $2p_z$ atomic orbitals) produces π molecular orbital, and the resulting bond is termed a π bond. Again, in the

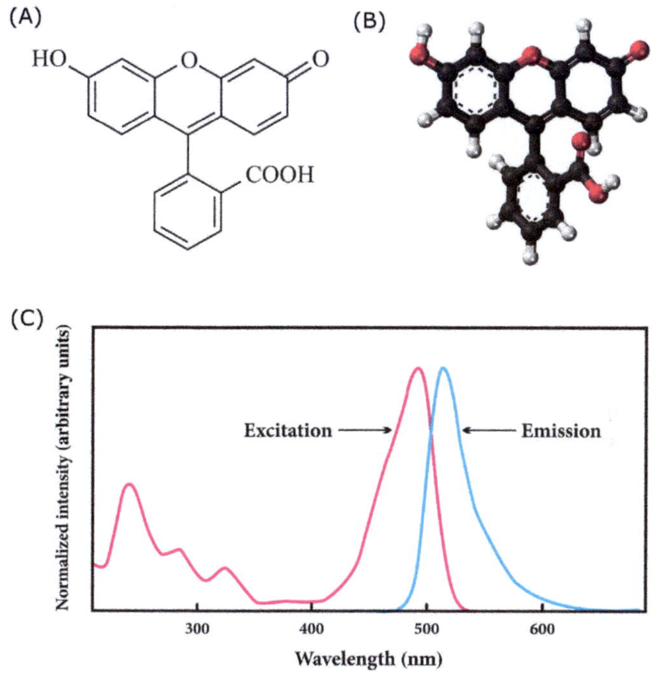

FIGURE 8.1 Fluorescein (A) chemical formula, (B) three-dimensional chemical structure, and (C) excitation (*magenta color*) and emission (*cyan color*) spectra.

ground state of a molecule the π orbital of lower energy is populated by two electrons (bonding orbital), while the molecular orbital of higher energy (antibonding π* orbital) remains unoccupied. Nonbonding electrons situated on heteroatoms such as oxygen or nitrogen may give rise to molecular orbitals called n-orbitals. Electronic transitions from the highest occupied molecular orbital (HOMO) to the lowest unoccupied molecular orbital (LUMO) (see Fig. 8.2 for an example of the molecular orbitals of formaldehyde) have energies, in principle, in the following order:

$$\sigma \rightarrow \sigma^* > \sigma \rightarrow \pi^* > \mathrm{n} \rightarrow \sigma^* > \pi \rightarrow \pi^* > \mathrm{n} \rightarrow \pi^*.$$

The same order applies for emission, i.e., for the radiative transitions from LUMO to HOMO.

Energies of σ→σ*, σ→π*, and n→σ* electronic transitions fall into the deep-ultraviolet and ultraviolet spectral regions. The color of a dye is thus dependent on the π→π* and n→π* transitions and their probabilities. The π→π* transition in organic dyes which entail a considerable charge transfer in donor/acceptor chromogens are highly probable and produce intense colors (i.e., produce a strong absorption of light). The widths of absorption and emission bands

TABLE 8.1 Optical Properties of Some Fluorescent Dyes

Dye	Solvent	λ_{exc} (nm)	λ_{em} (nm)	ε (L $mol^{-1}cm^{-1}$)	QY	τ (ns)	References
Rhodamine B	Water	554	573	9.810^4	0.31	1.75	[4, 5]
	Acidic ethanol	554	576	–	0.50	2.45	[6, 7]
Rhodamine 6G	Water	526	551	–	0.59	3.95	[5]
	Ethanol	530	556	10.5×10^4	0.95	3.99	[6, 8]
Rhodamine 101	Acidic ethanol	564	588	11.9×10^4	0.91	4.34	[9]
Rhodamine 110	Water	497	520				[10]
	Ethanol	510	535	8.99×10^4			[8]
Fluorescein	Ethanol	482	515	9.23×10^4	0.79		[11]
Perylene	Cyclohexane	436	468	4.3×10^4	0.94		[12]
Pyrene	Cyclohexane	335	385	5.4×10^4	0.65	450	[13]
Acridine orange	Water	450	535		0.25	1.7	[9, 14]
Coumarin 2	Ethanol	366	435	2.15×10^4	0.68		[15, 16]
Cy5	Water	650	667	25×10^4		0.27	[17]
Cy7	Water	750	777	20×10^4			[17]
IR 140	Ethanol	804	844	17×10^4	0.17		[9]
POPOP	Cyclohexane	358	387	4.2×10^4	0.97	1.12	[4]

λ_{exc}, Excitation maximum; λ_{em}, emission maximum; ε, molar extinction coefficient; *QY*, quantum yield; τ, lifetime.
Source: Data extracted from Valeur B, Berberan-Santos MN. Molecular fluorescence: principles and applications. Weinheim: Wiley-VCH; 2013. 569 p., with permission from Wiley.

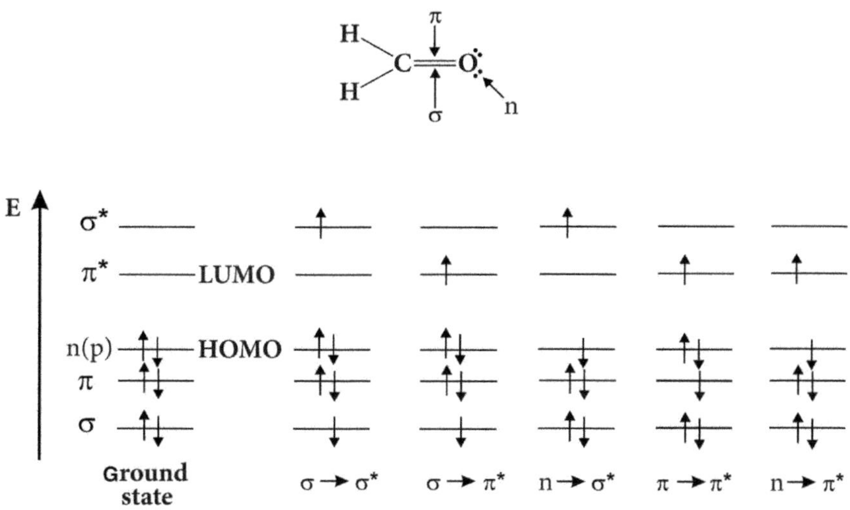

FIGURE 8.2 Energy levels of molecular orbitals in formaldehyde and the possible electronic transitions. *(Adapted from Valeur B, Berberan-Santos MN. Molecular fluorescence: principles and applications. Weinheim: Wiley-VCH; 2013. 569 p., with permission from Wiley).*

are determined by the distribution of vibrational energy levels superimposed on the electronic energy levels. Broader bands are generally observed with molecules which have a large spread of energies from bond vibrations.

The discussion so far has been related to the case of localized molecular orbitals between pairs of atoms. In conjugated molecules, i.e., molecules that possess a structure of alternating single and double carbon–carbon bonds, the π electron system can be regarded as independent of the σ bonds (delocalized) since there is no overlap between the σ and π orbitals. In such case, π electrons are delocalized over the whole conjugation length (the length of the carbon chain), their energy is lowered, and the energy gap between two successive π orbitals is decreased (see Fig. 8.3). The effect becomes larger as the conjugation length increases and causes a shift of the absorption band toward a longer wavelength in the visible region. The effect is well known in both aromatic molecules (cyclic conjugated systems) and polyenes (linear conjugated systems).

Regarding luminescent thermometry, fluorescent organic molecules are important since they are relatively inexpensive, can be handled with ease, and can be efficiently excited since they have intense and relatively wide absorption bands. In many cases, their fluorescence shows a strong temperature dependence and provides versatile temperature readings using all the techniques described in Chapter 4: Schemes for Temperature Read-out from Luminescence. Also, they can be easily combined with other organic or inorganic materials to form sensitive hybrid temperature probes. For thermometry applications, in addition to a strong dependence of fluorescence on temperature, organic

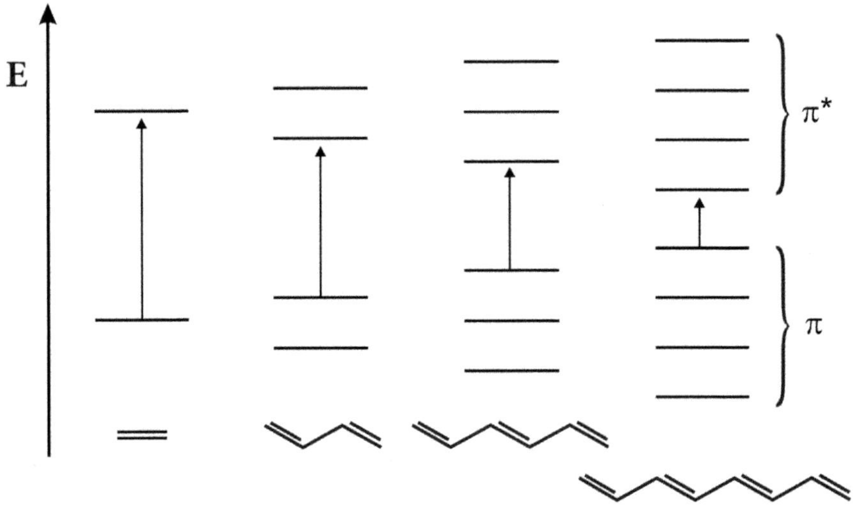

FIGURE 8.3 Lowering of energy of π electrons with an increase of conjugation length. Note the decrease in energy gap between π and π* levels with the increase of conjugation length.

molecules should have good thermostability and photostability, as well as an ability to undergo repeated excitation/emission cycles.

In addition to wide use as an efficient laser gain media and as a bioimaging dye, the rhodamine family of xanthene dyes has found a notable application in luminescence thermometry. Because of their rigid structure, rhodamines are highly fluorescent dyes whose efficiency depends on the substitution pattern of the amino groups [18]. When the amino groups are completely alkylated the quantum yield is strongly temperature dependent. This is the case with rhodamine B (RH B, $C_{28}H_{31}ClN_2O_3$; see Fig. 8.4A) whose quantum yield is highly temperature dependent above 220 K, and becomes 5-fold reduced with an increase in temperature to 373 K (see Fig. 8.5) [19, 20]. The quenching effect is reflected through a decrease of the emission intensity and a shortening of lifetime with temperature. On the contrary, when amino groups are only partially substituted by alkyl groups, like in rhodamine 101 (RH 101; $C_{32}H_{31}N_2O_3$) and rhodamine 110 (RH 110; $C_{20}H_{15}ClN_2O_3$; see Fig. 8.4B) the quantum yield is almost temperature independent and its value approaches unity [21].

Snare et al. [22] argued that strong fluorescence quenching of rhodamine B (when compared to almost negligible quenching of more rigid rhodamines) is caused by the torsional motion of the diethyl amino groups which narrow the S_1–S_0 gap, thus facilitating internal conversion. In fact, the fluorescence of Pyronin B ($C_{42}H_{54}Cl_8Fe_2N_4O_2$), a dye which has diethyl amino groups, but no carboxylphenyl substituent, also shows strong temperature dependency [23].

Because of the symmetry of the rhodamine chromophore, there is no dipole moment parallel to its long axis. For this reason, absorption and emission bands of rhodamines show considerable overlap and a small Stokes shift

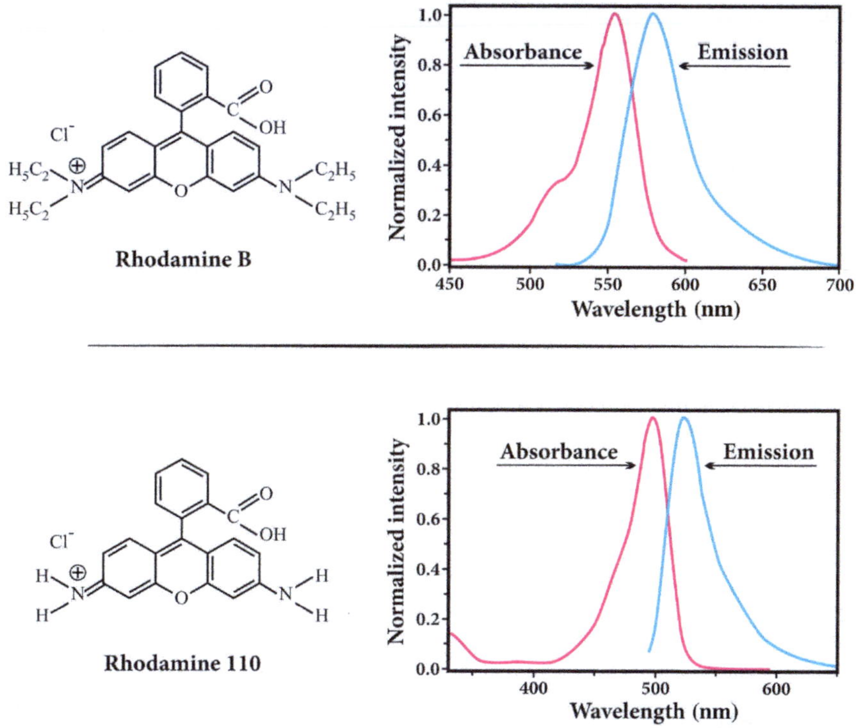

FIGURE 8.4 Structural formula (*left*) and normalized absorption and emission spectra (*right*) of rhodamine B and rhodamine 110.

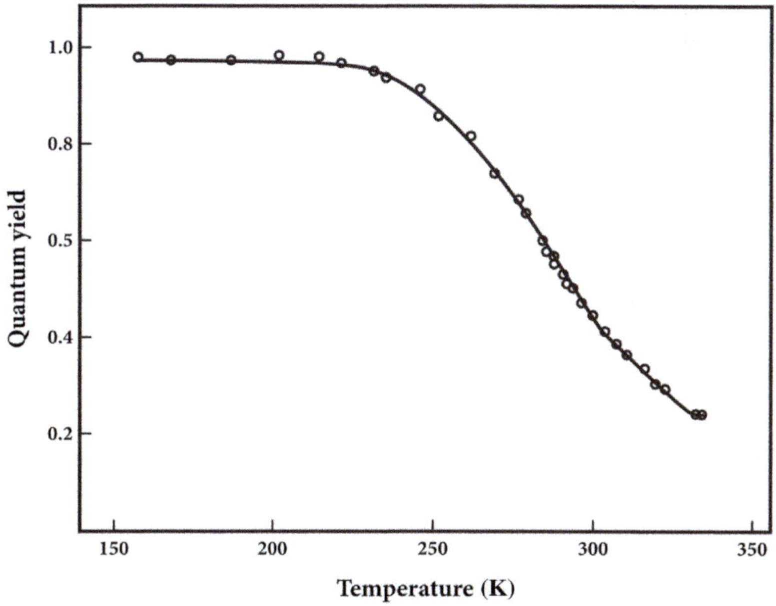

FIGURE 8.5 The temperature dependence of quantum yield of Rhodamine B in ethanol. *(Adapted from Ferguson J, Mau AWH. Spontaneous and stimulated emission from dyes. Spectroscopy of the neutral molecules of acridine orange, proflavine, and rhodamine B. Aust J Chem 1973;26(8):1617–24, with permission from CSIRO Publishing).*

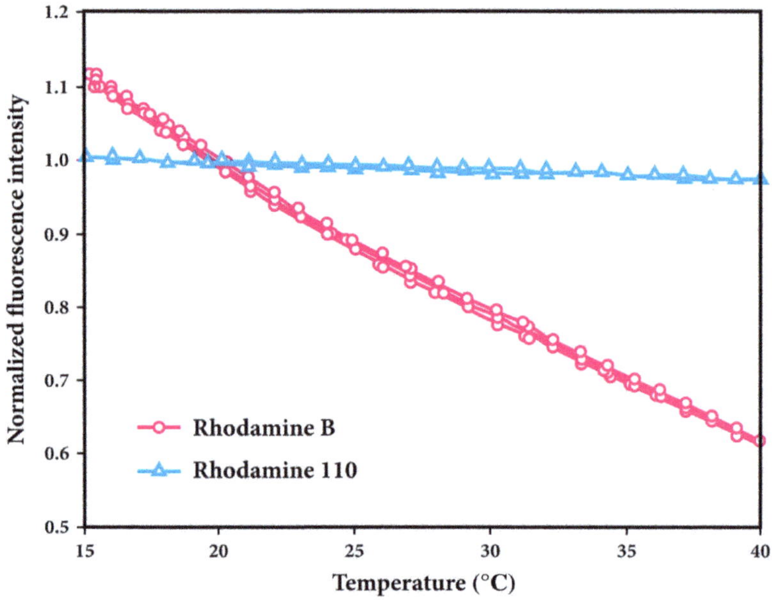

FIGURE 8.6 Temperature dependence of the fluorescence intensity of rhodamine B (measured at 575 nm) and rhodamine 110 (measured at 520 nm) normalized to values of fluorescence intensity measured at 20°C. *(Adapted from Sakakibara J, Adrian RJ. Whole field measurement of temperature in water using two-color laser induced fluorescence. Exp Fluids 1999;26:7–15, with permission from Springer).*

(see Fig. 8.4B). Also, as in all xanthene dyes, the intersystem crossing from S_1 level is quite low. Thus, luminescence thermometry can exploit the temperature-dependent fluorescence emission (emission intensity and lifetime) of rhodamine B and alike rhodamine derivatives (see Fig. 8.6). Temperature-independent emissions of rhodamine 101 and rhodamine 110 (Fig. 8.6) and alike derivatives may serve as excellent reference emissions for ratiometric temperature sensing.

Rhodamine dyes are chemically stable and soluble in water, methanol, and ethanol. However, they are toxic and harmful in contact with the skin and if swallowed. On the other hand, fluorescein, another well-known xanthene dye, is not considered hazardous. One should be aware of the potential hazards when using organic dyes for thermometry applications related to living organisms.

So far, for luminescent thermometry, fluorescent dyes have been mostly used as hybrid probes, attached or incorporated to organic (like polymers) or inorganic materials. To a lesser extent, they are used solely as solutions or buffer solutions. For example, rhodamine B solution (0.1 mM) in a carbonate buffer was used to monitor temperature in a microfluidic system [24] with achieved precision ranging from 2.4°C near room temperature to 3.5°C at high temperatures (see Fig. 8.7A). Also, rhodamine B solution has been used to measure the temperature fields of a stable Z thermal-stratified pipe flow [25]. Ratiometric thermal sensing from fluorescence of a mixed solution of 0.25 mM of rhodamine B and 0.05 mM of

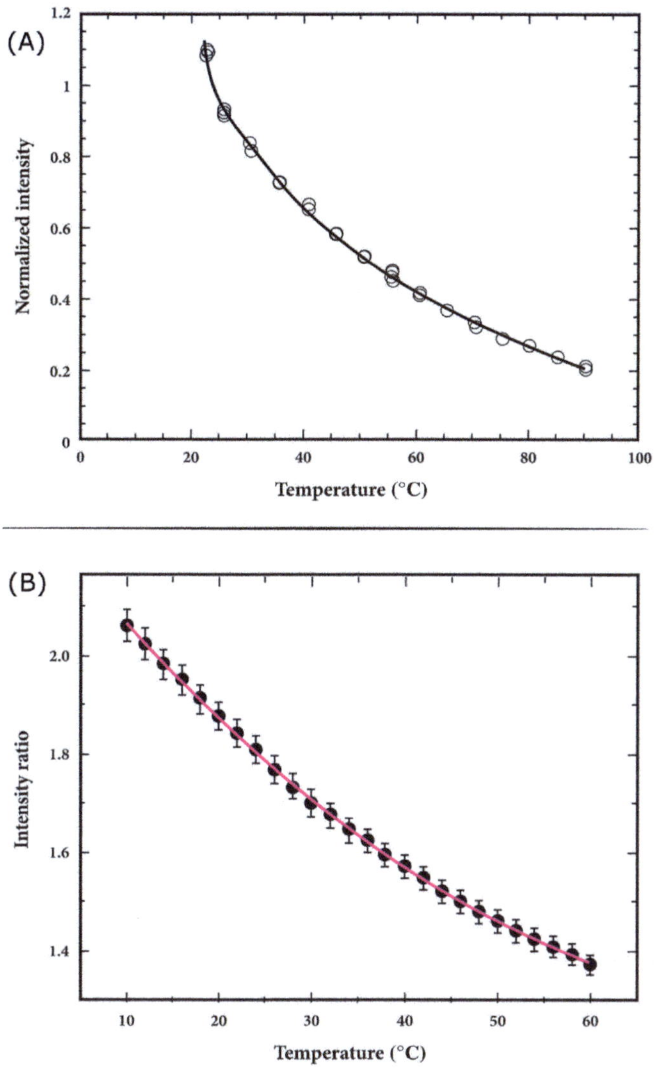

FIGURE 8.7 (A) Fluorescence intensity of rhodamine B dye as a function of temperature; intensities are normalized to the intensity measured at 22°C. (B) Intensity ratio of the fluorescence of rhodamine B and rhodamine 110 in a mixed solution. *((A) adapted from Ross D, Gaitan M, Locascio LE. Temperature measurement in microfluidic systems using a temperature-dependent fluorescent dye. Anal Chem 2001;73(17):4117–23 and (B) reprinted from del Rosal B, Sun C, Yan Y, Mackenzie MD, Lu C, Bettiol AA, et al. Flow effects in the laser-induced thermal loading of optical traps and optofluidic devices. Opt Express 2014;22(20):23938–54, with permission from OSA).*

rhodamine 110 has been used for temperature monitoring in optical traps and optofluidic devices [26] (see Fig. 8.7B). Emission of rhodamine 110 in this solution has been exploited as a reference emission. The intensity of rhodamine B fluorescence was quenched with temperature, so the ratio of emission was significantly decreased over the 10–60°C temperature range (see Fig. 8.7B). For thermometry of microfluidic systems, water and ethylene solutions of rhodamine B and sulforhodamine B have been exploited over the 10–60°C range [27]. Sulforhodamine B emission intensity slightly increases with temperature (0.14% K^{-1}), i.e., it shows a negative quenching effect. When sulforhodamine B emission was selected as a reference, ratiometric sensing using rhodamine B emission intensity had a relative sensitivity of 1%–1.6 % K^{-1} for the water solution and 1.4%–2% K^{-1} for the ethanol solution. This difference arises due to the weaker temperature sensitivity of rhodamine B in ethanol compared to when water is the solvent [27]. The aqueous solution of two dyes with strong temperature dependence, fluorescein 27 and rhodamine B, offers quite a high relative sensitivity of ratiometric measurements of 6%–10% K^{-1} depending on the combinations of wavelengths chosen for excitation and emission detection [28].

Besides solutions, fluorescent dyes have been used for luminescent thermometry in dried form, such as rhodamine B in the study performed by Löw et al. [29]. They mapped the temperature of the surface of resistively heated micro/nanowires in the 25–70°C range with a sensitivity of 0.7% K^{-1}. Jung et al. [30] demonstrated good thermographic properties of a thin layer of rhodamine B incorporated into optically transparent epoxy-based negative resist over the 25–70°C range. For potential use in luminescence thermometry, rhodamine B has also been entrapped in sol–gel matrix and silica gel [31]. Similar to rhodamine B, fluorescein and its derivatives have been combined with poly(vinyl alcohol) [32], chitosan [33], starch [34], and sulforhodamine [35] for use in a fluorescence thermometry.

8.3 DYES INCORPORATED IN FUNCTIONAL POLYMERS

Hybrid luminescent temperature probes can be formed, for instance, from a fluorescent dye and a functional polymer (or macromolecule when speaking in a general sense). In such case, the polymer provides the constitution for a hybrid probe and offers some additional functions which facilitate temperature sensing. For example, these specific functions can be fluorescence in the case of semiconducting polymers or a temperature-induced, solution-phase transition of a temperature responsive polymer which affects the fluorescence of dyes [36].

The semiconducting properties of polymers results from conjugation which facilitate electron delocalization, as discussed earlier in the text. They are promising materials for the preparation of luminescence thermometry probes since they are lightweight, flexible, cheap, and can be manufactured by various methods, including printing techniques. The intrinsic emission of semiconducting polymers may be combined with dye emissions for the preparation of sensitive probes. For example, Ye et al. [37] prepared luminescent temperature nanoprobes from semiconducting polymers (poly[(9,9-dioctylfluorenyl-2,7-diyl)-co-(1,4-benzo-{2,10,3}-thiadiazole)—PFBT and poly-[{9,9-dioctyl-2,7-divinylene-fluorenylene}-alt-co-{2-methoxy- 5-(2-ethylhexyloxy)-1,4-phenylene}]—PFPV) and rhodamine B. The PFBT and PFPV present almost temperature-insensitive emissions with maximal intensity at about 510 nm, while the intensity of the rhodamine B emission decreases with temperature (see Fig. 8.8). In both nanohybrids, with PFBT and PFPV, the ratio of emission intensities of polymer and dye is linear and decreases with temperature at a rate of ~1% K^{-1}.

Changes in the chemical structures and physical properties upon external stimuli, either reversible or irreversible, is a determining characteristic of stimuli-responsive polymers. They can be sensitive to many types of stimuli, such

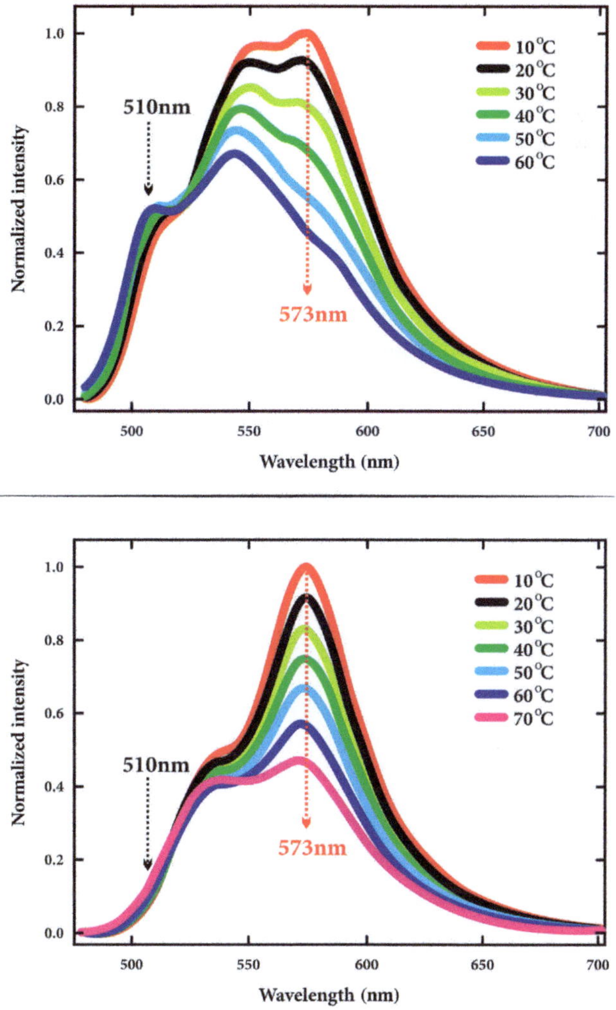

FIGURE 8.8 Fluorescence of PFBT–rhodamine B (*upper graph*) and PFPV–rhodamine B (*lower graph*) at different temperatures. *(Adapted from Ye F, Wu C, Jin Y, Chan Y-H, Zhang X, Chiu DT. Ratiometric temperature sensing with semiconducting polymer dots. J Am Chem Soc 2011;133(21):8146–9, copyright (2011) American Chemical Society).*

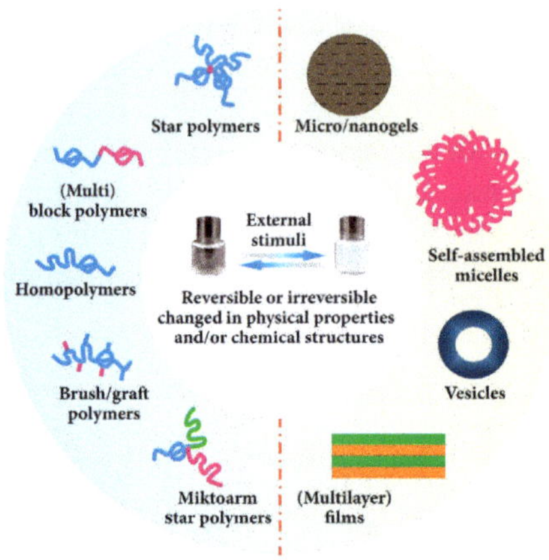

FIGURE 8.9 Different forms of stimuli-responsive polymers. *(Reprinted from Hu J, Liu S. Responsive polymers for detection and sensing applications: current status and future developments. Macromolecules 2010;43(20):8315–30, copyright (2010) American Chemical Society).*

as environmental temperature and pH, electric and magnetic fields, mechanical force, chemical or biochemical analytes, etc. This important group of materials may exist in the form of solutions, gels, self-assembled nanoparticles, (multilayer) films, and bulk solids [38] (Fig. 8.9).

Polarity-sensitive fluorescent dyes are weakly fluorescent in the hydrophilic microenvironment and strongly fluorescent in hydrophobic domains. When such dye is attached to a thermoresponsive polymer, its fluorescence becomes strongly dependent on polymer aggregation properties, since thermoresponsive polymers in aqueous solution normally form aggregates with decreased polarity above their lower critical solution temperatures (LCSTs) [38]. Table 8.2 lists some of the polymerizable monomers suitable for the preparation of temperature-responsive polymers of different LCSTs.

TABLE 8.2 Some Polymerizable Monomers Used for Preparations of Temperature-Responsive Polymers

Abbreviation	Chemical name	Structural formula
NIPAM	*N*-isopropylacrylamide	$CH_2CHCONHCH(CH_3)_2$
NIPMAM	*N*-isopropylmethacrylamide	$CH_2C(CH_3)CONHCH(CH_3)_2$
NNPAM	*N*-*n*-propylacrylamide	$CH_2CHCONH(CH_2)_2CH_3$
NTBAM	*N*-tert-butylacrylamide	$CH_2CHCONHC(CH_3)_3$
DEGMA	Di(ethylene glycol) methyl ether methacrylate	$CH_2C(CH_3)$ $COO(CH_2CH_2O)_2CH_3$
MMA	Methyl methacrylate	$CH_2C(CH_3)COOCH_3$

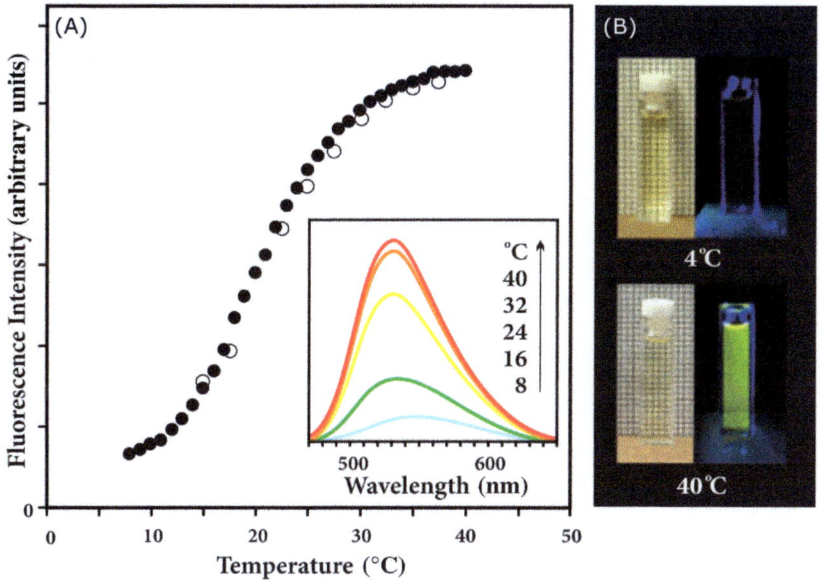

FIGURE 8.10 (A) Temperature dependence of fluorescence intensity of poly(DBD-*co*-DMAPAM-*co*-NTBAM) in buffer (pH 7) under 444 nm excitation; *full circles* represent values measured during heating while *empty circles* those measured during cooling; fluorescence emission spectra are given in the inset. (B) Visual and fluorescent images of the solution taken at 4 and 40°C. *(Adapted from Uchiyama S, Kawai N, de Silva AP, Iwai K. Fluorescent polymeric AND logic gate with temperature and pH as inputs. J Am Chem Soc 2004;126(10):3032–3, copyright (2004) American Chemical Society).*

The classical example of such a structure is poly(DBD-AE-co-DMAPAM-co-NTBAM) (of 0.1/20/80 copolymer composition) whose temperature-dependent fluorescence has been analyzed by Uchiyama et al [39] (see Fig. 8.10). *N,N*-Dimethylaminopro- pylacrylamide (DMAPAM) and *N*-*t*-butylacrylamide (NTBAM) were selected as *N*-alkylacrylamide monomers, and fluorescence was achieved by the 4-*N*-(2-acryloyloxy-ethyl)-*N*-methylamino-7-*N,N*-dimethylaminosulfonyl-2,1,3-benzoxa-diazole (DBD-AE) comonomer of the copolymer. Fig. 8.10A shows that the emission intensity of the copolymer is considerably increased when temperature rises from 8 to 40°C. At the same time, the position of the emission band changed from 550 to 530 nm (Fig. 8.10B, inset), indicating that the polarity near the DBD-AE unit was decreased. The strong rise in emission intensity with temperature can be observed in the images of Fig. 8.10B.

In a different approach, dye can be attached to chain terminals of pluronic triblock copolymers, as in the case of poly(ethylene oxide)-poly(propylene oxide)-poly(ethylene oxide) (PEO-PPO-PEO; commercial name: Pluronic F127) and Cy5.5 near-infrared dye ($C_{39}H_{42}N_2O_4P$) [40]. Below the critical micelle temperature (CMT), block copolymers exist in linear state in solution as dissolved monomers. Above the CMT, block copolymers gradually self-assemble into micelles, owing to an increase in their hydrophobic nature [40]. Emission of the Cy5.5 dye has maximal intensity around 695 nm and is reduced upon heating (see Fig. 8.11A). Since the dye is chemically conjugated to pluronic triblock copolymer, quenching of emission occurs while the copolymer changes its structure from the sol-state, through micelles to large aggregates (see Fig. 8.11B). Considering that Cy5.5 is attached to a hydrophilic PEO block, it appears that the formation of PPO-core micelles does not affect the trend of a decreasing fluorescence quantum yield with increasing temperatures.

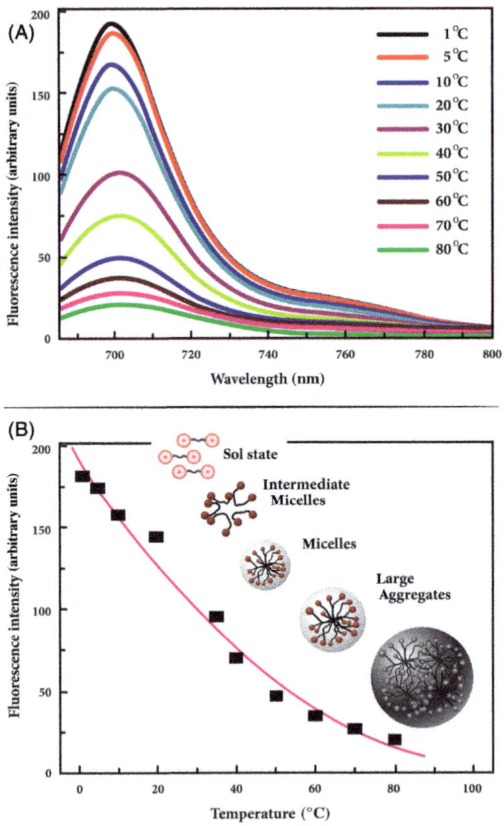

FIGURE 8.11 (A) The near-infrared emission spectra of the Pluronic F127–Cy5.5 hybrid probe measured at different temperatures under 675 nm excitation. (B) Changes of 695 nm emission intensity and probe structure with an increase in temperature. *(Adapted from Lee S-Y, Lee S, Youn I-C, Yi DK, Lim YT, Chung BH, et al. A near-infrared fluorescence-based optical thermosensor. Chem A Eur J 2009;15(25):6103–6, with permission with permission from Wiley).*

8.4 EXCIPLEX-TYPE PROBES

The term *exciplex* (*exci*ted com*plex*) describes an electronically excited molecular complex of definite stoichiometry. These complexes include excimers (excited dimers)—complexes formed between identical atoms or molecules; heteroexcimers—complexes formed between two nonidentical atoms or molecules; and excited charge-transfer complexes—produced by excitation of the ground-state molecular complexes for which there is conclusive evidence for association in the ground state [41]. Generally, excited fluorescent species lower their energy by forming a complex with another molecular species in its ground state. Normally, the exciplex produces an emission band that does not emerge in the emission spectra of its components [42]. Exciplex fluorescence occurs between bound upper state and unbound ground state. No spectral limitations from Frank–Condon rules are imposed since there is no quantization for vibrational states of an unbound electronic state. Therefore, when compared to the emission of a monomer, an exciplex emission, either fluorescence or phosphorescence, shows redshifted, broad, and structureless emission bands. Note that excimers are only formed when one of the dimer components is in the excited state. Upon return of the excimer to its ground state, its components dissociate and repel each other. For this reason, the formation of excimers is stimulated by the high density of the constitutive monomers.

An example of exciplex fluorescence use for temperature sensing is the work of Chandrasekharan and Kelly [43] who prepared a film from the blue-emitting perylene (0.017 wt%) and nonemissive *N*-allyl-*N*-methylaniline (18.96 wt%) in a polystyrene matrix. The film exhibited green fluorescence due to the formation of exciplexes (Fig. 8.12) showing two emission bands from perylene (at 463 and 475 nm) and a broad band (centered at 551 nm) due to exciplex formation

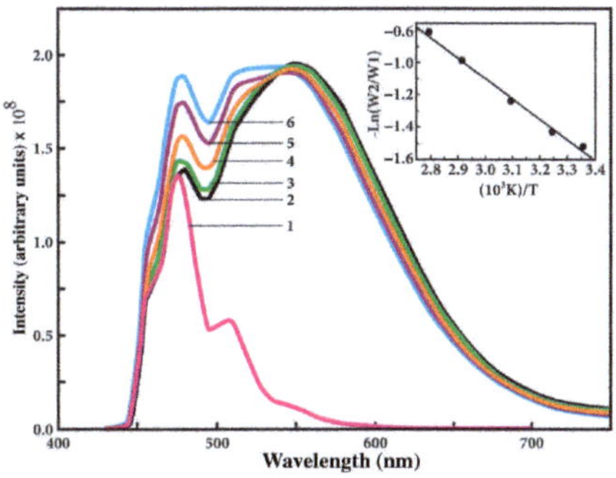

FIGURE 8.12 Emission spectra of a polystyrene film composed of perylene and *N*-allyl-*N*-methylaniline measured at 298 K (line 2), 304 K (line 3), 323 K (line 4), 343 K (line 5), and 358 K (line 6). Also shown is the emission spectrum of a film of perylene alone at 298 K (line 1). The inset shows the natural logarithm of the ratio of perylene and exciplex emission as a function of reciprocal temperature. *(Reprinted from Chandrasekharan N, Kelly LA. A Dual fluorescence temperature sensor based on perylene/exciplex interconversion. J Am Chem Soc 2001;123(40):9898–9, copyright (2001) American Chemical Society).*

between perylene and *N*-allyl-*N*-methylaniline. Upon heating, the intensity of the perylene emission increased while the emission due to the exciplex formation remained constant over the 25–85°C temperature range. The ratio of intensities of these emissions has a relative sensitivity ~1%K^{-1}.

A self-referencing luminescent thermometer based on a reversible, temperature-dependent monomer–excimer interconversion of 1,3-bis(1-pyrenyl)propane dissolved in an ionic liquid and operating in the 25–140°C range has been reported by Baker et al. [44]. The relatively high upper operational range of this probe, in respect to ones achieved with organic dyes, is governed by the thermal stability of the fluorophore itself, contrary to the majority of exciplex molecular thermometers for which the upper limit is determined by the boiling point of the organic solvents. Fig. 8.13 illustrates that the emission of 1-butyl-1-methyl pyrrolidinium bis(trifluoromethylsulfonyl)imide, [C_4mpy][Tf_2N] is temperature dependent. Spectra are composed of the monomer emission band that peaks at 376 nm and excimer emission around 476 nm. Under heating, the excited-state equilibrium of the system is shifted in favor of the excimer formation as a result of reduced bulk viscosity.

8.5 DISCRETE METAL–ORGANIC COMPLEXES

Metal–organic complexes are used today in many emerging and critical technologies, such as in highly efficient OLEDs and transistors (OLETs), photon upconversion and photovoltaics devices, etc. They are also often used for luminescence temperature sensing and are one of the largest group of materials investigated so far for this purpose. As mentioned earlier in this chapter, one may consider, as a rule of thumb, that the more rigid the molecule is, the more efficient its fluorescence will be. This is because the probability of nonradiative

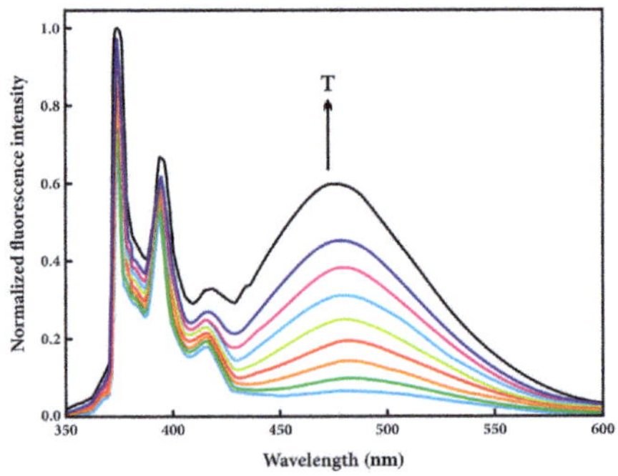

FIGURE 8.13 Emission spectra normalized to the intensity of the monomer band (emission intensity at 376 nm equals 1.00) for 5 mM py(3)py in [C_4mpy][Tf_2N] measured at different temperatures (excitation wavelength: 325 nm). *(Reproduced from Baker GA, Baker SN, McCleskey TM. Noncontact two-color luminescence thermometry based on intramolecular luminophore cyclization within an ionic liquid. Chem Commun 2003;(23):2932–3, with permission from The Royal Society of Chemistry).*

transition is suppressed by decreasing vibrations, i.e., by increasing the molecular rigidity. The complexation of metal ions by conjugated organic molecules can improve the structural rigidity, and, consequently, metal–organic complexes can exhibit enhanced fluorescence [45]. Also, new photophysical properties of organic compounds have emerged due to the participation of metal orbitals in excited states. One of them, for example, is an efficient phosphorescence accomplished by the addition of heavy metal ions that promote intersystem crossing via spin–orbital coupling. Note that pure organic compounds quite rarely exhibit room-temperature phosphorescence. In this sense, one can consider that the type of metal ions predominantly influence novel properties of metal–organic compounds, and the variety of metal ions that can be complexed with organics (such as heavy metals, transition metals, lanthanides, noble metals, etc.) provides a great variety of possible optical features.

Considering the advantageous characteristics for their use as luminescence thermometry probes, phosphorescent metal–organic complexes demonstrate relatively long lifetimes, absorption in the visible range, and moderate brightness. They usually have large Stokes shifts which prevent self-quenching and allow easy resolution of the emission and excitation light [46]. They are convenient probes for bioimaging due to long emission lifetimes, much longer than lifetimes of organic dyes, which permit discrimination of their emissions from emissions of endogenous fluorophores from biological environments (autofluorescence), and because of very small size (nanoscale dimensions) which permits an imaging with a high spatial resolution. Therefore, phosphorescent metal–organic complexes can provide monitoring of the temperature of biological objects along with imaging. Table 8.3 presents information on several metal–organic complexes that have been used for sensing temperature (the chemical structures of these complexes are given in Fig. 8.14).

Emissions of trivalent lanthanide ions (Ln^{3+}) are widely exploited for luminescence thermometry, as already discussed in Chapter 6, Lanthanide and Transition Metal Ion Doped Materials for Luminescence Temperature Sensing. These emissions are characterized by a narrow "line-like" bands and quite long lifetimes of excited states (of the order of microseconds and milliseconds) due to parity and spin forbidden *4f–4f* electronic transitions. The forbidden nature of 4f transitions presents some drawbacks for Ln^{3+} use since it causes a weak ability to directly absorb incident light. Complexation of Ln^{3+} ions with chromophore-containing organic groups circumvents the low absorption problem. The chromophore group acts as an "antenna", it strongly absorbs incident light and then transfers excitation to the Ln^{3+} ion, which then deexcites by its characteristic emission (see Fig. 8.15). For use in luminescent thermometry, mostly complexes prepared from Eu^{3+} and Tb^{3+} have been used, while β-diketone (1,3-diketones) and thenoyltrifluoroacetonate chelates were the organics of choice. Note that the practical application of Ln^{3+} complexes is difficult due to their poor thermal stability and processibility [58, 59]. The problem is often solved by the introduction of lanthanide complexes into a stable inorganic matrix.

TABLE 8.3 Working Ranges (in Kelvin), Spectral Maxima of Excitation and Emission (in Nanometers) of Selected Luminescent Metal–Organic Probes for Sensing Temperature

Complex	Range (K)	Excitation/Emission	Properties	References
$Ru(bpy)_3$	50–290	458/575	Water-soluble; can be excited with blue LEDs; resolution up to 0.1 K; fair photo-stability	[48]
$Ru(phen)_3$	273–393	470/580	High sensitivity to *T*; can be excited with blue LEDs; wide working range; photo-stable	[49–52]
$Ir(ppy)_2(carbac)$	274–323	405/519	High sensitivity and brightness; linear relationship between lifetime and *T*; poor photostability	[53, 54]
Brph-thq Pt	273–333	405/501	Blue LED excitable; good quantum yield	[55, 56]
PtOEP in polystyrene	300–450	387/650	Lifetime based thermal imaging; accuracy of about 0.25 K; enables imaging of thermal pulse induced *T* change	[57]

Source: Adapted from Wang X, Wolfbeis OS, Meier RJ. Luminescent probes and sensors for temperature. Chem Soc Rev 2013;42(19):7834–69, with permission from The Royal Society of Chemistry.

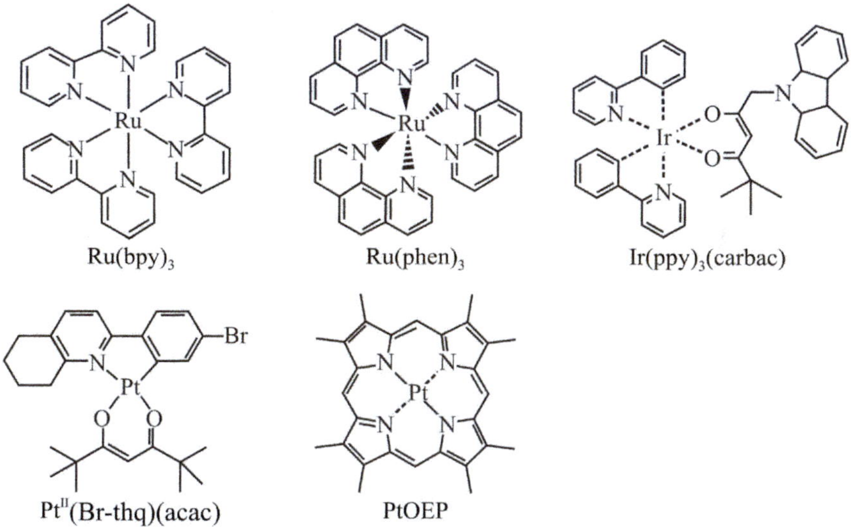

FIGURE 8.14 Chemical structures of metal–organic complexes from Table 8.3. *(Reproduced from Wang X, Wolfbeis OS, Meier RJ. Luminescent probes and sensors for temperature. Chem Soc Rev 2013;42(19):7834–69, with permission from The Royal Society of Chemistry).*

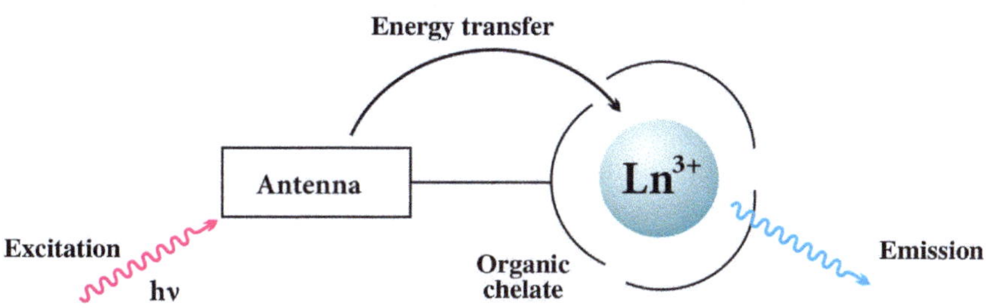

FIGURE 8.15 An "antenna" effect: an organic chromophore strongly absorbs incident radiation and transfers excitation energy to the Ln^{3+} ion which deexcites by its characteristic emission.

8.6 METAL–ORGANIC FRAMEWORKS

The power of combining organic and inorganic chemistry is reflected through MOFs, a special class of coordination polymers. These crystalline compounds form one-, two-, or three-dimensional structures from metal ions or clusters coordinated to organic ligands. The term framework is used to highlight the framework nature of MOFs, which mimics the porous nature of zeolite materials and differentiate them from other organic–inorganic structures. The astonishing degree of variability for both the organic and inorganic components make MOFs of significance for many applications, such as in clean energy, as storage media for gasses, as high-capacity adsorbents, for catalysis, for bioimaging,

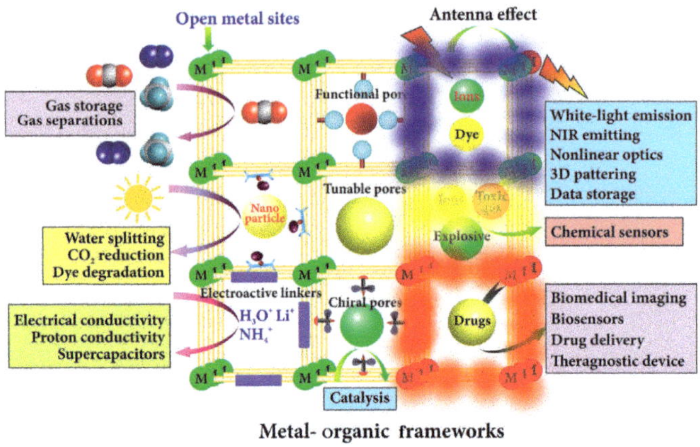

FIGURE 8.16 Graphic illustration of pore and function engineering to develop multifunctional MOFs. *(Reproduced from Li B, Wen H-M, Cui Y, Zhou W, Qian G, Chen B. Emerging multifunctional metal–organic framework materials. Adv Mater 2016;28(40):8819–60, with permission from Wiley).*

white and near-infrared light emission, chemical and biological sensing, etc. [60, 61] (see Fig. 8.16).

In recent times, numerous luminescent MOFs have been developed showing a wide range of emissive phenomena found in few other classes of materials. The variety of emission features comes from the unique properties of hybrid networks in which both the inorganic and organic moieties may be the source of the luminescence [62]. Also, emissions related to the metal–ligand charge transfer within MOFs can further enrich spectral features [63]. MOFs may also serve as hosts for the incorporation of luminescent species, such as fluorescent dyes, quantum dots (QDs), lanthanide or transition metal ions, etc.

Luminescence of MOFs shows strong temperature dependence. MOF luminescence temperature probes can be constructed to utilize intrinsic MOF luminescence or the combination of MOF luminescence and luminescence from intentionally incorporated moieties. The later approach facilitates the use of ratiometric temperature read-out (see Chapter 4: Schemes for Temperature Read-out from Luminescence) from the dual- or even multiple-emitting MOFs (see Fig. 8.17A). Cui et al. [63] proposed three designs of dual-emitting MOFs for ratiometric temperature sensing:

1. MOFs incorporating two distinct, electronically independent luminescence centers, such as metal ions and guest molecules (Fig. 8.17B). In the absence of an energy transfer between luminescence centers, the temperature caused changes of luminescence features occur due to the different quenching behavior of each luminescent center. To achieve strong sensitivity on temperature in ratiometric measurements, one center should have a strongly temperature-dependent emission, while the emission from other center should be temperature independent or, at least, only slightly sensitive to temperature.

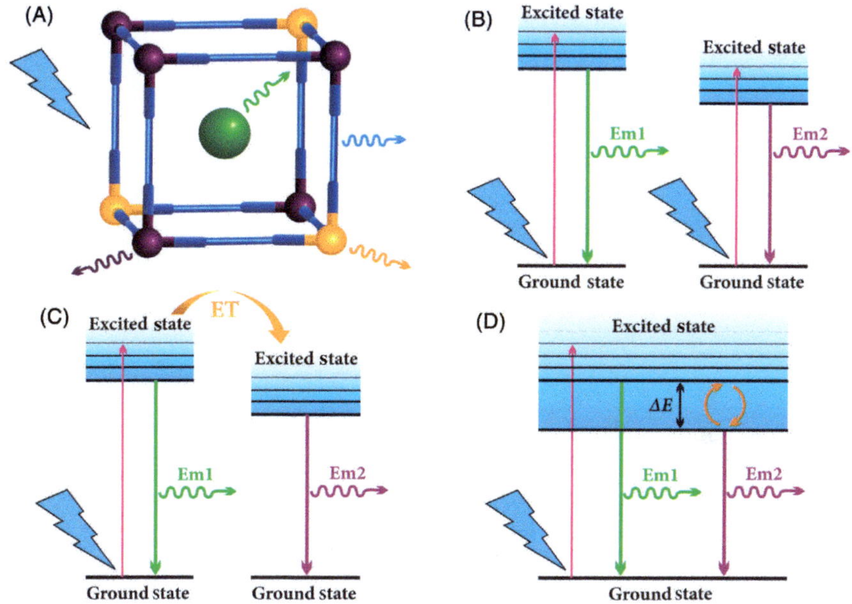

FIGURE 8.17 (A) Schematic representation of the possible luminescence centers in MOFs. (B–D) Three different energy level structures of dual-emitting MOFs for ratiometric luminescence temperature sensing. *ET*, energy transfer; *Em*, emission. *(From Cui Y, Zhu F, Chen B, Qian G. Metal-organic frameworks for luminescence thermometry. Chem Commun 2015;51(35):7420–31, with permission from The Royal Society of Chemistry).*

2. MOFs incorporating closely spaced donor–acceptor moieties, where the donor sensitizes the acceptor via an energy transfer, such as between ligands and lanthanides (see Fig. 8.17C). This design frequently leads to probes with quite high sensitivity to temperature.
3. MOFs incorporating a single emission center, showing at least two emission bands from closely spaced (thermally coupled) excited states (see Fig. 8.17D). This Boltzmann-type ratiometric measurement is discussed in detail in Chapter 4, Schemes for Temperature Read-out from Luminescence, and Chapter 6, Lanthanide and Transition Metal Ion Doped Materials for Luminescence Temperature Sensing.

An example of a MOF-type luminescence probe for ratiometric temperature sensing is $Eu_{0.0069}Tb_{0.9931}$-2,5-dimethoxy-1,4-benzenedicarboxylate (DMBDC), which has been investigated by Cui et al. [64]. It presents intense emissions of both Eu^{3+} and Tb^{3+} (Fig. 8.18A) which have an opposite temperature dependence. At 10 K, the emission bands from the $^5D_0 \rightarrow ^7F_2$ electronic transition of Eu^{3+} (centered around 613 nm) and the $^5D_4 \rightarrow ^7F_5$ electronic transition of Tb^{3+} (centered around 545 nm) have comparable intensity. Upon increasing temperature, the intensity of the Eu^{3+} emission band increases while the emission of Tb^{3+} decreases until the emission of Eu^{3+} almost dominates the

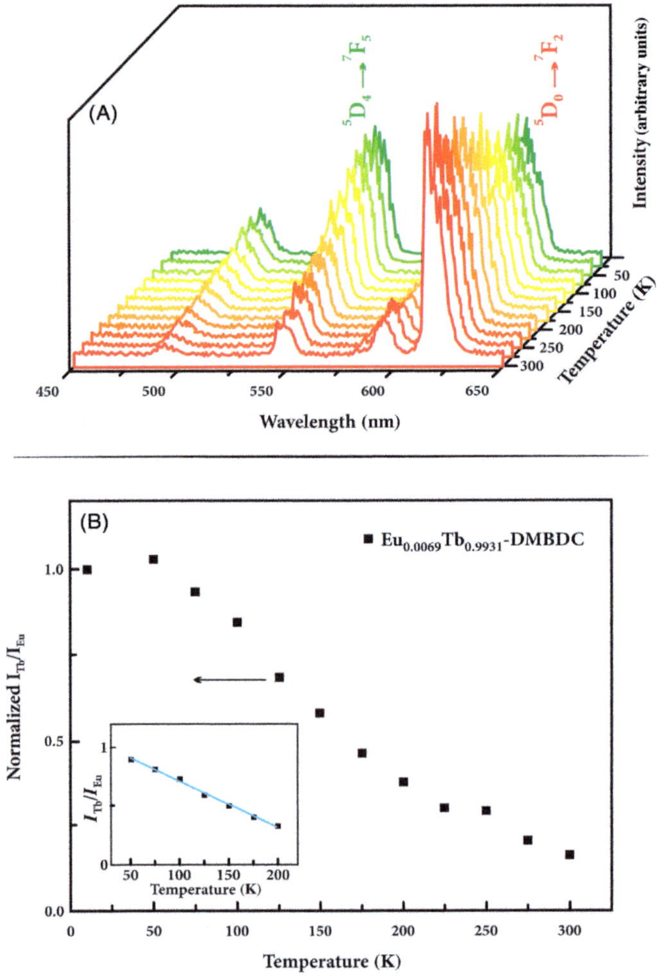

FIGURE 8.18 Luminescence of the $Eu_{0.0069}Tb_{0.9931}$-DMBDC (DMBDC = 2,5-dimethoxy-1,4-benzenedicarboxylate) MOF. (A) Temperature dependence of emission in the 10–300 K range. (B) The change with temperature of the ratio of integrated intensity of the Tb^{3+} ($^5D_4 \rightarrow {}^7F_5$, I_{Tb}) and Eu^{3+} ($^5D_0 \rightarrow {}^7F_2$, I_{Eu}) emissions. *(Adapted from Cui Y, Xu H, Yue Y, Guo Z, Yu J, Chen Z, et al. A luminescent mixed-lanthanide metal–organic framework thermometer. J Am Chem Soc 2012;134(9):3979–82, copyright (2012) American Chemical Society).*

whole spectrum at 300 K. Thus, the ratio of integrated intensity of the Tb^{3+} (I_{Tb}) and Eu^{3+} (I_{Eu}) emissions decreases with temperature and can be linearly related to temperature as $T = 287.09 - 263.85 \times I_{Tb} / I_{Eu}$ in the 50–200 K range (see Fig. 8.18B).

By using 5-(pyridin-4-yl)isophthalate, an organic ligand with higher triplet state energy than in the case of DMBDC, the ratiometric thermometry from the Tb^{3+} and Eu^{3+} emissions in $Tb_{0.9}Eu_{0.1}$PIA has been achieved [65]. This probe provided a significantly higher relative temperature sensitivity of 3.53% K^{-1} compared to 0.38% K^{-1} obtained with in $Eu_{0.0069}Tb_{0.9931}$-DMBDC. The highly efficient emission from the triplet state of the disulfobenzoic ligand in the Gd-MOF, along with the exponential temperature dependence of the ratio of intensities of triplet and lanthanide emissions (Eu, Gd, and Tb), provided a ratiometric temperature sensing with sensitivities up to 32% K^{-1} over the 10–300 K range [66]. This value of relative sensitivity is among the highest ever reported in the literature.

As we reach the end of this chapter, it is important to mention some other types of organic materials that may find potential applications in luminescence thermometry. These include intrinsically fluorescent polymers (such as polydiacetylene supramolecules [67]), fluorescent proteins (such as green fluorescent protein [68]), and carbon nanostructures (such as C_{70} [69]).

REFERENCES

[1] Herschel JFW. On a case of superficial colour presented by a homogeneous liquid internally colourless. Philos Trans R Soc Lond 1845;135:147–53.

[2] Booth G, Zollinger H, McLaren K, Sharples WG, Westwell A. Dyes, General Survey. Ullmann's encyclopedia of industrial chemistry. Wiley-VCH Verlag GmbH & Co. KGaA; 2000.

[3] Valeur B, Berberan-Santos MN. Molecular fluorescence: principles and applications. Weinheim: Wiley-VCH; 2013. p. 569.

[4] Boens N, Qin W, Basarić N, Hofkens J, Ameloot M, Pouget J, et al. Fluorescence lifetime standards for time and frequency domain fluorescence spectroscopy. Anal Chem 2007;79(5):2137–49.

[5] López Arbeloa T, López Arbeloa F, Hernández Bartolomé P, López Arbeloa I. On the mechanism of radiationless deactivation of rhodamines. Chem Phys 1992;160(1):123–30.

[6] Magde D, Rojas GE, Seybold PG. Solvent dependence of the fluorescence lifetimes of xanthene dyes. Photochem Photobiol 1999;70(5):737–44.

[7] Reija B, Al-Soufi W, Novo M, Vázquez Tato J. Specific interactions in the inclusion complexes of pyronines Y and B with β-Cyclodextrin. J Phys Chem B 2005;109(4):1364–70.

[8] Brackmann U. Lambdachrome® laser dyes. Goettingen; 2000.

[9] Rurack K, Spieles M. Fluorescence quantum yields of a series of red and near-infrared dyes emitting at 600 − 1000 nm. Anal Chem 2011;83(4):1232–42.

[10] Invitrogen. (http://www.invitrogen.com/).

[11] Kellogg RE, Bennett RG. Radiationless intermolecular energy transfer. III. Determination of phosphorescence efficiencies. J Chem Phys 1964;41(10):3042–5.

[12] Berlman IB. Handbook of fluorescence spectra of aromatic molecules, London; 1965.

[13] Birks JB. Photophysics of aromatic molecules. London: Wiley-Interscience; 1970.

[14] Kubota Y, Steiner RF. Fluorescence decay and quantum yield characteristics of acridine orange and proflavine bound to DNA. Biophys Chem [Internet] 1977;6(3):279–89.

[15] Reynolds GA, Drexhage KH. New coumarin dyes with rigidized structure for flashlamp-pumped dye lasers. Opt Commun [Internet] 1975;13(3):222–5.

[16] Jones G, Jackson WR, Choi CY, Bergmark WR. Solvent effects on emission yield and lifetime for coumarin laser dyes. Requirements for a rotatory decay mechanism. J Phys Chem 1985;89(2):294–300.

[17] Mujumdar RB, Ernst LA, Mujumdar SR, Lewis CJ, Waggoner AS. Cyanine dye labeling reagents: sulfoindocyanine succinimidyl esters. Bioconjug Chem 1993;4(2):105–11.
[18] Arden J, Deltau G, Huth V, Kringel U, Peros D, Drexhage KH. Fluorescence and lasing properties of rhodamine dyes. J Lumin 1991;48:352–8.
[19] Ferguson J, Mau AWH. Spontaneous and stimulated emission from dyes. Spectroscopy of the neutral molecules of acridine orange, proflavine, and rhodamine B. Aust J Chem 1973;26(8):1617–24.
[20] Chen YY, Wood AW. Application of a temperature-dependent fluorescent dye (Rhodamine B) to the measurement of radiofrequency radiation-induced temperature changes in biological samples. Bioelectromagnetics 2009;30(7):583–90.
[21] Karstens T, Kobs K. Rhodamine B and rhodamine 101 as reference substances for fluorescence quantum yield measurements. J Phys Chem 1980;84(14):1871–2.
[22] Snare MJ, Treloar FE, Ghiggino KP, Thistlethwaite PJ. The photophysics of rhodamine B. J Photochem 1982;18(4):335–46.
[23] Sakakibara J, Adrian RJ. Whole field measurement of temperature in water using two-color laser induced fluorescence. Exp Fluids 1999;26:7–15.
[24] Ross D, Gaitan M, Locascio LE. Temperature measurement in microfluidic systems using a temperature-dependent fluorescent dye. Anal Chem 2001;73(17):4117–23.
[25] Sakakibara J, Hishida K, Maeda M. Measurements of thermally stratified pipe flow using image-processing techniques. Exp Fluids 1993;16(2):82–96.
[26] del Rosal B, Sun C, Yan Y, Mackenzie MD, Lu C, Bettiol AA, et al. Flow effects in the laser-induced thermal loading of optical traps and optofluidic devices. Opt Express 2014;22(20):23938–54.
[27] Christensen VKN, KT. Two-color laser-induced fluorescent thermometry for microfluidic systems. Meas Sci Technol 2009;20(1):15401.
[28] Estrada-Pérez CE, Hassan YA, Tan S. Experimental characterization of temperature sensitive dyes for laser induced fluorescence thermometry. Rev Sci Instrum 2011;82(7):74901.
[29] Löw P, Kim B, Takama N, Bergaud C. High-spatial-resolution surface-temperature mapping using fluorescent thermometry. Small 2008;4(7):908–14.
[30] Jung W, Kim YW, Yim D, Yoo JY. Microscale surface thermometry using SU8/Rhodamine-B thin layer. Sens Actuat A Phys 2011;171(2):228–32.
[31] Duong HD, Rhee JIl. Exploitation of thermo-effect of rhodamine B entrapped in sol–gel matrix and silica gel for temperature detection. Sens Actuat B Chem 2007;124(1):18–23.
[32] Guan X, Liu X, Su Z, Liu P. The preparation and photophysical behaviors of temperature/pH-sensitive polymer materials bearing fluorescein. React Funct Polym 2006;66(11):1227–39.
[33] Guan X, Liu X, Su Z. Preparation and photophysical behaviors of fluorescent chitosan bearing fluorescein: potential biomaterial as temperature/pH probes. J Appl Polym Sci 2007;104(6):3960–6.
[34] Guan X, Su Z. Synthesis and characterization of fluorescent starch using fluorescein as fluorophore: potential polymeric temperature/pH indicators. Polym Adv Technol 2008;19(5):385–92.
[35] Kim M, Yoda M. Dual-tracer fluorescence thermometry measurements in a heated channel. Exp Fluids 2010;49(1):257–66.
[36] Vancoillie G, Zhang Q, Hoogenboom R. Polymeric temperature sensors. In: Carlos LD, Palacio F, editors. Thermometry at the nanoscale: techniques and selected applications. The Royal Society of Chemistry; 2016. p. 190–236.
[37] Ye F, Wu C, Jin Y, Chan Y-H, Zhang X, Chiu DT. Ratiometric temperature sensing with semiconducting polymer dots. J Am Chem Soc 2011;133(21):8146–9.

[38] Hu J, Liu S. Responsive polymers for detection and sensing applications: current status and future developments. Macromolecules 2010;43(20):8315–30.
[39] Uchiyama S, Kawai N, de Silva AP, Iwai K. Fluorescent polymeric AND logic gate with temperature and pH as inputs. J Am Chem Soc 2004;126(10):3032–3.
[40] Lee S-Y, Lee S, Youn I-C, Yi DK, Lim YT, Chung BH, et al. A near-infrared fluorescence-based optical thermosensor. Chem A Eur J 2009;15(25):6103–6.
[41] Davidson RS. The chemistry of excited complexes: a survey of reactions. Adv Phys Org Chem 1983;19:1–130.
[42] Lou J, Finegan TM, Mohsen P, Hatton AT, Laibinis PE. Fluorescence-based thermometry: principles and applications. Reviews in analytical chemistry. vol. 18, pp. 235-84.
[43] Chandrasekharan N, Kelly LA. A dual fluorescence temperature sensor based on perylene/exciplex interconversion. J Am Chem Soc 2001;123(40):9898–9.
[44] Baker GA, Baker SN, McCleskey TM. Noncontact two-color luminescence thermometry based on intramolecular luminophore cyclization within an ionic liquid. Chem Commun 2003;(23):2932–3.
[45] Xu H, Chen R, Sun Q, Lai W, Su Q, Huang W, et al. Recent progress in metal–organic complexes for optoelectronic applications. Chem Soc Rev 2014;43(10):3259–302.
[46] Ma D-L, He H-Z, Leung K-H, Chan DS-H, Leung C-H. Bioactive luminescent transition-metal complexes for biomedical applications. Angew Chemie Int Ed 2013;52(30):7666–82.
[47] Wang X, Wolfbeis OS, Meier RJ. Luminescent probes and sensors for temperature. Chem Soc Rev 2013;42(19):7834–69.
[48] Maruszewski K, Andrzejewski D, Strek W. Thermal sensor based on luminescence of Ru(bpy) 3^{2+} entrapped in sol-gel glasses. J Lumin 1997;72:226–8.
[49] Liebsch G, Klimant I, Wolfbeis OS. Luminescence lifetime temperature sensing based on sol-gels and poly(acrylonitrile)s dyed with ruthenium metal–ligand complexes. Adv Mater 1999;11(15):1296–9.
[50] Kocincova AS, Borisov SM, Krause C, Wolfbeis OS. Fiber-optic microsensors for simultaneous sensing of oxygen and pH, and of oxygen and temperature. Anal Chem 2007;79(22): 8486–93.
[51] Baleizão C, Nagl S, Schäferling M, Berberan-Santos MN, Wolfbeis OS. Dual fluorescence sensor for trace oxygen and temperature with unmatched range and sensitivity. Anal Chem 2008;80(16):6449–57.
[52] Borisov SM, Vasylevska AS, Krause C, Wolfbeis OS. Composite luminescent material for dual sensing of oxygen and temperature. Adv Funct Mater 2006;16(12):1536–42.
[53] Fischer LH, Borisov SM, Schaeferling M, Klimant I, Wolfbeis OS. Dual sensing of pO_2 and temperature using a water-based and sprayable fluorescent paint. Analyst 2010;135(6): 1224–9.
[54] Fischer LH, Stich MIJ, Wolfbeis OS, Tian N, Holder E, Schäferling M. Red- and green-emitting iridium(III) complexes for a dual barometric and temperature-sensitive paint. Chem A Eur J 2009;15(41):10857–63.
[55] Fischer LH, Karakus C, Meier RJ, Risch N, Wolfbeis OS, Holder E, et al. Referenced dual pressure- and temperature-sensitive paint for digital color camera read out. Chem A Eur J 2012;18(49):15706–13.
[56] Karakus C, Fischer LH, Schmeding S, Hummel J, Risch N, Schaferling M, et al. Oxygen and temperature sensitivity of blue to green to yellow light-emitting Pt(ii) complexes. Dalt Trans 2012;41(32):9623–32.
[57] Stehr J, Lupton JM, Reufer M, Raschke G, Klar TA, Feldmann J. Sub-microsecond molecular thermometry using thermal spin flips. Adv Mater 2004;16(23-24):2170–4.

[58] Lin J, Feng C, He X, Wang W, Fang Y, Liu Z, et al. Europium (III) organic complexes in porous boron nitride microfibers: efficient hybrid luminescent material. Sci Rep 2016;6:34576.
[59] Binnemans K. Lanthanide-based luminescent hybrid materials. Chem Rev 2009;109(9): 4283–374.
[60] Zhou H-C, Long JR, Yaghi OM. Introduction to metal–organic frameworks. Chem Rev 2012;112(2):673–4.
[61] Li B, Wen H-M, Cui Y, Zhou W, Qian G, Chen B. Emerging multifunctional metal–organic framework materials. Adv Mater 2016;28(40):8819–60.
[62] Millan A, Carlos LD, Brites CDS, Silva NJO, Pinol R, Palacio F. Organic–inorganic hybrids thermometry. In: Carlos LD, Palacio F, editors. Thermometry at the nanoscale: techniques and selected applications. The Royal Society of Chemistry; 2016. p. 237–72.
[63] Cui Y, Zhu F, Chen B, Qian G. Metal–organic frameworks for luminescence thermometry. Chem Commun 2015;51(35):7420–31.
[64] Cui Y, Xu H, Yue Y, Guo Z, Yu J, Chen Z, et al. A luminescent mixed-lanthanide metal–organic framework thermometer. J Am Chem Soc 2012;134(9):3979–82.
[65] Rao X, Song T, Gao J, Cui Y, Yang Y, Wu C, et al. A highly sensitive mixed lanthanide metal–organic framework self-calibrated luminescent thermometer. J Am Chem Soc 2013;135(41):15559–64.
[66] D'Vries RF, Alvarez-Garcia S, Snejko N, Bausa LE, Gutierrez-Puebla E, de Andres A, et al. Multimetal rare earth MOFs for lighting and thermometry: tailoring color and optimal temperature range through enhanced disulfobenzoic triplet phosphorescence. J Mater Chem C 2013;1(39):6316–24.
[67] Yarimaga O, Lee S, Ham D-Y, Choi J-M, Kwon SG, Im M, et al. Thermofluorescent conjugated polymer sensors for nano- and microscale temperature monitoring. Macromol Chem Phys 2011;212(12):1211–20.
[68] Donner JS, Thompson SA, Kreuzer MP, Baffou G, Quidant R. Mapping intracellular temperature using green fluorescent protein. Nano Lett 2012;12(4):2107–11.
[69] Baleizão C, Nagl S, Borisov SM, Schäferling M, Wolfbeis OS, Berberan-Santos MN. An optical thermometer based on the delayed fluorescence of C70. Chem A Eur J 2007;13(13): 3643–51.

Chapter 9

Applications of Luminescence Thermometry in Engineering

9.1 INTRODUCTION

Temperature measurements in engineering encompass a great variety of applications and needs. Measurement ranges exist between cryogenic temperatures and high temperatures. Environments can be extremely harsh and objects which need to be measured may be stationary or moving. In mechanical engineering, for example, temperature measurement practices range from simple monitoring of water temperature within an engine to complex measurements such as measuring temperatures of a weld, or smoke stack gas temperatures from a power station, or the distribution of temperature on the surfaces inside internal combustion engines. Similar wide ranges of temperature measurement applications can be found in other engineering disciplines. In response to this, a great number of methods and devices have been developed over the years. Among them, luminescence thermometry is gaining attention, in particular for measurements at spots which are difficult to access by traditional measuring devices as well as in harsh environments. In fact, engineering applications represent by far the largest fraction of all luminescence thermometry applications, and it would take a complete book to review them all in detail.

In this chapter, an emphasis is given to several engineering applications in which luminescence can be used for the real-time, two-dimensional temperature mapping of different objects. Temperature distribution data are quite useful to mechanical engineers in thermodynamics and fluid mechanics, while civil engineers may benefit from a similar knowledge on construction elements and the combustion of materials. The reliable detection of hot-spots in electrical devices is becoming increasingly important since the size of devices has been rapidly decreasing, making them more sensitive to thermal overstress. The measurement of heat flux is related to the measurement of temperature. Temperature gradients may be estimated from thermal images obtained in the real-time, providing additional benefit to the engineering practice of luminescence thermometry. This chapter presents the applications of luminescence for temperature measurements and imaging of electronic circuits and electrical machines, fluid flows, thermal and environmental barrier coatings, aerospace and automotive testing in wind tunnels, and surfaces inside internal combustion engines. It is

Luminescence Thermometry. http://dx.doi.org/10.1016/B978-0-08-102029-6.00009-9

impossible to cover all the engineering applications of luminescence thermometry in a single chapter. For this reason, some important applications, such as the use of fiber-optic luminescence probes for point temperature measurements, are not covered; readers interested in the additional applications of luminescence thermometry may find information in review articles [1,2] listed in the reference section.

9.2 THERMAL IMAGING OF ELECTRONIC CIRCUITS AND ELECTRICAL MACHINES VIA LUMINESCENCE

Over the years, electronic equipment has becoming more complex, rapidly miniaturized, and increasingly powerful, thereby generating more heat. Current electronic designs are characterized by extremely small component packages and high assembly design densities (component to component spacing is smaller than 0.02 inch with short copper interconnections). Consequently, electronic equipment is also becoming more sensitive to thermal overstress (i.e., excess heat). The defects causing overheating of either the circuitry or the components are more challenging to locate. In circuit boards, finding and fixing “hotspots” is done by a physical inspection; traditionally, by touching the circuit board surface (this may be a dangerous undertaking if a high voltage is present). Alternatively, hotspots can be located using thermocouples or spot pyrometers. These methods, however, provide values of temperature only at certain points in the circuitry, and, thus, some hotspots might go unnoticed. Circuit inspection by thermal-imaging infrared cameras is superior to the previously mentioned methods since it provides temperature readings of complete circuits, both large or micro-sized. Also, cameras can be used alone for thermal imaging or as a component part of an intricate testing system. Nevertheless, infrared cameras are quite expensive devices, and temperature measurements with them have some specific drawbacks, including relatively poor spatial resolution (see Chapter 2) which is, according to Rayleigh’s criterion, proportional to the wavelength λ of radiation utilized for the imaging:

$$\text{Spatial resolution} = 0.61 \times \frac{\lambda}{\text{NA}}, \tag{9.1}$$

where NA is optical system numerical aperture. Since infrared detectors operate under micrometer wavelengths the spatial resolution of infrared thermal imaging is limited to a few micrometers.

A promising alternative may be luminescence thermometry, especially in the case of testing extremely small electronic systems where the required spatial resolution of the temperature measurements comes near to the optical resolution limit, or even drops below it. Two approaches which differ in the positioning of the luminescent probe have been tested so far. In the first approach, the luminescent temperature probe is attached (glued) at the end of the tip of a scanning thermal microscope (SThM) and then used for surface thermal scanning [3–5]. This type of measurement provides spatial resolutions of thermal imaging beyond an optical resolution limit, which is exceptionally significant for applications at the nanoscale and cellular levels (see Chapters 10 and 11). In the second approach, luminescence temperature probes are dispersed and attached to the surface of interest [6,7], and thermal images are derived from traditional fluorescence imaging techniques. In this later case, the spatial resolution is worse than using a SThM and near-field optical imaging. On the other hand, this approach can be used for real-time thermal imaging of large surfaces and facilitates the use of a wide range of different types of luminescence temperature probes and almost all temperature read-out schemes (see Chapter 4). Depending on the measurement configuration, this approach yields better temporal resolution than SThM.

A typical luminescence SThM experimental set-up (see Fig. 9.1A) consist of a piezoelectric scanning stage where the sample is mounted, a tip with the luminescence probe, a laser for luminescence excitation, an objective, optical lenses and filters, and photodetectors. The most convenient method used to attach the luminescent particle at the end of the tip is careful gluing [3,8]. An example of fixing a luminescent probe at the end of the tip of a SThM is described by Aigouy et al. [3]. First, a thin layer of adhesive is deposited on the tip, and the tip is brought toward the luminescent probe so that the probe attaches to the tip. By using a second tip, attached to piezoelectric scanners, the luminescent probe is positioned at the end of the tip. Finally, the probe size is reduced by focused ion beam milling.

To obtain scanning thermal images of the resistor stripe in an operating integrated circuit (see Fig. 9.1B—image on the left) Aigouy et al. [4] used the upconversion luminescent probe made from a Yb,Er codoped Zbyban glass fragment (probe composition: $(ZrF_4)_{45.5}$, $(BaF_2)_{23}$, $(YbF_3)_{11}$, $(ErF_3)_3$, $(AlF_3)_3$, $(InF_3)_{0.5}$, $(NaF)_{14}$). Temperature was derived using the luminescence intensity ratio method (see Chapters 4 and 6) by exploiting Er^{3+} emissions centered at 527 and 550 nm under laser excitation (975 nm) into the Yb^{3+} level. As discussed in detail in Chapter 6, these emissions correspond to the $^2H_{11/2} \rightarrow {}^4I_{15/2}$ and $^4S_{3/2} \rightarrow {}^4I_{15/2}$ f–f electronic transitions of the Er^{3+} ions. The ratio of their intensities is sensitive to temperature change due to a thermal coupling of closely spaced $^2H_{11/2}$ and $^4S_{3/2}$ excited electronic levels. To obtain a thermal image, two upconversion emission images were taken while scanning the microchip's surface with a luminescent tip. During scanning, upconversion radiation from the tip was passed through an objective, then divided into two beams by a beam splitter, and finally passed to two photomultiplier tubes (PMTs) through interference filters (one centered at 520 nm and a second at 550 nm; bandwidth used was approximately 10 nm). After the scan, a thermal image was obtained as the ratio of respective pixel intensities in 520 nm and 550 nm images. Along with thermal imaging, this experimental set-up allowed measurements of the surface topography by conventional atomic force microscopy (AFM) tapping mode. The middle and right-hand images in Fig. 9.1B show topography (*upper images*) and thermal images of the resistive stripe when no current is passing

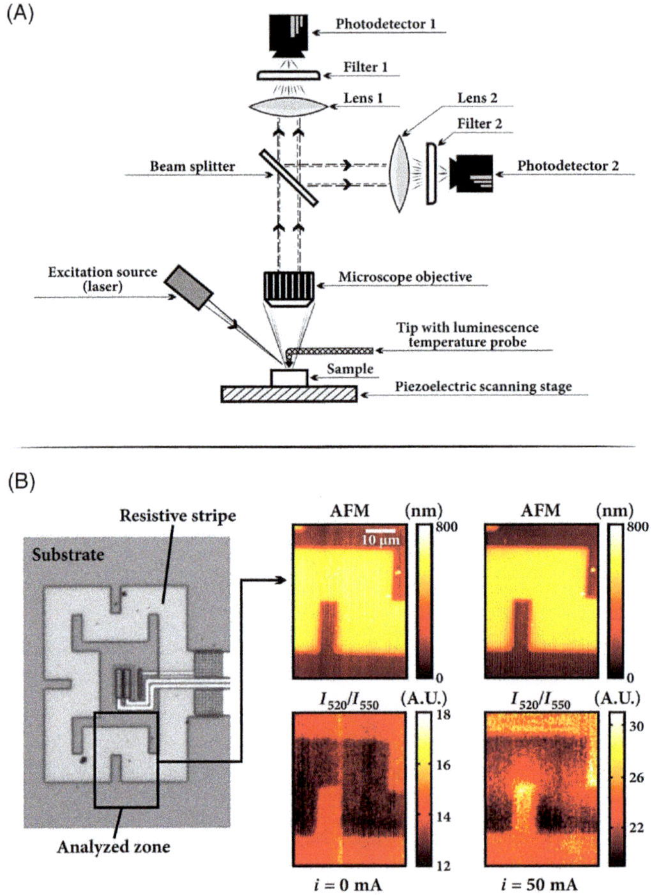

FIGURE 9.1 (A) Schematic of the luminescence scanning thermal microscope experimental set-up. (B) Optical micrograph of the microelectronic circuit (*left*); topography and fluorescence ratio images of the structure when no current is passing through the stripe (*images in the middle*); and images when a current of 50 mA is passing through the structure (*images on the right*). The bright hot zones are clearly visible on the stripe. The image size is 45 × 60 µm^2. *(Reprinted from Aigouy L, Tessier G, Mortier M, Charlot B. Scanning thermal imaging of microelectronic circuits with a fluorescent nanoprobe. Appl Phys Lett 2005;87(18):184105 3 pp., with the permission from AIP Publishing).*

through the stripe and when a current of 50 mA is present (*bottom images*). Hot zones in the stripe can be easily identified as bright spots on the image. The main drawback of this method is its extremely slow scanning speed (10 surface points per second) which is needed to facilitate the proper thermalization of a luminescence probe. Therefore, the method is not suited to real-time thermal imaging since many processes may appear or disappear while measuring one thermal image.

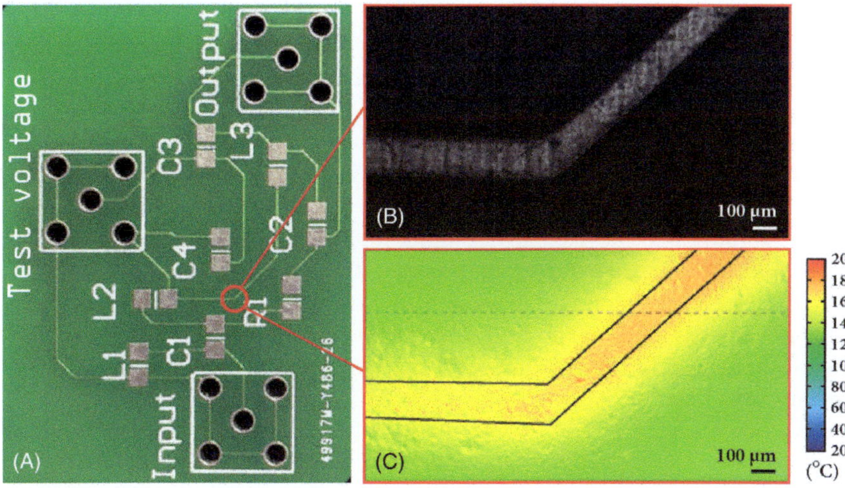

FIGURE 9.2 (A) Image of the printed circuit board with a *circle* marking the imaged area; (B) the optical monochrome image of the thermally imaged area; (C) thermal image of the marked circuit area operating with a DC current of 5.3 A; the scale bars represent 100 μm. *(Reprinted from Xiong J, Zhao M, Han X, Cao Z, Wei X, Chen Y, et al. Real-time micro-scale temperature imaging at low cost based on fluorescent intensity ratio. Sci Rep 2017;7:41311 8 pp., under a Creative Commons Attribution 4.0 license).*

Much faster, real-time thermal imaging can be obtained when the complete surface of interest is covered with luminescent temperature probes, affording in this way luminescence information from each surface point. Then, steady-state or time-resolved luminescence images of the surface can be captured and converted into thermal images in accordance with the chosen temperature read-out scheme (luminescence intensity ratio, lifetime, etc.; see Chapter 4). For example, Xiong et al. [6] used a luminescence microscope to thermally image a 140 μm wide area on a printed circuit board (PCB) operating at a DC current of 5.3 A (see Fig. 9.2A) that was previously covered with an Sm^{2+} doped SrB_4O_7 fine luminescent powder. Sm^{2+} doped SrB_4O_7 is well known because of its pressure and (high-)temperature sensitivity. Its emission spectra are characterized by narrow emission bands which appear due to the $^5D_0 \rightarrow {}^7F_J$ (J = 0,1,...,6) $^4f_6$ electronic transitions (at wavelengths of 680 nm and higher), low intensity narrow bands between 630 and 690 nm due to the $^5D_1 \rightarrow {}^7F_J$ transitions, and broadband from the interconfigurational transition from the low-energy $4f^55d^1$ excited configuration to the $^4f_6$ ground configuration. The energy difference between $4f^55d^1$ and $^5D_0$ ($^4f_6$) is relatively high (about 3800 cm^{-1}), however, thermal coupling between these excited levels is possible due to the strong electron–phonon coupling involving $4f^{(n-1)}5d^1$ levels [9]. Then, the Boltzmann-type luminescence intensity ratio read-out of temperature is possible by utilizing emissions from $4f^55d^1$ and $^5D_0$ ($^4f_6$) excited levels expecting, due to a large energy difference between emitting levels, a relative sensitivity much higher than in the case of trivalent lanthanide ions (see Chapter 6). Indeed, Cao et al. [9] obtained a relative sensitivity of 2.16% K^{-1} at 500K with this material. To make a thermal image of the part of the operating PCB covered with luminescent powder (see Fig. 9.2B and C), Xiong et al. [6] illuminated the PCB with 365 nm radiation and recorded its luminescence using two monochrome CMOS cameras (1.3 million pixels; pixel size of 5.2 μm × 5.2 μm) after splitting the emission beam into two and passing one of them through a shortwave-pass filter (590 nm), and the second through a longwave-pass filter (650 nm). The thermal image was then created by rationing the respective pixel intensities of two images and calculating the temperature using calibration data. Two major problems arise with this type of measurement. The first comes from the slight spatial difference in the two images which occurs due to a difference in the optical paths from the sample to the two CMOS cameras. The problem can be resolved with a geometrical transformation of images using image processing to adjust images to reflect the same point of origination. The second problem is that the low-temperature measurement bound lies at a relatively high temperature, about 373K. This occurs because the thermal coupling between the $4f^55d^1$ and $^5D_0$ ($^4f_6$) levels is quite weak at temperatures below 373K, due to the large energy gap between levels, so the $4f^55d^1$emission is barely visible and cannot be utilized for sensing. For temperatures lower than 373K, the ratio of emission intensities from the $^5D_1$ and $^5D_0$ levels can be used; however, with a lower relative sensitivity.

One promising avenue for thermal imaging of electronic devices and circuits may be based on the use of π-conjugated polydiacetylene (PDA) supramolecules embedded in transparent polymer films. When compared to other conjugated polymers, PDAs present a range of exceptionally interesting and useful optical properties, such as thermochromism, mechanochromism, solvatochromism, and affinochromism (change of color in response to temperature, mechanical stress, solvent, and ligand interactions, respectively), as well as the change from a nonfluorescent to a fluorescent state [10] (see Fig. 9.3A). The changes in color (absorption) and fluorescence of PDA's conjugated backbones caused by the PDA's interaction with the environment are much investigated for sensing applications [11]. A great number of different forms can be produced by the self-assembly of diacetylene amphiphiles, and further incorporation in various host materials, such as polymers [(poly(vinylpyrrolidone), poly(vinyl alcohol) (PVA), etc.], carbohydrates, sol–gel materials. Of interest here is PDA incorporated into a PVA (see Fig. 9.3B).

These films can be used for the simple, low-cost evaluation of the surface temperature of electronic devices with submicrometer spatial resolution. For example, Yarimaga et al. [7] used this to observe the temperature distribution on a conventional integrated circuit (IC). They prepared the 50 μm thick film by drop-casting the PDA/PVA composite solution on the IC surface and induced the 18 h long, room-temperature curing of the film. When the IC was in operation, the color of the film changed from blue to red, and the film became

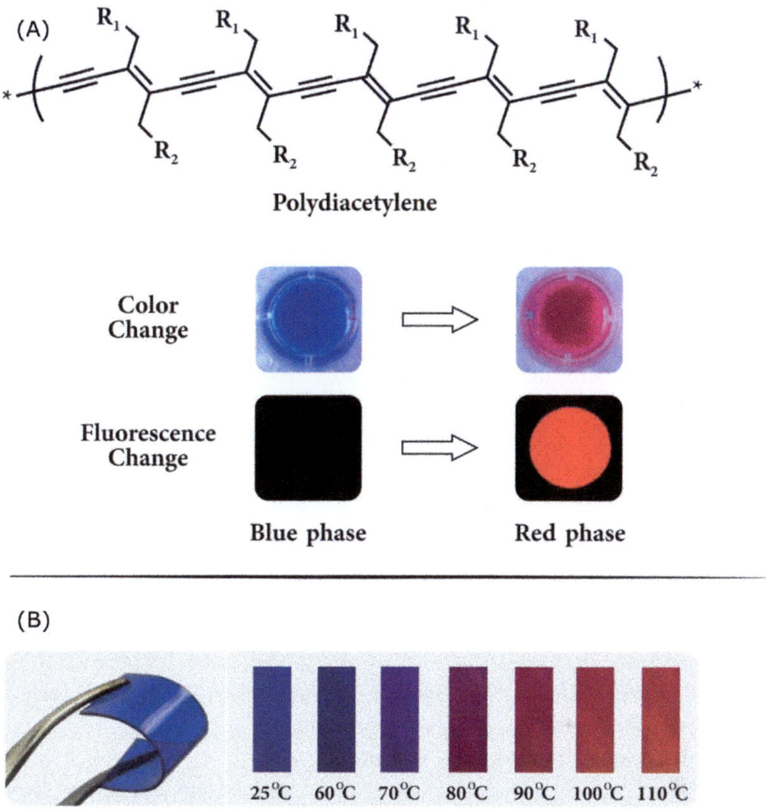

FIGURE 9.3 (A) The structure of polydiacetylenes (PDAs) and their typical color/fluorescence changes upon stimulation. (B) photographs of a PDA-embedded poly(vinyl alcohol) film during the heating process. *(Adapted from Yoon B, Lee S, Kim J.-M. Recent conceptual and technological advances in polydiacetylene-based supramolecular chemosensors. Chem Soc Rev 2009;38(7):1958–68, with permission from The Royal Society of Chemistry).*

fluorescent, all due to heating. Then, it was possible to remove the film from the IC, and analyze the temperature distribution of the surface from fluorescent intensities measured at different film spots. Using a similar procedure, Choi et al. [12] were able to detect nanosized hot spots on the surface of a poly-Si thin film transistor from the fluorescence of PDAs in a poly(vinylpyrrolidone) film.

Besides applications in thermal imaging of electrical circuits, luminescence thermometry can be used for temperature measurements of moving surfaces in electrical machinery. In fact, traditional contact methods are difficult to use where one needs to obtain temperature distributions on rotating elements. Pyrometry, on other hand, does not produce visible radiation at low temperatures (close to room temperature) and suffers when there is a high blackbody radiation background present in a measurement environment. Infrared thermography can be used in such conditions, but only to measure the spatially averaged temperature. In these situations, luminescence thermometry may be the method of choice. For example, Allison et al. [13] demonstrated the luminescence thermometry of a rotor in an electric motor which was constructed from high-permeability magnets. A mixture of epoxy paint and $La_2O_2S:Eu^{3+}$ phosphor particles was used to cover the underside of the rotor. Ultraviolet radiation for excitation and the radiation from emitting phosphor particles were carried by two optical fibers with the help of a small mirror which was mounted on the stator. This approach yielded temperature measurements of the rotor's surface as a function of its angular position, with a spatial resolution of about 2 mm and a precision of about 0.3°C. In another example, Mannik et al. [14,15] used a mixture of silicon binder and phosphor particles ($La_2O_2S:Eu^{3+}$) to paint a stripe around the circumference of the rotor in a high-power electrical generator. By using the emission decay time temperature read-out, they determined the temperature at 24 positions around the rotor surface while the generator was operating at 500 MW. The measurements had a 5 mm spatial resolution, and a temperature accuracy of about 2°C.

9.3 LUMINESCENCE THERMOMETRY OF FLUID FLOWS

A combination of luminescence thermometry with laser Doppler velocimetry or particle image velocimetry (PIV) allows combined measurements of velocity and high-temperature fields in gas flows. When compared to point-measurements, field measurements offer more advantages since they provide two-dimensional information about both velocity and temperature (or velocity and pressure, for example), thus enabling analyses of complex phenomena related to fluid mechanics. PIV is a nonintrusive optical technique which indirectly determines the velocity field of fluid components by measuring the velocity of particles (so-called tracer particles) that are, in most cases, intentionally added to the flow. Details on the PIV technique's principles and applications can be found in an extensive literature, for example in Ref. [16]. In a typical experiment, tracer particles are seeded into the gas flow (the same applies to a liquid flow) and are illuminated by a light sheet generated from high-energy laser pulses. The displacement of particles due to the local flow is determined by imaging their positions on account of the scattered light for two consecutive laser pulses (see Fig. 9.4). With known values of imaging magnification and knowledge of the time lag between the two laser pulses, vectors of particle velocities in different subareas of the flow (the so-called interrogation areas) can be calculated using cross-correlation or autocorrelation functions. One should be aware that tracer particles do not follow fluid elements exactly (a phenomenon known as velocity lag). Generally, in order to have smaller velocity lags one should use smaller particles; however, the intensity of scattered light considerably decreases with increased particle size [17].

Microspheres of similar densities to the investigated fluid are used as tracer particles in PIV. They are usually prepared from glass beads, polystyrene, or

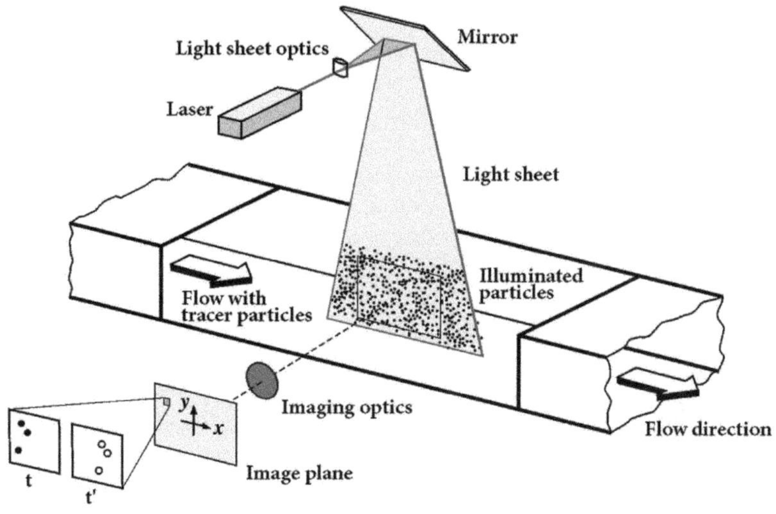

FIGURE 9.4 The experimental arrangement for particle image velocimetry in a wind tunnel. *(Reprinted from Raffel M, Willert CE, Wereley ST, Kompenhans J. Particle image velocimetry. 2nd ed. Berlin, Heidelberg: Springer Berlin Heidelberg; 2007, with permission from Springer).*

polyethylene, but also from ceramic materials (such as Al_2O_3, TiO_2, and ZrO_2) when high-temperature flows are to be investigated. By using luminescent tracer particles it is possible to measure both velocity and temperature fields in one experiment [18–22]. Such particles should be designed to facilitate PIV, to rapidly assume fluid temperature, and to have a sufficient luminescence sensitivity on temperature so that some of the well-established luminescence read-outs of temperature can be employed (see Chapter 4).

An example of the usefulness of this approach may be drawn from the recent work of Fond et al. [23] who have demonstrated the technique for the simultaneous measurement of temperature/velocity/gas-mixture-fraction fields in a turbulent jet using 2 μm Eu^{2+} doped BAM ($BaMgAl_{10}O_{17}$) tracer particles (see Fig. 9.5). After seeding phosphorescent particles into the flow, the velocity field was derived using traditional PIV methodology, while the temperature field was evaluated using emission from Eu^{2+} integrated over two spectral regions (see Fig. 9.5A). The precision of the temperature measurements was estimated to be between 2 and 5°C. The luminescence of tracer particles was also used for the determination of the mixture fraction of two gas streams (see Fig. 9.5B). For this purpose, particles were introduced only in one stream, and the mixture fraction was evaluated from the emission intensity after temperature conditioning.

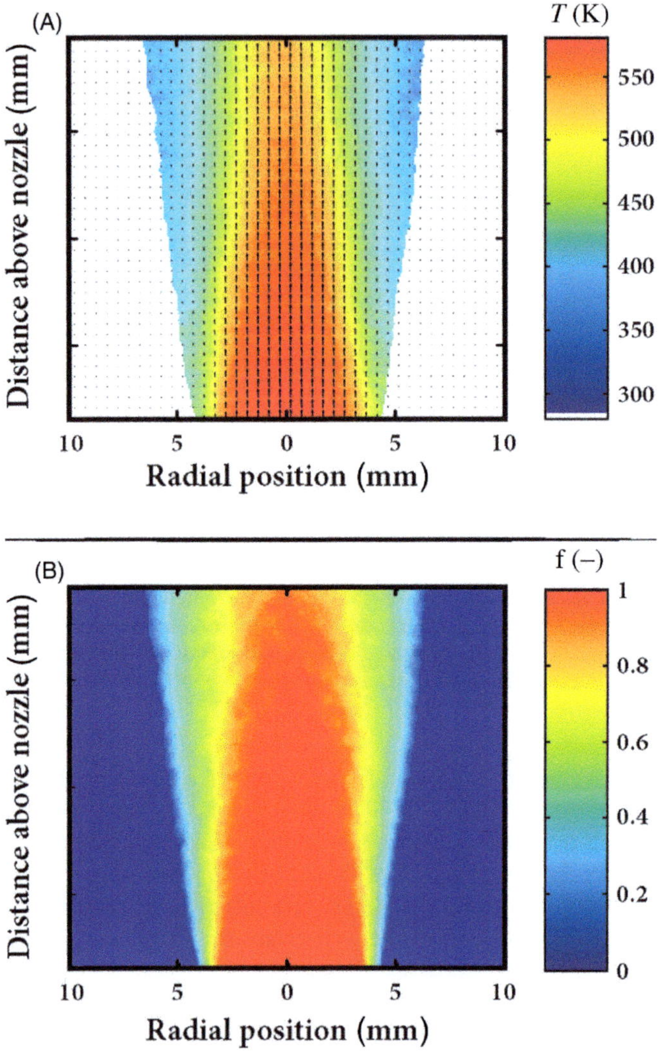

FIGURE 9.5 (A) Average temperature and velocity fields for a jet temperature of 530K (compiled from 100 single shots). (B) Average mixture fraction image which corresponds to the same velocity–pressure image in (A). *(Reprinted from Fond B, Abram C, Heyes AL, Kempf AM, Beyrau F. Simultaneous temperature, mixture fraction and velocity imaging in turbulent flows using thermographic phosphor tracer particles. Opt Express 2012;20(20):22118–22133, with permission from OSA Publishing).*

9.4 THERMAL AND ENVIRONMENTAL BARRIER COATINGS

Thermal barrier and environmental barrier coatings (TBCs and EBCs, respectively) are used to protect structural materials when they are operating in high temperatures, high heat fluxes, and thermal oxidation environments [24,25]. They are of special interest for aerospace (components of gas turbine engines), energy (components of industrial power generators and nuclear reactors), automotive (components of engine exhaust systems and turbochargers), and oil and gas (pumps, compressors) technologies and industries. Not only do TBCs and EBCs protect structural materials but they also facilitate operations at higher temperatures, thus increasing the efficiencies of engines. It is generally assumed that a 1% gain in engine efficiency may be achieved with a 50°C increase in firing temperature (the 1% enhancement of engine efficiency reduces fuel costs by about $20 million over the lifecycle of a 400–500 MW combined-cycle, gas-fired industrial power plant [26]).

The current industry standard for TBCs is a prime-tetragonal yttria-stabilized zirconia (YSZ), but some other oxides are of interest, such as pyrochlore $A_2^{3+}B_2^{4+}O_7$ oxides (particularly rare earth zirconate pyrochlores) and fluorite HfO_2, CeO_2, and ThO_2 oxides. Rare earth silicates (Re_2SiO_5 and $Re_2Si_2O_7$ Re = Y, Gd, Lu) and $(Ba_{0.75}Sr_{0.25})Al_2Si_2O_8$ are used for EBCs. All these oxides can be easily doped with a small amount of trivalent lanthanide ions to provide luminescence suitable for temperature sensing, as discussed in detail in Chapter 6 and Refs. [27,28]. Doping of coatings does not disturb their materials properties, which is of utmost importance for the performance of TBCs and EBCs. To activate luminescence in TBCs and EBCs, mainly Eu^{3+}, Dy^{3+}, and Tb^{3+} are used, with Sm^{3+} to a lesser extent. These ions have a large energy difference between their excited and ground states, so their luminescence is quenched only at the high temperatures at which the excited state lifetimes undergo a steep decrease in value, as can be seen in Fig. 9.6 for the case of the lifetime of Eu^{3+} doped YSZ ($Y_{0.06}Eu_{0.01}Zr_{0.93}O_{1.965}$) $^5D_0$ level [29]. For both bulk materials and coatings prepared by electron-beam physical vapor deposition (EB-PVD), the lifetime value does not change with temperatures up to 400 °C, but decreases rapidly (for 4 orders of magnitude) over the 400–1200°C range. Such great temperature dependence of lifetime suggests that decay time read-outs can be effectively used for the evaluation of temperature of TBCs with good measurement sensitivity. One should note that traditional thermometry methods cannot be implemented straightforwardly for measurements of TBC and EBC temperatures. The high-temperature emissivity of many materials is not yet precisely known, which causes problems for pyrometry use. What is more, pyrometry provides information about the surface temperature of TBCs and EBCs, while of real interest is the temperature of the structural material beneath (alloy, SiC, etc.) which is substantially lower (>200°C) due to the high-temperature gradient over the coating. Taking into account that TBCs protect superalloys from temperatures that may exceed their melting points and that the temperature at the superalloy–TBC interface crucially influences the duration of the TBC, it is of primary interest to measure the temperature at the interface [30]. Gentleman et al. [29] have shown that an actual temperature of a superalloy at an interface can be measured when a layer of the thermographic phosphor (10 μm thick Eu doped YSZ) is introduced between superalloy and YSZ–TBC, as shown in Fig. 9.6B.

The red luminescence of the phosphor layer can be easily observed from the cross-sectional image obtained under ultraviolet radiation on the

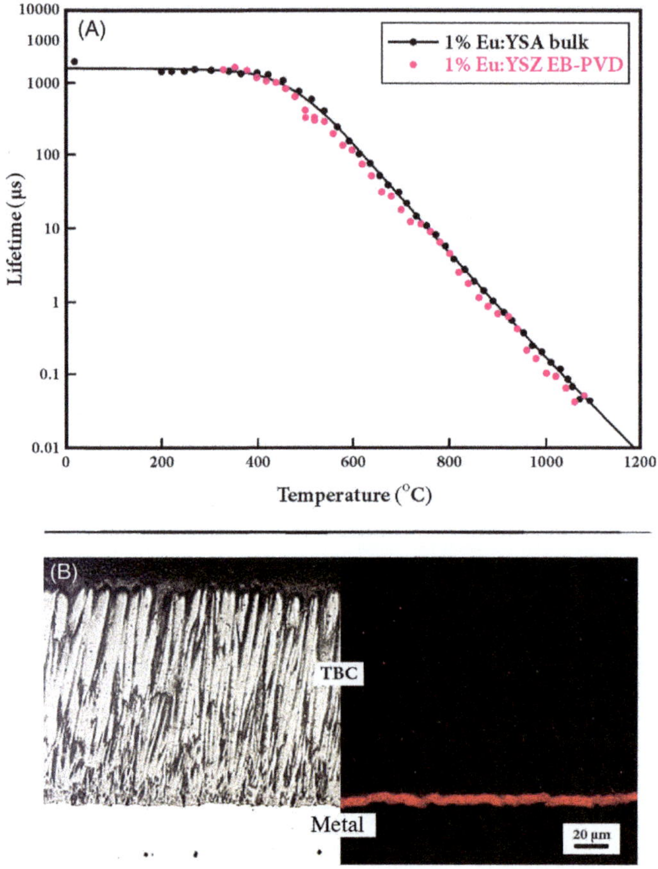

FIGURE 9.6 (A) Temperature dependence of the Eu^{3+} $^5D_0$ excited state lifetime in the yttria-stabilized zirconia (YSZ) bulk and coated materials. (B) Cross section of the YSZ thermal barrier coating with a layer of Eu^{3+} doped YSZ introduced between a superalloy and its coating: (*left*) image obtained under white light, (*right*) image obtained under UV radiation. *(Reprinted from Gentleman MM, Eldridge JI, Zhu DM, Murphy KS, Clarke DR. Non-contact sensing of TBC/BC interface temperature in a thermal gradient. Surf Coatings Technol 2006;201(7):3937–41, copyright (2006), with permission from Elsevier).*

right-hand side of Fig. 9.6B. A white light image of the coating is shown on the left-hand side. Recently, Nada et al. [31] tested a similar approach in measuring the superalloy temperature beneath a YSZ–TBC using an Eu^{3+} doped Y_2O_3 buried layer. In addition, they analyzed the possibility of using both lifetime and intensity ratio methods for measuring the temperature of the TBC after YSZ doping with Dy^{3+}. Also recently, a comparison between thermometric performances of Dy^{3+} and Eu^{3+} doped YSZ layers deposited between a bond coat and YSZ–TBC has been made by Yang et al. [32]. They showed that Dy^{3+} doped layers give stronger emission intensities at high temperatures than Eu^{3+} doped ones, thus having higher measurement limits and a slightly better sensing performance.

Thermally grown oxide at the interference of the bond and top coats of a TBC is a frequent cause of TBC failure. This oxide growth can be suppressed with the introduction of an intermediate layer prepared from a material that shows a low oxygen diffusivity. The material with such a property is yttrium–aluminum garnet ($Y_3Al_5O_{12}$—YAG) which also has a low thermal conductivity, only slightly higher than YSZ [32]. YAG is also one of the most important host materials for the preparation of rare earth phosphors, and Dy^{3+} doped YAG has been used for the luminescence temperature measurement of the highest ever temperature of 2000K [33]. Steenbakker et al. [26] showed that it is possible to measure the temperature of YAG:Dy^{3+} layers in TBC structures over the 1080–1500°C range.

It is worthy of mentioning that doping TBC and EBC oxides with lanthanide ions provides luminescence which can be used to observe defects and cracks in coating structures, in addition to luminescence thermometry, as shown in a recent literature report [34].

9.5 SURFACE TEMPERATURE MEASUREMENTS USING TEMPERATURE-SENSITIVE PAINTS AND PHOSPHOR COATINGS

As discussed at the beginning of this chapter, electronic circuits and many electrical devices can be successfully thermally imaged when covered with a coating or thin film that contains a temperature-responsive luminescent species. If sufficiently thin, coatings and films promptly equilibrate with their surroundings, and local temperature can be monitored in real-time over the complete surface of interest. However, the spatial resolution of such thermal imaging is limited by the wavelength of the emitted radiation (Rayleigh's limit, see Eq. (9.1)), and by a heat transfer which tends to average temperature over the surface and perpendicularly inside the film or coating. For this reason, luminescence coatings and films pose difficulties for luminescence thermal imaging of very small surfaces ($<10\ \mu m$), and cannot be used at the nanoscale. Nevertheless, this does not impose problems for many engineering applications, such as aerodynamics and combustion engine testing, where spatial resolutions of luminescence thermal imaging with films and coatings are sufficient or even exceed the spatial resolutions offered by traditional thermometry methods. What is more, thermal imaging using luminescence coatings and films can be combined with luminescence imaging of other surface physical properties for achieving multiphysics imaging technologies [35]. One well-known example of such an approach can be found in aerodynamic testing where combined imaging of temperature and pressure of surfaces in wind tunnels is conducted using temperature and pressure sensitive paints (TSP and PSP, respectively; see Chapter 12).

Thermographic coatings and films can be prepared from almost all luminescence thermographic materials (see Chapters 6–8) mixed with appropriate binders. The most frequent formulations contain lanthanide or transition metal ion doped phosphors for ceramic coatings, and organic dyes or metal–organic complexes in the case of films. To avoid quenching of luminescence by oxygen, metal–organic complexes and dyes should be incorporated into polymers that are not permeable to oxygen (for example, poly(methyl methacrylate), polyacrylonitrile, or poly(vinyl methyl-ketone)). Surfaces of interest may be covered with thermographic coatings by traditional physical and chemical techniques (see Chapter 6). Also, a variety of convenient and fast solution-processing techniques can be used for the same purpose, such as spin coating, doctor blade method, slot-die coating, spray coating (airbrush, electrospray, ultrasonic), dip-coating. They can also be printed onto the surface to achieve full or patterned surface coverage, for example, by screen, gravure, flexographic, or inkjet printing [36]. The details on applications of coating and printing methods can be found in [37], as well as in an extensive literature on this subject.

Thermal and pressure imaging with luminescent paints has been used in the field of aerodynamic science and technology for more than 50 years. Mainly, TSPs and PSPs are utilized for testing model designs in wind tunnels. Thermocouples and small-diameter gauges are traditionally used as heat flux sensors for surface measurements in wind tunnels. However, they lack spatial resolution (since they are in fact point measurements) and impose geometrical constraints making them sometimes difficult to install on models having complicated geometries. In addition, they are much more costly than TSPs. The transition point between laminar and turbulent flow, characterized by a step change of surface temperature, can be detected by TSP instead of using the traditional acenaphthene sublimation technique. Several types of TSPs have been used so far for aerodynamics testing [38,39], with single or dual emission bands, providing temperature read-outs from both emission intensity and decay time. For example, the luminescence of Ru(phen) (Ruthenium tris(1,10-phenanthroline) dichloride) incorporated into a polyacrylic acid binder has been successfully used for the assessment of the heat-flux distribution on a slender 7 degree half-angle cone in a hypersonic shock tunnel [40]. In the Mach-6 Quiet Tunnel at Purdue University, the heat-flux distribution on a nylon cone has been measured by a luminescent coating consisting of a luminescent Ru(bpy) (tris(2,2-bipyridyl) ruthenium) incorporated into automobile clear coat (DuPont ChromaClear) binder [41]. In addition to these TSP formulations, EuTTA (europium(III) thenoyltrifluoroacetonate), in model airplane dope, is one of the most utilized paints for luminescence applications in aerodynamics.

TSPs composed of organic luminophores and polymer binders decompose at high temperatures and may suffer from photobleaching. Therefore, such TSPs are not suitable for long-duration testing in high-temperature ($> 200°C$) hypersonic wind tunnels. For such conditions, Crafton et al. [42] suggest the use of inorganic phosphor-based TSPs having high-temperature binders which do not degrade at elevated temperatures and which can be removed from the surface after measurements and reused several times. An anodized aluminum may be a promising alternative to polymer binders as a supporting matrix for the preparation of dye-based and quantum dot–based luminescence thermometers [43,44]. It can be used at cryogenic temperatures where polymers suffer stiffening and cracking, and, also, has a much higher melting point than polymers. Claucherty and Sakaue [43] have recently demonstrated surface temperature measurements in the 150–500K range on account of the luminescence of rhodamine B adsorbed onto an anodized aluminum surface or incorporated into voids.

Due to the high temporal resolution and contactless nature of measurements, TSPs are invaluable for the thermal imaging of moving surfaces, for example, moving rotor blades. By using fast-response PSP and TSP, Jiao et al. [45] simultaneously imaged pressure and temperature fields on turbocharger compressor blades that had rotation speeds as high as 110,000 rpm. The blades were separately painted with PSP and TSP. While the compressor was operating, the temperature and pressure of the surface were evaluated using single-shot decays of emission excited by a short-pulsed 532 nm laser and detected using a CCD camera.

PSPs and TSPs are also used for automotive model testing. Grenier et al. [46] used TSP (Mg-tetra(pentafluorophenyl)porphine metal–organic complex in a copolymer of heptafluoro-n-butyl methacrylate and hexafluoroisopropyl methacrylate) for compensation of the temperature sensitivity of PSP (Pt-tetra(pentafluorophenyl)porpholactone metal-organic complex in co-polymer of heptafluoro-n-butyl methacrylate and hexafluoroisopropyl methacrylate) in testing passenger car models in a low-speed continuous atmospheric wind tunnel. The temperature compensation of the PSP's response is essential in auto model testing since at low wind speeds the lift pressure is only a small fraction of the static pressure [47]. When selected paints have similar temperature sensitivity, the ratio of emissions from two paints is almost temperature insensitive. In this case, the temperature sensitivity of PSP was $-0.32\%K^{-1}$ and TSP was $-0.25\%K^{-1}$, yielding pressure imaging with a sensitivity of about 65% per bar, with only a small error due to the temperature sensitivity of $-0.07\%K^{-1}$.

Thermal imaging via luminescence of surfaces in automotive science and the automotive industry is not limited only to model testing in wind tunnels. It is of importance for any surface where noninvasive temperature measurements are required, especially at surfaces which are difficult to access, such as inside an internal combustion engine, where taking surface temperature measurements is a sophisticated task due to very fast temperature changes (of the order of 10,000K per second and even faster in some cases) [48,49]. Temperature point-measurements by luminescence thermometry have been demonstrated on pistons in single-cylinder diesel engines [48,49] and cylinder walls [50–52]. Recently, coinciding measurements of the piston surface temperature and flame impingement in an optically reachable spark ignition engine has been facilitated by coating the piston's surface with $Gd_3Ga_5O_{12}$:Cr,Ce phosphor [53]. By using a high-speed CMOS camera, Someya et al. [54] performed two-dimensional surface temperature measurements inside an optical engine. A mixture of Y_2O_2S:Eu,Sm phosphor (3%) and ceramic binder (97%) was coated on the surface of an engine wall. While the engine was operating, the coating was excited by a pulsed Nd:YAG laser (355 nm) and imaged using a high-speed CMOS camera. The temperature was derived from emission decays. It was necessary to apply very short measurement times (250 μs) since, under standard conditions, the crank in an optical engine operating at 1500 rpm rotates by 2 degree in 250 μs. To meet this demand, the surface of interest was imaged at 40,000 fps and the temperature was derived from 10 images.

As can be seen from previous examples, and as expected due to the high-temperature and volatile environmental conditions inside internal combustion engines, inorganic lanthanide or transition metal–based phosphors are the materials of choice for luminescence thermometry. So far, the majority of applications have been conducted in research-type optical engines. They are valuable devices since they give access to physical processes that occur in combustion engines by facilitating the use of nonintrusive optical diagnostic techniques for studies of in-cylinder flow, mixing, combustion, and emissions. Nevertheless, experimental studies [55] have revealed that differences (mainly in heat transfer properties) between optical and all-metal engines influence mixing, combustion, and emission processes, and that there is a significant difference in the temperature across combustion chamber walls. To measure the two-dimensional distribution of temperature on surfaces in an all-metal engine by luminescence thermometry, Fuhrmann et al. [48] developed an optical system which employs a rigid borescope. The use of this optical system inside an engine enabled thermal imaging (using a high-speed CMOS camera) of its surfaces (coated with $Gd_3Ga_5O_{12}:Cr^{3+}$) under fired and motored conditions.

REFERENCES

[1] Aldén M, Omrane A, Richter M, Särner G. Thermographic phosphors for thermometry: a survey of combustion applications. Prog Energy Combust Sci 2011;37(4):422–61.

[2] Khalid AH, Kontis K. Thermographic phosphors for high temperature measurements: principles, current state of the art and recent applications. Sensors 2008;8(9):5673–744.

[3] Aigouy L, De Wilde Y, Mortier M, Giérak J, Bourhis E. Fabrication and characterization of fluorescent rare-earth-doped glass-particle-based tips for near-field optical imaging applications. Appl Opt 2004;43(19):3829–37.

[4] Aigouy L, Tessier G, Mortier M, Charlot B. Scanning thermal imaging of microelectronic circuits with a fluorescent nanoprobe. Appl Phys Lett 2005;87(18):184105.

[5] Samson B, Aigouy L, Löw P, Bergaud C, Kim BJ, Mortier M. AC thermal imaging of nanoheaters using a scanning fluorescent probe. Appl Phys Lett 2008;92(2):23101.

[6] Xiong J, Zhao M, Han X, Cao Z, Wei X, Chen Y, et al. Real-time micro-scale temperature imaging at low cost based on fluorescent intensity ratio. Sci Rep 2017;7:41311.

[7] Yarimaga O, Lee S, Ham D-Y, Choi J-M, Kwon SG, Im M, et al. Thermofluorescent conjugated polymer sensors for nano- and microscale temperature monitoring. Macromol Chem Phys 2011;212(12):1211–20.

[8] Michaelis J, Hettich C, Mlynek J, Sandoghdar V. Optical microscopy using a single-molecule light source. Nature 2000;405(6784):325–8.

[9] Cao Z, Wei X, Zhao L, Chen Y, Yin M. Investigation of $SrB_4O_7:Sm^{2+}$ as a multimode temperature sensor with high sensitivity. ACS Appl Mater Interfaces 2016;8(50):34546–51.

[10] Yoon B, Lee S, Kim J-M. Recent conceptual and technological advances in polydiacetylene-based supramolecular chemosensors. Chem Soc Rev 2009;38(7):1958–68.

[11] Reppy MA, Pindzola BA. Biosensing with polydiacetylene materials: structures, optical properties and applications. Chem Commun 2007;42:4317–38.

[12] Choi JM, Choi SJ, Yarimaga O, Yoon B, Kim JM, Choi YK. Detection of a nanoscale hot spot by hot carriers in a Poly-Si TFT using polydiacetylene-based thermoresponsive fluorometry. IEEE Trans Electron Devices 2011;58:1570–4.

[13] Allison SW, Cates MR, Noel BW, Gillies GT. Monitoring permanent-magnet motor heating with phosphor thermometry. IEEE Trans Instrum Meas 1988;37:637–41.

[14] Mannik L, Brown SK, Campbell SR. Phosphor-based thermometry of rotating surfaces. Appl Opt 1987;26(18):4014–7.

[15] Mannik L, Brown SK, Campbell SR. Phosphor-based thermometry of the rotor in a high-power electrical generator. Appl Opt 1991;30(19):2670–2.

[16] Raffel M, Willert CE, Wereley ST, Kompenhans J. Particle image velocimetry. 2nd ed. Berlin, Heidelberg: Springer Berlin Heidelberg; 2007.

[17] Melling A. Tracer particles and seeding for particle image velocimetry. Meas Sci Technol 1997;8(12):1406.

[18] Omrane A, Petersson P, Aldén M, Linne MA. Simultaneous 2D flow velocity and gas temperature measurements using thermographic phosphors. Appl Phys B 2008;92(1):99–102.

[19] Yin Z, Fond B, Eckel G, Abram C, Meier W, Boxx I, et al. Investigation of $BAM:Eu^{2+}$ particles as a tracer for temperature imaging in flames. Combust Flame 2017;184:249–51.

[20] Lawrence M, Zhao H, Ganippa L. Gas phase thermometry of hot turbulent jets using laser induced phosphorescence. Opt Express 2013;21(10):12260–81.

[21] Lee H, Böhm B, Sadiki A, Dreizler A. Turbulent heat flux measurement in a non-reacting round jet, using $BAM:Eu^{2+}$ phosphor thermography and particle image velocimetry. Appl Phys B 2016;122(7):209.

[22] Kim D, Yi SJ, Kim HD, Kim KC. Simultaneous measurement of temperature and velocity fields using thermographic phosphor tracer particles. J Vis 2017;20(2):305–19.

[23] Fond B, Abram C, Heyes AL, Kempf AM, Beyrau F. Simultaneous temperature, mixture fraction and velocity imaging in turbulent flows using thermographic phosphor tracer particles. Opt Express 2012;20(20):22118–33.

[24] Clarke DR, Phillpot SR. Thermal barrier coating materials. Mater Today 2005;8(6):22–9.

[25] Xu Y, Hu X, Xu F, Li K. Rare earth silicate environmental barrier coatings: present status and prospective. Ceram Int 2017;43(8):5847–55.

[26] Steenbakker RJL, Feist JP, Wellman RG, Nicholls JR. Sensor thermal barrier coatings: remote in situ condition monitoring of EB-PVD coatings at elevated temperatures. J Eng Gas Turbines Power 2009;131(4):41301–9.

[27] Chambers MD, Clarke DR. Doped oxides for high-temperature luminescence and lifetime thermometry. Annu Rev Mater Res 2009;39(1):325–59.

[28] Dramićanin MD. Sensing temperature via downshifting emissions of lanthanide-doped metal oxides and salts. A review. Methods Appl Fluoresc 2016;4(4):42001.

[29] Gentleman MM, Eldridge JI, Zhu DM, Murphy KS, Clarke DR. Non-contact sensing of TBC/BC interface temperature in a thermal gradient. Surf Coatings Technol 2006;201(7):3937–41.

[30] Pin L, Pilgrim C, Feist J, Le Maoult Y, Ansart F, Lours P. Characterisation of thermal barrier sensor coatings synthesised by sol–gel route. Sensors Actuators A Phys 2013;199:289–96.

[31] Abou Nada F, Lantz A, Larfeldt J, Markocsan N, Aldén M, Richter M. Remote temperature sensing on and beneath atmospheric plasma sprayed thermal barrier coatings using thermographic phosphors. Surf Coatings Technol 2016;302:359–67.
[32] Yang L, Peng D, Zhao C, Xing C, Guo F, Yao Z, et al. Evaluation of the in-depth temperature sensing performance of Eu- and Dy-doped YSZ in air plasma sprayed thermal barrier coatings. Surf Coatings Technol 2017;316:210–8.
[33] Cates MR, Allison SW, Jaiswal SL, Beshears DL. YAG: Dy and YAG:Tm fluorescence to 1700°C. ISA Int Instrum Symp 49th (Orlando); 2003.
[34] Eldridge JI, Singh J, Wolfe DE. Erosion-indicating thermal barrier coatings using luminescent sublayers. J Am Ceram Soc 2006;89(10):3252–4.
[35] Wolfbeis OS. Sensor paints. Adv Mater 2008;20(19):3759–63.
[36] Matsuda Y, Kameya T, Suzuki Y, Yoshida Y, Egami Y, Yamaguchi H, et al. Fine printing of pressure- and temperature-sensitive paints using commercial inkjet printer. Sensors Actuators B Chem 2017;250:563–8.
[37] Sankir ND, Sankir M, editors. Printable solar cells. Hoboken, NJ, USA: John Wiley & Sons, Inc.; 2017.
[38] Liu T, Campbell BT, Burns SP, Sullivan JP. Temperature- and pressure-sensitive luminescent paints in aerodynamics. Appl Mech Rev 1997;50(4):227–46.
[39] Liu T. Pressure- and temperature-sensitive paints. Encyclopedia of aerospace engineering. John Wiley & Sons, Ltd; 2010.
[40] Ozawa H, Laurence SJ, Schramm JM, Wagner A, Hannemann K. Fast-response temperature-sensitive-paint measurements on a hypersonic transition cone. Exp Fluids 2014;56(1): 1853.
[41] Liu T, Cai Z, Lai J, Rubal J, Sullivan JP. Analytical method for determining heat flux from temperature-sensitive-paint measurements in hypersonic tunnels. J Thermophys Heat Transf 2010;24(1):85–94.
[42] Crafton JW, Stanfield S, Palluconi S, Lui T, Montefort J. Heat flux measurements on the impingement surface of a jet operating at an inclined angle. 55th AIAA Aerospace Sciences Meeting. American Institute of Aeronautics and Astronautics; 2017.
[43] Claucherty S, Sakaue H. An optical-chemical sensor using rhodamine B on anodized-aluminum for surface temperature measurement from 150 to 500K. Sensors Actuators B Chem 2017;240:956–61.
[44] Sakaue H, Aikawa A, Iijima Y. Anodized-aluminum as quantum dot support for global temperature sensing from 100 to 500K. Sensors Actuators B Chem 2010;150(2):569–73.
[45] Jiao L, Peng D, Yu Y, Liu Y, Oshio T, Oouchida S, et al. Single-shot lifetime-based PSP and TSP measurements on rotating blades – challenges at high RPMs. 33rd AIAA Aerodynamic Measurement Technology and Ground Testing Conference, 2017. American Institute of Aeronautics and Astronautics Inc, AIAA; 2017. p. Code 192599 (14 pp.).
[46] Gouterman M, Callis J, Dalton L, Khalil G, Mébarki Y, Cooper KR, et al. Dual luminophor pressure-sensitive paint: III. Application to automotive model testing. Meas Sci Technol 2004;15(10):1986.
[47] Liu T, Guille M, Sullivan JP. Accuracy of pressure-sensitive paint. AIAA J 2001;39(1): 103–12.
[48] Fuhrmann N, Schneider M, Ding C-P, Brübach J, Dreizler A. Two-dimensional surface temperature diagnostics in a full-metal engine using thermographic phosphors. Meas Sci Technol 2013;24(9):95203.
[49] Kashdan JT, Bruneaux G. Laser-induced phosphorescence measurements of combustion chamber surface temperature on a single-cylinder diesel engine. SAE Technical Paper; 2011.

[50] Särner G, Richter M, Aldén M, Vressner A, Johansson B. Cycle resolved wall temperature measurements using laser-induced phosphorescence in an HCCI engine. SAE Technical Paper; 2005.

[51] Knappe C, Andersson P, Algotsson M, Richter M, Linden J, Alden M, et al. Laser-induced phosphorescence and the impact of phosphor coating thickness on crank-angle resolved cylinder wall temperatures. SAE Int J Engines 2011;4(1). 2011-01-1292.

[52] Abou Nada F, Hult J, Knappe C, Richter M, Mayer S, Aldén M. A first application of thermographic phosphors in a marine two-stroke diesel engine for surface temperature measurement. ASME. Internal combustion engine division fall technical Conference, volume 1: Large bore engines; fuels; advanced combustion; emissions control systems; 2014. p. V001T01A001.

[53] Ding C-P, Honza R, Böhm B, Dreizler A. Simultaneous measurement of flame impingement and piston surface temperatures in an optically accessible spark ignition engine. Appl Phys B 2017;123(4):110.

[54] Someya S, Uchida M, Tominaga K, Terunuma H, Li Y, Okamoto K. Lifetime-based phosphor thermometry of an optical engine using a high-speed CMOS camera. Int J Heat Mass Transf 2011;54(17):3927–32.

[55] Kashdan J, Thirouard B. Optical engines as representative tools in the development of new combustion engine concepts. Oil Gas Sci Technol 2011;66(5):759–77.

Chapter 10

Biomedical Applications of Luminescence Thermometry

10.1 INTRODUCTION

Temperature plays an essential role in biological systems, affecting a variety of their properties. For example, cell division rate, and consequently tissue growth, are both critically influenced by temperature [1]. The list of examples is long. It is not strange then that temperature measurements in biomedicine, especially those at intracellular and intercellular levels, are of paramount importance. The use of luminescent nanothermometers opens the way for accurate and even real-time temperature monitoring in many biomedical applications, as has been demonstrated by several research groups recently.

One should note though that choosing a luminescent thermal probe in biomedical applications is considerably more rigorous than in other fields. Materials must be biocompatible, photostable and chemically stable, and bright. Their sizes and properties should allow complete removal from the host organism, but only after sufficient time has passed after their introduction in order to permit temperature measurements to be completed. Also, these probes should be dispersible within the biological media.

The beginning of this chapter deals with the optical properties of tissues, with an emphasis on absorption, scattering, and fluorescence of tissue components in different spectral regions. This knowledge is required when selecting the spectral ranges of excitation and emission of luminescence thermometers aimed at biomedical uses. Several applications are then depicted: intracellular and intercellular temperature measurements, in-tumor temperature monitoring during thermal therapy, and diagnosis of ischemia by thermal imaging.

10.2 OPTICAL PROPERTIES OF TISSUE; BIOLOGICAL WINDOWS

To understand light propagation and interactions in biological tissues one must consider tissue as a complex system consisting of many different components which have diverse optical properties, sizes, and shapes. The fundamental processes that affect light propagation are refraction, reflection, scattering, absorption, and fluorescence (as outlined in Chapter 3), and these processes are

Luminescence Thermometry. http://dx.doi.org/10.1016/B978-0-08-102029-6.00010-5

governed by the optical properties of the constituents of tissues. In biomedicine, for example, knowledge of the optical properties of tissues is required for both diagnostic and therapeutic applications [2]. In diagnostics, both diffuse reflectance and fluorescence spectra of tissues may be regarded as a spectral fingerprint of the tissue chromophores, scatters, and fluorophores. When tissue progresses from the normal physiological condition, concentrations and microenvironments of chromophores, scatters, and fluorophores alter, some new types may emerge, and diffuse reflectance and fluorescence spectra change indicating a change in tissue condition. This phenomenon is used, for example, for cancer diagnostics via fluorescence [3,4]. In therapeutic applications, scattering signals are used to establish the best light dosimetry for therapy, while absorption is the main mechanism that facilitates light-induced physical effects on tissue for treatment purposes, such as heat in photodynamic therapy. In food chemistry, small fluctuations in concentrations of the main food fluorophores are reflected in fluorescence spectra, which then may be used for resolving the botanical or geographical origins of a food product [5,6]. Regarding luminescence thermometry, knowledge of the optical properties of tissues is mainly needed when considering the ability of light to excite a luminescence thermometer inside a biological object, to detect the luminescence of the thermometer, and to discriminate its luminescence for the luminescence of the surrounding tissue.

Light transport in tissues can be considered two ways: by using the optical properties of various tissue components in calculations, or by considering the tissue as a whole with an "averaged" value of the properties of its components [7]. The situation is further complicated by the fact that even tissues of the same type vary in composition, for example, in the contents of blood (human and animal tissue) or chlorophyll (plant tissue), meaning their optical properties are not identical.

The largest constituent portion of most tissues is water, which has a refractive index of 1.33. Thus, human and animal body fluids and soft-tissues have refraction indexes only slightly higher than water. For example, extracellular fluids and intercellular cytoplasm has a refractive index of ~1.35 to 1.38, whole tissues such as parts of the lung, kidney, brain, and bladder have refractive indexes of ~1.36 to 1.40, fatty tissues have a refractive index of 1.45, tooth enamel has 1.62, melanin has an index higher than 1.6 [7], and plant cell walls have a refractive index of about 1.42 [8]. Generally, plant cells have larger refractive indexes than human and animal cells. The nonuniform spatial/temporal variation of the refractive index and the presence of heterogeneities within a tissue (random thermal motion caused fluctuations in dielectric constant and physical inclusions) are responsible for light scattering. The cells themselves (especially at low incident angles), that is, the cell nuclei, mitochondria, lysosomes, the Golgi apparatus, striations in collagen fibrils, macromolecular aggregates, and membranes, are the main causes of scattering [9]. Due to different sizes of tissue components, scattering is composed of both Mie and Rayleigh scattering.

Tissue absorption and fluorescence depend on molecular composition. In human and animal tissues, the main absorbers in visible and near-infrared spectral regions are hemoglobin and lipid (fat). In the ultraviolet–visible region, light is also absorbed by DNA, coenzymes ($NAD(P)^+/NAD(P)H$ and $FAD/FADH_2$), structural proteins (collagen and elastin), amino acids (phenylalanine, tyrosine, and tryptophan), bilirubin, and melanin pigment (in skin), while water contributes to absorption in the near-infrared. In plant tissues, visible light is absorbed by pigments, mainly by chlorophylls a and b, carotenoids (carotenes and xanthophylls), and by some other pigments depending on plant species (such as, for example, anthocyanins, flavons, and chlorophyll c). Water absorption becomes significant in the near-infrared and mid-infrared spectral region.

So, which wavelengths of light are suitable for the excitation of a luminescence thermometer inside biological matter and for the detection of its luminescence? Regarding light's penetration through tissue, the lower energy photons (higher wavelengths), such as those in the near-infrared spectral region, can penetrate tissues deeper than high-energy ones, since tissue scatters and absorbs lower-energy photons less. This can be simply evidenced by the naked eye while watching a hand in sunlight glow red due to highly preferential transmission of red and near-infrared light [10]. For example, the approximate depth of penetration of optical radiation in fair Caucasian skin, for a value of $1/e$ (37%) of the incident energy density, is 6 μm for a 300 nm ultraviolet radiation, 230 μm for 500 nm visible radiation, and 2200 μm for 1200 nm near-infrared radiation, as shown by Anderson et al. [11]. One can see in Fig. 10.1 that the penetration depth of radiation D_{pen} (in μm) through skin strongly increases with

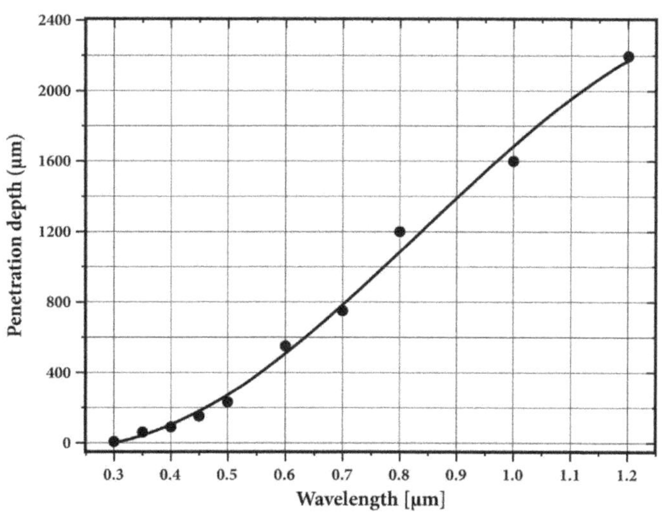

FIGURE 10.1 The penetration depth of radiation through fair Caucasian skin. Symbols represent data extracted from [11] and the *solid line* the symmetric spline regression.

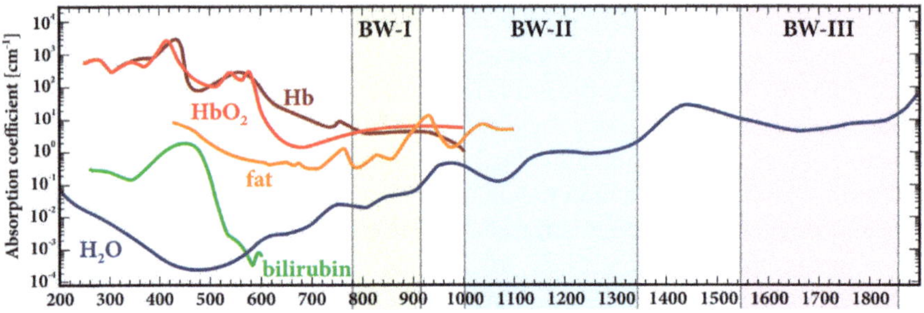

FIGURE 10.2 Wavelength dependence of absorption coefficients of H_2O, HbO_2, Hb, fat, and bilirubin. BW-I, BW-II, and BW-III show spectral ranges of the first, second, and third biological window, respectively.

the wavelength of radiation λ (in μm), and this dependence can be approximated (for the case of fair Caucasian skin) with cubic or symmetric spline functions:

$$D_{pen} = 372.1525 - 3214.68 \times \lambda + 7545.393 \times \lambda^2 - 3012.379 \times \lambda^3,$$
$$D_{pen} = 3193.178 - \frac{58.85767 + 3193.178}{1 + \left(\frac{\lambda}{0.9542909}\right)^{3.36247}} \tag{10.1}$$

Propagation of light in the near-infrared is, however, attenuated by increased absorption of water and lipids. Nonetheless, there are three spectral regions in near-infrared where these absorptions are minimal. These regions are frequently termed biological windows and span the following wavelength ranges: I, 750–950 nm; II, 1000–1350 nm; and III, 1550–1870 nm (see Fig. 10.2). While luminescence bioimaging and thermometry in the first and second biological windows are well-established techniques, the use of the third biological window for these applications is currently at the research stage [12, 13].

10.3 TISSUE FLUORESCENCE

Tissue fluorescence comes from the combined contributions of aromatic amino acids, proteins, vitamins, coenzymes, porphyrins, and chlorophyll. *Tryptophan and phenylalanine* are nonpolar and *tyrosine* is a polar essential amino acid which participates in the synthesis of human proteins. They contain an α amino group, a carboxylic group, and an indole (tryptophan) or phenyl group (tyrosine and phenylalanine), and can be excited by near-ultraviolet light via $\pi \rightarrow \pi^*$ and $n \rightarrow \pi^*$ transitions. Unpaired n-electrons originate from the p orbital of oxygen (C═O group). They show ultraviolet and visible fluorescence (emission maxima around 350 nm for tryptophan, 300 nm for tyrosine, and 280 nm for phenylalanine) due to $\pi^* \rightarrow \pi$ and $\pi^* \rightarrow n$ radiative transitions [14].

Among *vitamins*, notable fluorescence is shown by vitamin A (retinol), vitamin D (calciferol), vitamin K (vitamin K_1-phylloquinone, vitamin K_2-menaquinone), vitamin B_6 (pyridoxine), and vitamin B_2 (riboflavin). Vitamin A is part of a family of unsaturated organic compounds that include retinol, retinal, and retinoic acid [15]. Chemically, they have a cyclic ring and unsaturated chain. Provitamin A (β carotene) is an inactive form of vitamin A, and via the action of *beta-carotene 15,15'-monooxygenase* forms retinol. It is an red–orange pigment abundant in plants, algae, and several bacteria and fungi, in contrast to vitamin A which is not present in plants. Five conjugated π bonds (π electrons) are responsible for the luminescent properties of vitamin A. The active form of vitamin B_6, pyridoxal phosphate, is a coenzyme in many enzymatic reactions. Pyridoxine is the most stable form of all, and it is contained in plant tissues, while pyridoxal and pyridoxamine are found in animal tissues. All forms of vitamin B_6 are fluorescent, with an emission maximum near 420 nm [16]. Riboflavin (vitamin B_2) is formed of tricyclic heterocycle isoalloxazine and ribose. It is a yellow or orange–yellow pigment used for the coloring and the improvement of flavor of food. Riboflavin is a precursor of the important coenzymes flavin mononucleotide (FMN) and flavin adenine dinucleotide (FAD). Absorption (around 450 nm) and emission (around 550 nm) of riboflavin are in the visible spectral range.

FAD and FMN coenzymes belong to the huge family of flavins and are the chromophores of biological blue-light photoreceptors. Like riboflavin, FAD and FMN have an isoalloxazine ring and ribityl side chain [17]. The absorption bands of FAD are centered at 390 and 450 nm; the later band comes from the $\pi \rightarrow \pi^*$ transition of isoalloxazine. The broadband emission of FAD spans the 500–600 nm spectral range and is due to the $\pi^* \rightarrow \pi$ radiative transition of isoalloxazine. While FAD molecules (redox form) are fluorescent, $FADH_2$ (oxidize form) does not fluoresce. Generally, fluorescence properties of this coenzyme and flavins are caused by conjugation of π bonds and the free electron pair of the N atom in the heterocycle ring. Emissions from FAD and flavins are usually quenched upon binding to proteins. *Nicotinamide adenine dinucleotide (phosphate) (NAD(P)H)* is a dinucleotide in which one nucleotide contains an adenine base and the other nicotinamide, connected via two phosphate groups. Nicotinamide adenine dinucleotide exists in two forms, an oxidized ($NAD(P)^+$) and reduced NAD(P)H) form, which show different fluorescence. Both forms absorb in the ultraviolet spectral region. In solution, NADH absorbs ultraviolet light with a maximum at 340 nm and has a broadband blue emission centered at 460 m, while NAD^+ does not fluoresce.

Porphyrins are a group of heterocyclic tetrapyrrole derivatives having a basic structure made of a porphyrin ring. All of them are fluorescent. The most interesting member of the family is heme, which has a Fe^{2+} ion located in the center of a porphyrin. Heme is a prosthetic group of some proteins well known as hemoproteins. The best-known hemoprotein is hemoglobin, the red pigment in blood, which transports O_2 via heme to every cell in the body. The absorption spectra of porphyrins exhibit a band centered at 400 nm due to an intense $\pi \rightarrow \pi^*$ transition to the second excited state (the so-called Soret band or B band) and four bands in the 450–650 nm region due to weaker $\pi \rightarrow \pi^*$ transitions to the first excited state (the so-called Q bands). The fluorescence emission spectrum of porphyrin consists of two strong emission bands (between 600 and 800 nm) [18]. *Chlorophyll* is a green pigment present in all green plants, algae, and cyanobacteria. Chlorophylls a and b are responsible for the absorption of light to provide energy for photosynthesis. Chemically, chlorophyll is a heterocyclic aromatic ring termed chlorin with an Mg^{2+} ion in the ring's center. As a porphyrin derivative, chlorophyll has a similar structure to it. There are six different naturally occurring forms of chlorophyll (a, b, c1, c2, d, and f); the most widely distributed form is chlorophyll a. Chlorophyll strongly absorbs energy in the blue and red regions of the visible spectrum and emits close to 676 and 681 nm.

Collagen is the main structural protein in the extracellular space in various connective tissues and is the most abundant protein in mammals [19]. The collagen protein is composed of a triple helix, which consists of two identical chains and an additional chain that differs slightly in its chemical composition. Its excitation band has a maximum around 325 nm and an emission band around 400 nm. *Elastin* is an elastic protein in connective tissue and allows many tissues in the body to resume their shape after stretching or contracting. The excitation band of elastin is centered at 320 nm and its emission band at 400 nm.

Tissue fluorescence (sometimes called autofluorescence) can significantly interfere with the luminescence signal from luminescence thermometers incorporated into biological matter. Since most of the tissue fluorophores are excited by high-energy ultraviolet or visible radiation, the luminescence of thermometers that emit in near-infrared is much easier to discriminate from tissue fluorescence. However, recent research [20] has revealed tissue autofluorescence in near-infrared whose origin and spectral properties have yet to be determined.

10.4 INTRACELLULAR AND INTERCELLULAR TEMPERATURE MEASUREMENTS VIA LUMINESCENCE

Intracellular and intercellular temperature measurements in biomedicine may be among the most exciting applications of luminescence thermometry. A typical example is the incubation of water-dispersible $NaYF_4:Er^{3+},Yb^{3+}$ nanoparticles into HeLa cancer cells for the monitoring of cell temperature during apoptosis [21] (see Fig. 10.3). Under near-infrared 980 nm excitation, $NaYF_4:Er^{3+},Yb^{3+}$ nanoparticles show green upconversion emission that is typical for Er^{3+} ions. As discussed in detail in Chapter 6, the ratio of intensities of the two emission bands centered at 525 and 545 nm provides a measure of temperature of the environment around the nanoparticles, with good sensitivity. The authors used a metallic platform connected to a resistor to change the cell temperature by varying the voltage across the resistor while simultaneously observing the temperature via luminescence (see bottom of Fig. 10.3). Transmission optical images

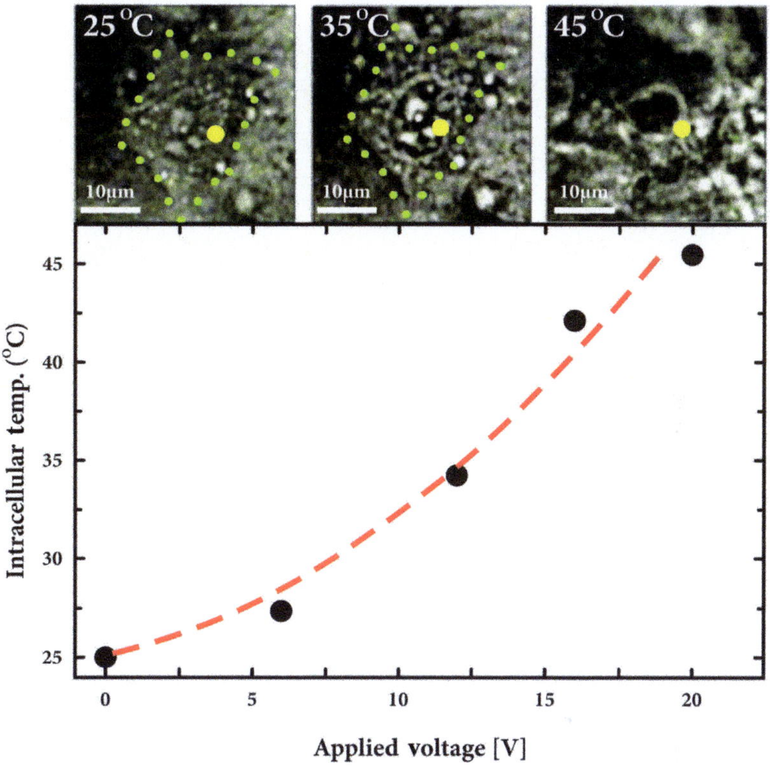

FIGURE 10.3 (Top) Optical transmission images of an individual HeLa cell at three inner temperatures. Cell death is observed at 45°C. (Bottom) Temperature of the HeLa cell determined from the Er^{3+} ion luminescence in $NaYF_4$:Er^{3+},Yb^{3+} nanoparticles as a function of applied voltage. *(Reprinted from Vetrone F, Naccache R, Zamarrón A, De La Fuente AJ, Sanz-Rodríguez F, Maestro LM, et al. Temperature sensing using fluorescent nanothermometers. ACS Nano 2010;4(6):3254–8. [21], copyright (2010) American Chemical Society).*

were recorded at different temperatures (see top of Fig. 10.3). Images that were taken at 25°C and 35°C have revealed typical morphology of the HeLa cancer cells. An image measured at 45°C showed a small membrane fragment which points to cell death caused by external heating.

Luminescence thermometry also facilitates intracellular thermal mapping as shown by Okabe et al. [22] by using fluorescent polymeric nanothermometers (with an average hydrodynamic diameter of 8.9 nm at 40°C) incorporated into COS7 living cells. This research disclosed that the temperature distribution inside living cells, obtained from nanothermometer fluorescence lifetimes, is not uniform. Temperatures of nucleus and centrosome of a cell were higher than that of the cytoplasm—the temperature difference changed depending on cell cycle. The centrosome was warmer than the cytoplasm (see Fig. 10.4) about 0.75°C on average, with significant cell-to-cell variation (see Fig. 10.4B).

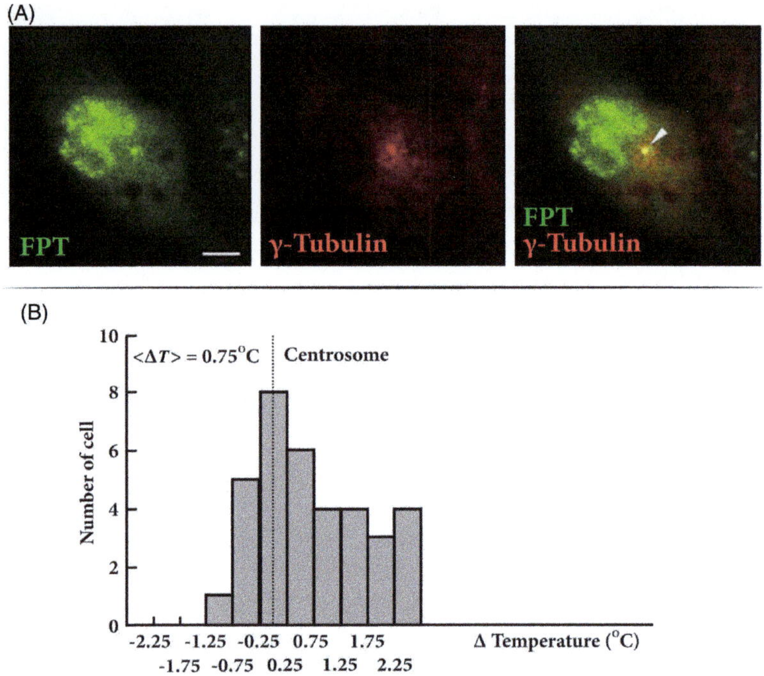

FIGURE 10.4 **Heat production from the centrosome in living COS7 cells.** (A) Heat production near the centrosome (*arrowhead*). Epifluorescence images of fluorescence polymeric thermometer–FPT (*left*), the centrosome visualized with γ-tubulin antibody (*middle*), and their merged image (*right*). (B) Histogram of temperature differences between the centrosome and the cytoplasm (n = 35). ΔT (temperature difference) was calculated by subtracting the average temperature of the cytoplasm from that of the centrosome with a potential maximum error of 0.35°C. The temperature of the medium was maintained at 30°C. The scale bar represents 10 μm. *(Reprinted from Okabe K, Inada N, Gota C, Harada Y, Funatsu T, Uchiyama S. Intracellular temperature mapping with a fluorescent polymeric thermometer and fluorescence lifetime imaging microscopy. Nat Commun 2012;3:705 (9 pp.) [22], copyright (2012), with permission from Macmillan Publishers Ltd).*

In addition, authors were able to observe heat production from mitochondria via luminescence.

However, according to the discussion from earlier sections, luminescence thermometry with excitations and emission detection in spectral regions of biological windows is preferable over those conducted in the visible spectral range. Today, there are a number of available luminescent nanomaterials that operate in biological windows with emissions giving sufficient temperature sensitivity. Such nanothermometers can be made from a variety of Nd^{3+} doped nanomaterials, as recently discussed by del Rosal et al. [23]. Other types of nanothermometers include near-infrared–emitting quantum dots (PbS/CdS/ZnS, Ag_2S/Ag), Er^{3+}, Ho^{3+}, and Tm^{3+} doped fluorides, and poly(lactic-co-glycolic acid) (PLGA) (see Fig. 10.5). For example, Ag/Ag_2S near-infrared nanothermometers [24] provide

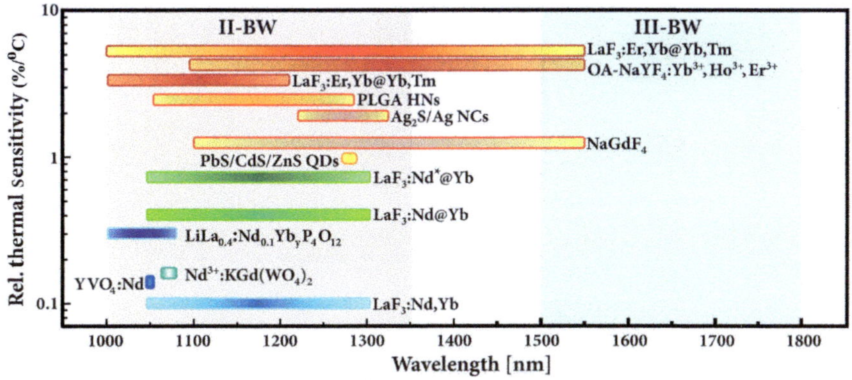

FIGURE 10.5 Thermal sensitivity of different luminescent nanothermometers and their spectral operating ranges. *(Reprinted from Ximendes EC, Rocha U, Sales TO, Fernández N, Sanz-Rodríguez F, Martín IR, et al. In Vivo Subcutaneous Thermal Video Recording by Supersensitive Infrared Nanothermometers. Adv Funct Mater. 2017;27(38):1702249 (10 pp.) [26], with permission from Wiley).*

temperature readings from both the temperature-induced shift of emission band (relative sensitivity of almost 2%/K) and emission intensity quenching (relative sensitivity of 4%/K) over the 15–50°C temperature range. More information on the materials available for *in vivo* luminescence nanothermometry can be found in [25].

Today, it is extremely difficult to perform noncontact, accurate subcutaneous *in vivo* two-dimensional temperature measurements. Imaging with thermographic cameras is limited to tissue surface, and luminescence thermometry with nanothermometers that emit in a visible spectral range is not feasible since visible radiation penetrates tissues to only extremely small depths (see Fig. 10.1). The possible solution to the problem may be in the use of luminescence thermometry with near-infrared–emitting nanoparticles that possess emission, which is strongly temperature dependent, in the physiological temperature range (20–50°C). Ximendes et al., [26] recently demonstrated *in vivo* subcutaneous thermal video recording using Er^{3+}, Tm^{3+}, and Yb^{3+} core–shell LaF_3nanoparticles. These nanoparticles under 690 nm excitation emit in the near-infrared spectral range with emission bands centered at 1000, 1230, and 1550 nm due to electronic f–f transition in Yb^{3+}, Tm^{3+}, and Er^{3+} ions, respectively. The authors found that in Er-Yb@Yb-Tm LaF_3 core–shell nanoparticles, the ratio between emission intensity at 1000 nm (Yb^{3+} emission) and 1230 nm (Tm^{3+} emission) showed a remarkable temperature sensitivity of 5%/K (at room temperature). These particles were then tested for the *in vivo* acquisition of two-dimensional thermal videos at the subcutaneous level. The procedure consisted of injecting of a phosphate-buffered saline dispersion of nanoparticles into a mouse, with confirmation of the subcutaneous presence of nanoparticles completed by simple luminescence imaging. A laser was used to induce a moderate

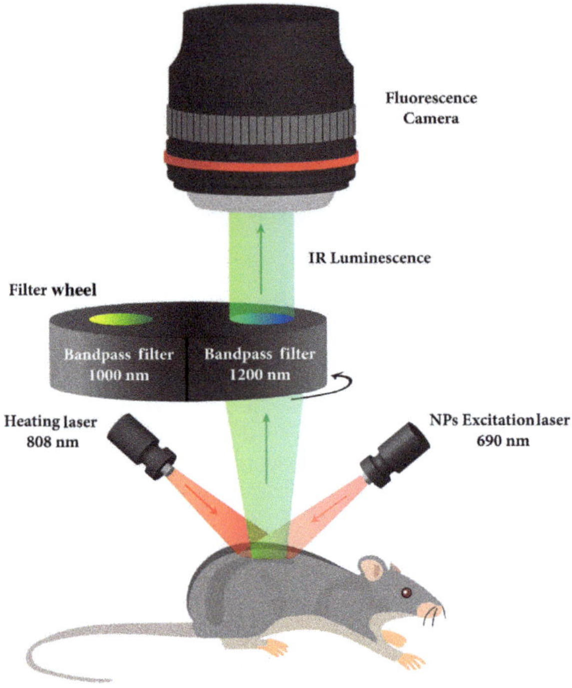

FIGURE 10.6 **The experimental set-up for the acquisition of two-dimensional thermal videos at a subcutaneous level based on the ratio of intensities of emissions at 1000 and 1230 nm of Er-Yb@Yb-Tm LaF_3 core–shell nanoparticles injected into a mouse.** *(Reprinted from Ximendes EC, Rocha U, Sales TO, Fernández N, Sanz-Rodríguez F, Martín IR, et al. In Vivo Subcutaneous Thermal Video Recording by Supersensitive Infrared Nanothermometers. Adv Funct Mater. 2017;27(38):1702249 (10 pp.) [26], with permission from Wiley).*

increase of temperature at the injection site. The heating and cooling dynamics of the tissue was performed with an experimental set-up (see Fig. 10.6) which consisted of 690 nm and 808 nm lasers, an InGaAs camera, and two bandpass filters (1000 and 1200 nm) that were placed in a motorized wheel with a 0.3 Hz switching frequency.

Two different methodologies were used for thermometry during heating and cooling cycles. For the heating cycle, the mouse was illuminated with 690 nm laser radiation to warm the tissue and excite luminescence of the nanoparticles. For the cooling cycle, the tissue was warmed using the 808 nm laser, with the 690 nm laser (of low power density) being used to excite luminescence of the nanoparticles. Thermal images derived from the intensity ratio of 1000 and 1200 nm emissions during the heating and cooling processes are shown in Fig. 10.7A and B, respectively (an optical image of the mouse is superimposed on the thermal images). The average temperature of the injection area as a function of time during heating and thermal relaxation processes is shown in Fig. 10.7C.

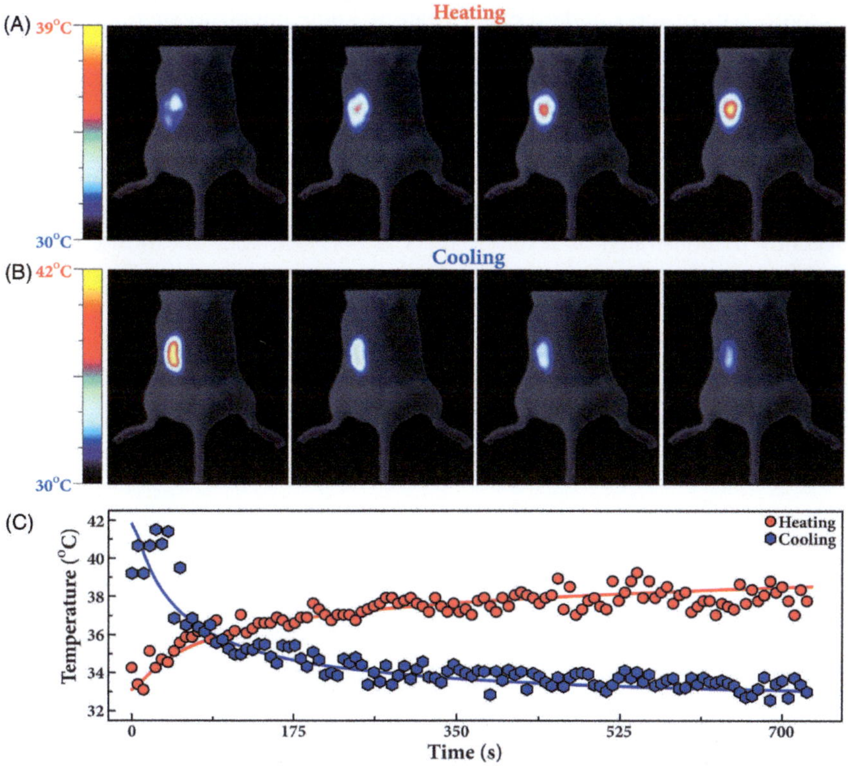

FIGURE 10.7 Thermal images obtained by dividing the luminescence images at 1000 and 1200 nm during (A) heating and (B) the relaxation processes. An optical figure of the anesthetized mouse was superimposed. (C) Time evolution of the average temperature of the injection area as measured by subcutaneous luminescence thermometry during the heating and thermal relaxation processes. *(Reprinted from Ximendes EC, Rocha U, Sales TO, Fernández N, Sanz-Rodríguez F, Martín IR, et al. In Vivo Subcutaneous Thermal Video Recording by Supersensitive Infrared Nanothermometers. Adv Funct Mater. 2017;27(38):1702249 (10 pp.) [26], with permission from Wiley).*

10.5 USING LUMINESCENCE FOR MEASUREMENTS OF IN-TUMOR TEMPERATURE DURING THERMAL THERAPY

The manipulation of tissue or body temperature is a basis of thermal therapy which, today, has a growing number of clinical applications in the treatment of a variety of diseases, including cancer. Among them, the most widely used are hyperthermia and thermal ablation. For cancer treatments, hyperthermia can be used as an adjuvant to radiation and chemotherapy or to activate the cytotoxic effects of chemotherapy within tumors. It is well known that within tumors temperatures in the 42–45 C range are cytotoxic, and many clinical trials conducted over the past 20 years have shown that these temperatures can sensitize tumors to radiation and/or chemotherapy. In fact, hyperthermia is today part of the standard cancer care in some countries such as Germany and the Netherlands [27].

The critical factor in hyperthermia applications is the accurate real-time monitoring of tissue temperature during therapy. Conventional methods, such as imaging by thermal cameras are able to measure skin temperatures which may be different to the actual tumor's temperature. The use of luminescent nanothermometers may facilitate accurate measurements of in-tumor temperatures as recently demonstrated in several papers [28–31]. In addition to thermometry, luminescence nanoparticles can enable luminescence imaging of a tumor and act as heating sources. Examples of luminescence temperature monitoring during hyperthermia treatment are discussed in detail in Chapter 12.

10.6 USING LUMINESCENCE THERMOMETRY TO DIAGNOSE ISCHEMIA

Ischemia is a serious, life-threatening medical condition characterized by a shortage of oxygen, glucose, and nutrients in cellular metabolism caused by a reduced blood supply to tissues. It is also characterized by inadequate removal of metabolic wastes. An insufficient blood supply to tissues and an accumulated metabolic waste can cause tissue damage which may progress to an irreversible state in just a few minutes in highly aerobic tissues. Considering the significant change in cellular metabolism, it is reasonable to expect that ischemic tissue shows altered physical behavior compared to healthy tissue, such as a tissues ability to transfer bioheat. The well-known Penne's equation [32] gives a simple description of bioheat transfer:

$$\begin{aligned} \rho_t c_t \frac{\partial T_t}{\partial_t} - \nabla(k_t \nabla T_t)\dot{q}_t + \dot{q}_p, \\ \dot{q}_p = \rho_b c_b \omega_b (T_b - T_t), \end{aligned} \tag{10.2}$$

where $\dot{q}_t$ is the metabolic heat generation rate per unit volume, $\dot{q}_p$ is the heat generation rate per unit volume due to blood perfusion, T_t is the tissue temperature, T_b is the blood temperature, k_t is the thermal conductivity of tissue, ρ_t is the density of tissue, ρ_b is density of blood, c_t is specific heat capacity of tissue, c_b is specific heat capacity of blood, and ω_b is the blood perfusion rate. This equation clearly displays that any change in tissue thermal properties and/or the blood perfusion rate affects heat transfer through the tissue. When tissue temperature is raised above the basal temperature by an external heat source, the temperature relaxes back to the basal value after termination of heating. Variations of temperature relaxation profiles between tissues may be exploited for the diagnosis of different medical conditions including ischemia by a method known as transient thermometry (TTh) [33].

TTh is traditionally conducted with infrared thermal cameras, so it is limited only to analyses of superficial tissues. However, Ximendes et al. [34] recently

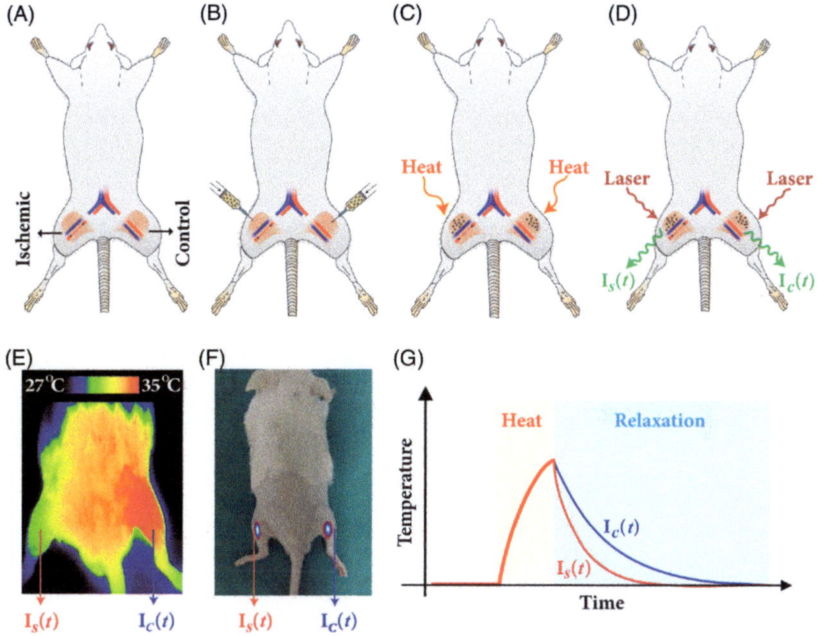

FIGURE 10.8 (A) Artery ligation. (B) Thermal image of an individual 24 hours after artery ligation in which the lower temperature in the ischemic leg is evidenced. (C) Injection of near-infrared quantum dots. (D) Near-infrared luminescence image of a characteristic individual as obtained just after the injection of the near-infrared quantum dots. (E) Soft heating. (F) Measurement of the thermal transient. (G) Schematic representation of the working principle of transient thermometry (C stands for the control and S for the ischemic condition). *(Reprinted from Ximendes EC, Rocha U, del Rosal B, Vaquero A, Sanz-Rodríguez F, Monge L, et al. In Vivo Ischemia Detection by Luminescent Nanothermometers. Adv Healthc Mater 2017;6(4):1601195 (8 pp.) [34], with permission from Wiley).*

demonstrated that it is possible to access the temperature in deep tissues using luminescence thermometry (see Fig. 10.8).

In their research, a murine model of hindlimb ischemia was surgically induced in mice by a procedure which consisted of a double ligation of the femoral artery in one limb (see Fig. 10.8A). The surgery resulted in a low blood perfusion rate in the entire hindlimb tissue of the mouse, as evidenced by a thermal image (see Fig. 10.8B). Then, 24 hours after surgery, 50 μL of a 0.01 mg/mL solution of near-infrared–emitting PbS/CdS/ZnS quantum dots in a phosphate buffered saline, were subcutaneously injected into the surae triceps muscles of both limbs of mouse (see Fig. 10.8C). The location of quantum dots in the limbs was confirmed by a deep-tissue 1200 nm luminescence image obtained under 808 nm excitation (see Fig. 10.8D). TTh was performed with two lasers operating at 808 nm. The first laser produced tissue heating (by 8–10°C), and the second one, of much lower photon density, provided an excitation of the quantum dot thermometers (see Fig. 10.8E and F). Temperature relaxation after

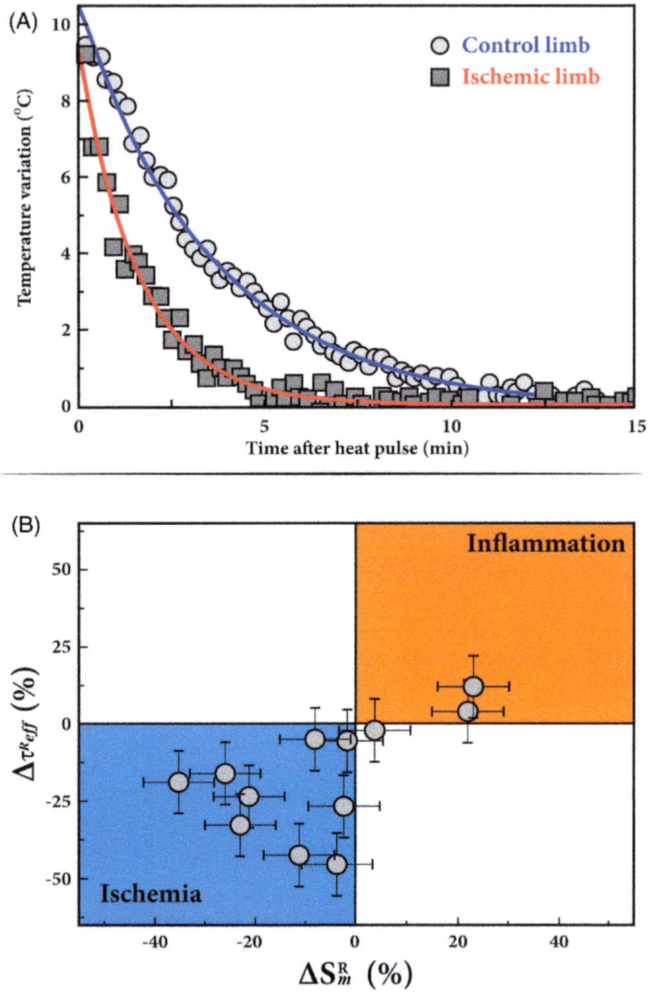

FIGURE 10.9 (A) Representative thermal transients obtained from an ischemic and healthy leg. The faster thermal dynamics of the ischemic leg is evidenced. (B) Statistical results of $\Delta\tau^{R}_{eff}$ versus ΔS^{R}_{m}, taken from 12 mice. *(Reprinted from Ximendes EC, Rocha U, del Rosal B, Vaquero A, Sanz-Rodríguez F, Monge L, et al. In Vivo Ischemia Detection by Luminescent Nanothermometers. Adv Healthc Mater 2017;6(4):1601195 (8 pp.) [34], with permission from Wiley).*

heating was monitored by luminescence thermometry in both the ischemic and the control limb—the obtained temperature relaxation profiles were unmistakably different (see Fig. 10.8G). After measurements and approximately 72 hour after surgery, the mice were euthanized so that their tissue could be analyzed to evaluate the relative modification of the distal area of their surae triceps muscles (ΔS^{R}_{m}). This parameter revealed whether the tissue was in ischemia (ΔS^{R}_{m}) or inflammatory phase ($\Delta S^{R}_{m} < 0$) (detailed explanation is given in [34]). Thermal relaxation times of healthy and ischemic mouse legs were evaluated from relaxation profiles (see Fig. 10.9A), and were $\Delta S_m^R > 0$, on average, respectively. Evidently, the characteristic thermal relaxation time of tissue was shorter after artery ligation. Ximendes et al. [34] systematically investigated the relative shortening of thermal relaxation time, $\tau_{rel}^C = 196 \pm 5\,\text{s}$ and $\tau_{rel}^i = 113 \pm 5\,\text{s}$ for 12 mice, and correlated values with ΔS_m^R (see Fig. 10.9B). They found that the $\Delta\tau_{eff}^R$ value can be used as an indicator of both ischemic and inflammatory conditions of the tissue.

REFERENCES

[1] Jaque D, Vetrone F. Luminescence nanothermometry. Nanoscale 2012;4(15):4301–26.

[2] Jacques SL. Optical properties of biological tissues: a review. Phys Med Biol 2013;58(11):R37–61.

[3] Dramićanin T, Dramićanin M. Using fluorescence spectroscopy to diagnose breast cancer. In: Stauffer MT, editor. Applications of molecular spectroscopy to current research in the chemical and biological sciences. Rijeka: InTech; 2016. p. 261–80.

[4] Dramićanin T, Dramićanin MD, Jokanović V, Nikolić-Vukosavljević D, Dimitrijević B. Three-dimensional total synchronous luminescence spectroscopy criteria for discrimination between normal and malignant breast tissues. Photochem Photobiol 2005;81(6):1554–8.

[5] Lenhardt L, Bro R, Zeković I, Dramićanin T, Dramićanin MD. Fluorescence spectroscopy coupled with PARAFAC and PLS DA for characterization and classification of honey. Food Chem 2015;175:284–91.

[6] Lenhardt L, Zeković I, Dramićanin T, Milićević B, Burojević J, Dramićanin MD. Characterization of cereal flours by fluorescence spectroscopy coupled with PARAFAC. Food Chem 2017;229(Suppl. C):165–71.

[7] Mobley J, Vo-Dinh T. Optical properties of tissue. In: Vo-Dinh T, editor. Biomedical photonics handbook. CRC Press; 2003. p. 2-1–2-75.

[8] Gausman HW, Allen WA, Escobar DE. Index of plant cell walls. Appl Opt 1974;13(1):109–11.

[9] Mourant JR, Canpolat M, Brocker C, Esponda-Ramos O, Johnson TM, Matanock A, et al. Light scattering from cells: the contribution of the nucleus and the effects of proliferative status. J Biomed Opt 2000;5(2):5–7.

[10] Smith AM, Mancini MC, Nie S. Second window for in vivo imaging. Nat Nanotechnol 2009;4:710.

[11] Anderson RR, Parrish JA. The optics of human skin. J Invest Dermatol 1981;77(1):13–9.

[12] Hemmer E, Benayas A, Legare F, Vetrone F. Exploiting the biological windows: current perspectives on fluorescent bioprobes emitting above 1000 nm. Nanoscale Horizons 2016;1(3):168–84.

[13] Skripka A, Benayas A, Marin R, Canton P, Hemmer E, Vetrone F. Double rare-earth nanothermometer in aqueous media: opening the third optical transparency window to temperature sensing. Nanoscale 2017;9(9):3079–85.

[14] Myhovich MI, Kelman VA. Theoretical and experimental study of spectroscopic characteristics of aromatic amino acids. Ukr J Phys 2014;59(6):581–8.

[15] Damodaran S, Parkin KL. . In: Damodaran S, Parkin KL, editors. Fennema's food chemistry. fifth ed. CRC Press/Taylor and Francis; 2017. 1107pp.

[16] Bridges JW, Davies DS, Williams RT. Fluorescence studies on some hydroxypyridines including compounds of the vitamin B6 group. Biochem J 1966;98(2):451–68.

[17] Islam SDM, Susdorf T, Penzkofer A, Hegemann P. Fluorescence quenching of flavin adenine dinucleotide in aqueous solution by pH dependent isomerisation and photo-induced electron transfer. Chem Phys 2003;295(2):137–49.
[18] Uttamlal M, Sheila Holmes-Smith A. The excitation wavelength dependent fluorescence of porphyrins. Chem Phys Lett 2008;454(4):223–8.
[19] Di Lullo GA, Sweeney SM, Korkko J, Ala-Kokko L, San Antonio JD. Mapping the ligand-binding sites and disease-associated mutations on the most abundant protein in the human, type I collagen. J Biol Chem 2002;277(6):4223–31.
[20] del Rosal B, Villa I, Jaque D, Sanz-Rodríguez F. In vivo autofluorescence in the biological windows: the role of pigmentation. J Biophotonics 2016;9(10):1059–67.
[21] Vetrone F, Naccache R, Zamarrón A, De La Fuente AJ, Sanz-Rodríguez F, Maestro LM, et al. Temperature sensing using fluorescent nanothermometers. ACS Nano 2010;4(6):3254–8.
[22] Okabe K, Inada N, Gota C, Harada Y, Funatsu T, Uchiyama S. Intracellular temperature mapping with a fluorescent polymeric thermometer and fluorescence lifetime imaging microscopy. Nat Commun 2012;3:705. 9pp.
[23] Del Rosal B, Perez-Delgado A, Misiak M, Bednarkiewicz A, Vanetsev AS, Orlovskii Y, et al. Neodymium-doped nanoparticles for infrared fluorescence bioimaging: the role of the host. J Appl Phys 2015;14(118):143104. 11pp.
[24] Ruiz D, del Rosal B, Acebrón M, Palencia C, Sun C, Cabanillas-González J, et al. Ag/Ag_2S nanocrystals for high sensitivity near-infrared luminescence nanothermometry. Adv Funct Mater 2017;27(6):1604629. 9pp.
[25] del Rosal B, Ximendes E, Rocha U, Jaque D. In vivo luminescence nanothermometry: from materials to applications. Adv Opt Mater 2017;5(1):1600508. 14pp.
[26] Ximendes EC, Rocha U, Sales TO, Fernández N, Sanz-Rodríguez F, Martín IR, et al. In vivo subcutaneous thermal video recording by supersensitive infrared nanothermometers. Adv Funct Mater 2017;27(38):1702249. 10pp.
[27] What Thermal Medicine? Web page. Society for Thermal Medicine. Available from: http://www.thermaltherapy.org/ebusSFTM/SOCIETYINFO/WhatisThermalMedicine.aspx.
[28] Carrasco E, del Rosal B, Sanz-Rodríguez F, de la Fuente ÁJ, Gonzalez PH, Rocha U, et al. Thermal reading during photo-thermal therapy by multifunctional fluorescent nanoparticles. Adv Funct Mater 2015;25(4):615–26.
[29] del Rosal B, Carrasco E, Ren F, Benayas A, Vetrone F, Sanz-Rodríguez F, et al. Infrared-emitting QDs for thermal therapy with real-time subcutaneous temperature feedback. Adv Funct Mater 2016;26(33):6060–8.
[30] del Rosal B, Pérez-Delgado A, Carrasco E, Jovanović DJ, Dramićanin MDG, et al. Neodymium-based stoichiometric ultrasmall nanoparticles for multifunctional deep-tissue photothermal therapy. Adv Opt Mater 2016;4(5):782–9.
[31] Zhu X, Feng W, Chang J, Tan Y-W, Li J, Chen M, et al. Temperature-feedback upconversion nanocomposite for accurate photothermal therapy at facile temperature. Nat. Commun 2016;7:10437. 10pp.
[32] Pennes HH. Analysis of tissue and arterial blood temperatures in the resting human forearm. 1948. J Appl Physiol 1998;85(1):5–34.
[33] Newman WH, Lele PP. A transient heating technique for the measurement of thermal properties of perfused biological tissue. J Biomech Eng 1985;107(3):219–27.
[34] Ximendes EC, Rocha U, del Rosal B, Vaquero A, Sanz-Rodríguez F, Monge L, et al. In vivo ischemia detection by luminescent nanothermometers. Adv Healthc Mater 2017;6(4):1601195. 8pp.

Chapter 11

Temperature Measurements at the Nanoscale

11.1 INTRODUCTION

When making an introduction to the field of nanoscience and nanotechnology two things are usually mentioned early on. The first is the length scale at which processes in nanoscience are observed. It is generally assumed nowadays that nanoscience and nanotechnology are dealing with objects that have at least one dimension smaller than 100 nm. The second thing is historical remarks on the use of gold nanoparticles in the Middle Ages as red pigments in glasses, and on the famous speech of Richard Feynman "There's plenty of room at the bottom," given at the annual meeting of the American Physical Society at the California Institute of Technology (Caltech) on December 29, 1959. This speech has opened the first window into a nanoworld, and has foreseen many outstanding benefits to be achieved with nanoscience.

Since then, scientists have put effort into trying to understand the fundamentals of the physical processes of the nanoscale—driven by the urge to find benefits to the world by exploiting the small. Today, nanoscience represents a leading research field in physics, chemistry, and biology. The possible applications of nanotechnology are countless, and range from nanoelectronics and energy to health. Continuous research in nanoscience has led to developments of the smallest structures ever made—just a few nanometers in size. These structures are the smallest structures humans will ever make since we now approach dimensions similar to just a few atoms.

At such small scales, the very definition of temperature (Chapter 2) comes into question since assumptions of the continuum and thermodynamic equilibrium may not hold anymore [1]. It is of fundamental importance, thus, to determine what is the minimal dimension an object can have for which a local temperature exists [2]. Besides interest from the fundamental point of thermodynamics, and physics in general, temperature measurements with nanoscale spatial resolution are of considerable interest for many existing and emerging nanotechnologies in which the performance of structures are strongly determined by temperature. These are, for example, nanoelectronics and integrated photonics [3]. We already pointed out in Chapter 9, that miniaturization of electronic circuits and devices, as well as escalated switching speeds, have provoked

Luminescence Thermometry. http://dx.doi.org/10.1016/B978-0-08-102029-6.00011-7

localized heating problems and frequent appearance of "hot-spots." Locations of hot-spots are difficult or even impossible to predict, and, therefore, it is of utmost importance to detect them in a convenient but reliable way. The refractive index is a physical property, the value of which is considerably influenced by temperature change. In integrated photonics, the submicrometric variations of the refractive index are utilized to form photonic devices [3].

Today, we still have no established reference methods for nanoscale temperature measurements, although pioneering work in this field has been made [1]. Some nanothermometry methods will be outlined in this chapter and discussed in terms of their performance. The rest of the chapter will present several applications of luminescence nanothermomometry in the fields of nanofluidics and nanoelectronics.

11.2 METHODS OF NANOSCALE TEMPERATURE MEASUREMENT

Devices used for temperature measurements at the nanoscale are either adapted from conventional instruments by miniaturization or are completely new designs which utilize some nanoscale phenomena or nanosized objects. For example, a carbon nanothermometer containing gallium [4] can be regarded as a nanoscale counterpart of the liquid-in-glass thermometer (Fig. 11.1). When a liquid gallium column is introduced inside a carbon nanotube its height changes with temperature (Fig. 11.1A–C). The height linearly changes over the 50–500°C range (Fig. 11.1D) at a rate of $0.095 \pm 0.006 \times 10^{-3}$/°C. The rate has a physical meaning in terms of the volumetric-expansion coefficient and its value is similar to that of liquid gallium at a microscopic level (0.1015×10^{-3}/°C) [4]. Several other material compositions have been used to successfully fabricate nanoscale counterparts of liquid-in-glass thermometer, such as Ga in MgO nanotubes [5], Pb in ZnO nanotubes [6] and in single-crystalline In_2O_3 nanotubes [7], Au(Si) in β-Ga_2O_3 nanotubes [8], and Bi in Bi_2O_3 nanotubes [9]. So far, among all the nanothermometers based on thermal expansion, the atomic layer deposition-capped Bi-Al_2O_3 nanocable has shown the widest operational range (300–1000°C) and the highest operating temperature of 1000°C [9].

Electrical thermometry devices can also be downscaled for high-resolution temperature measurements. This is achieved mainly with nanosized semiconductor components by observing temperature-induced changes in their operating characteristics or with patterned metal bridges and junctions by monitoring changes in their electrical resistance and Seebeck voltage [10]. For example, it has been demonstrated that a resonant tunneling diode (RTD) in sequence with a GaInNAs/GaAs quantum well shows a large temperature-dependent screening effect that can be exploited for sensitive temperature measurements over a wide range of temperatures (from 4.5 to 295 K) [11]. The current voltage characteristics of this nanothermometer shows a large linear shift of threshold voltages

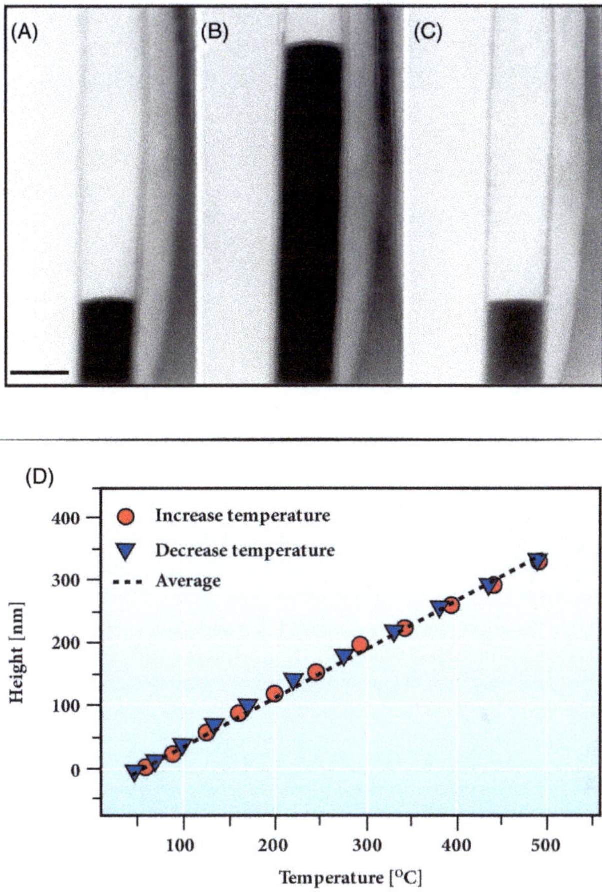

FIGURE 11.1 Expansion of gallium inside a carbon nanotube with increasing temperature. Changing level of the gallium meniscus at (A) 58°C, (B) 490°C, and (C) 45°C; scale bar, 75 nm. (D) Height of the gallium meniscus plotted against temperature during heating [light grey circles (*red circles in the web version*)] and cooling [dark grey triangles (*blue triangles in the web version*)]. *(Reprinted from Gao Y, Bando Y. Nanotechnology: carbon nanothermometer containing gallium. Nature 2002;415(6872):599, copyright (2002), with permission from Macmillan Publishers).*

with temperature of about 19.2 mV K^{-1}, which enables electrical read-out of temperature with a 5 mK uncertainty in the 40–295 K range, somewhat higher below 40 K. It is interesting to note that this nanothermometer also affords optical read-out of temperature from electroluminescence (Fig. 11.2A and B) which occurs over the GaAs band gap due to electron–hole recombination. Both the position of the spectral maximum and integrated intensity of the emission change with temperature (Fig. 11.2C), and both can be used for the determination of temperature with relative sensitivities of 0.035% and 4.35% K^{-1}, respectively [11].

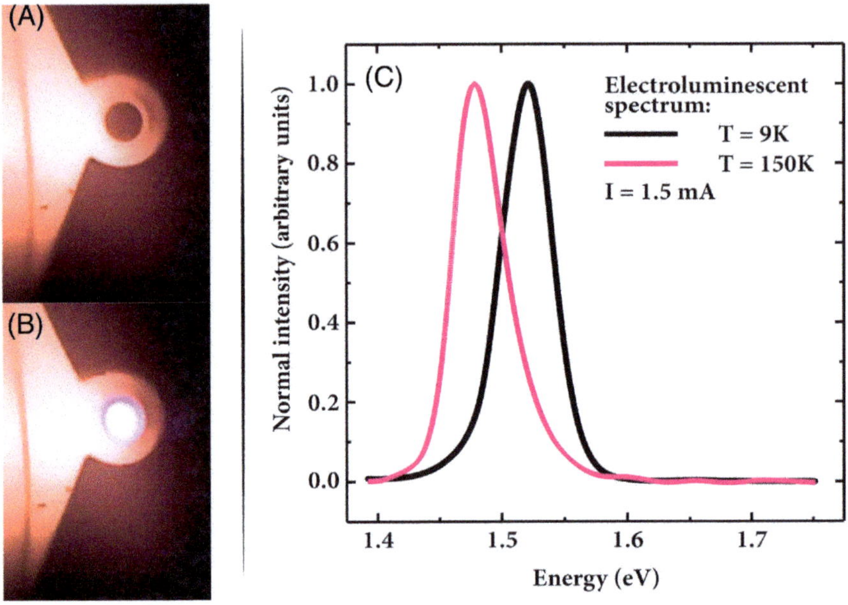

FIGURE 11.2 Electroluminescent read-out of temperature from the nanothermometer fabricated from a resonant tunneling diode (RTD) in sequence with a GaInNAs/GaAs quantum well. (A) Top-view photograph of the nanothermometer for zero bias voltage. The gold–ring contact is centered on top of the RTD mesa structure. (B) Top-view photograph of the nanothermometer bias above a threshold voltage: EL-signal is observed from the RTD-mesa. (C) Normalized electroluminescent spectra for temperatures of 9 K (*black*) and 150 K [light grey (*red in the web version*)]. The EL emission can be ascribed to the GaAs drain region of the RTD nanothermometer. *(Reprinted with permission from Pfenning A, Hartmann F, Rebello Sousa Dias M, Castelano LK, Süßmeier C, Langer F, et al. Nanothermometer based on resonant tunneling diodes: from cryogenic to room temperatures. ACS Nano 2015;9(6):6271–7, copyright (2015) American Chemical Society).*

The spatial resolution of patterned electrical nanothermometers depends on the size of their junction, meaning that it can be quite small. However, these devices can measure temperature only at certain, fixed points. An example of such a nanodevice is a thin film gold–nickel thermocouple with a junction area of 100 nm^2, deposited on a silicon [12]. The temperature mapping of surfaces is possible, however, with scanning thermal microscopy (SThM) techniques in which resistance temperature elements [13] or thermocouples [14], or even both, are attached to a probe tip. When the surface of a sample is scanned by the tip, the apex temperature of the tip changes due to heat transfer between the sample and tip. The temperature is read from changes in the voltage at the nanothermocouple ends or by changes in the electrical resistivity. Other thermal nanoprobes can also be used in the SThM. They can be based on the temperature-induced changes in the contact potential [15], the bimetal thermal expansion [16], or fluorescence [17]. More information regarding SThM and

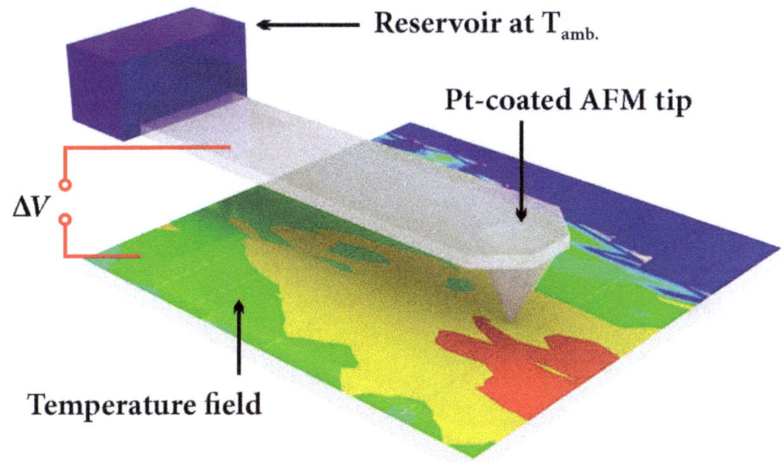

FIGURE 11.3 Schematic view of the nanoscale point contact thermocouple set-up. A Pt-coated AFM cantilever connected to a thermal reservoir at ambient temperature on one end is brought into soft mechanical contact with a metallic surface whose temperature field is to be determined. *(Reprinted with permission from Sadat S, Tan A, Chua YJ, Reddy P. Nanoscale thermometry using point contact thermocouples. Nano Lett 2010;10(7):2613–7, copyright (2010) American Chemical Society).*

temperature nanoprobes can be found in [18,19]. Temperature mapping of metallic surfaces with nanometric spatial resolution can be achieved even without fabrication of a dedicated temperature-sensitive probe being attached to the tip. Sadat et al. [20] used a platinum-coated atomic force microscopy (AFM) cantilever to measure the temperature distribution of a gold surface with ~10 mK temperature resolution (Fig. 11.3). When the tip was placed in soft contact with a gold surface, the temperature was retrieved from the thermoelectric voltage of the tip–gold point contact.

Regarding electrical nanothermometry methods, it is important to note that Crossno et al. [21] recently adopted the Johnson noise method, which is a primary thermometry method (Chapter 2) for nanoscale temperature measurements with a sensitivity of 5.5 mK (in 1 s of integration time) over the 3–300 K range. Besides electrical thermometry, optical thermometry methods are frequently used for temperature measurements at the nanoscale. They provide excellent spatial and temporal resolution and, more importantly, they facilitate thermal imaging. Optical phenomena of interest include Rayleigh and Raman light scattering, reflection, deflection, refraction, interference, birefringence, absorption, and luminescence. The use of luminescence for nanothermometry is motivated by the large sensitivity of luminescence on temperature and the ease with which luminescence can be detected in comparison to other optical signals. The typical temperature, spatial, and temporal resolutions of the most commonly used nanothermometry methods [10,21–23] are given in Table 11.1.

TABLE 11.1 The typical temperature, spatial, and temporal resolutions of selected nanothermometry methods

Thermometry method	Temperature resolution (mK)	Spatial resolution (nm)	Temporal resolution (μs)
Thermal expansion	10	100	1000
Seebeck effect	1	25	1
Electrical resistance	1	25	1
Johnson noise	5	100	100
SThM	100	100	100
Thermoreflectance	10	100	10
Interferometry	0.01	1000	0.001
Raman	100	1000	10^6
Luminescence	10	100	10

Source: Data adapted from Asheghi M, Yang Y. Micro- and nano-scale diagnostic techniques for thermometry and thermal imaging of microelectronic and data storage devices. In: Breuer KS, editor. Microscale diagnostic techniques, Berlin, Heidelberg: Springer Berlin Heidelberg; 2005. p. 155–96; Crossno J, Liu X, Ohki TA, Kim P, Fong KC. Development of high frequency and wide bandwidth Johnson noise thermometry. Appl Phys Lett 2015;106(2):23121; Christofferson J, Maize K, Ezzahri Y, Shabani J, Wang X, Shakouri A. Microscale and nanoscale thermal characterization techniques. J Electron Packag 2008;130(4):41101–6; Brites CDS, Lima PP, Silva NJO, Millan A, Amaral VS, Palacio F, et al. Thermometry at the nanoscale. Nanoscale 2012;4(16):4799–829.

11.3 TEMPERATURE MAPPING OF MICROFLUIDIC AND NANOFLUIDIC SYSTEMS

Microfluidics and nanofluidics deal with properties, behavior, control, and manipulation of extremely low quantities of fluids which are geometrically constrained to a small scale. In the case of nanofluids, this scale is less than 100 nm. As one may expect, the behavior and properties of such fluids drastically differ, in many ways, from those known at the macroscale. For example, the Reynolds number of microfluids and nanofluids is very low and their flow is more laminar than turbulent. Consequently, coflowing fluids do not mix in a conventional way. Since the first major developments in the 1980s, microfluidics and nanofluidics have found significant applications, such as in lab-on-a-chip technology, in inkjet printheads, in molecular biology (for enzymatic and DNA analysis), tracking cellular aging, analyzing antibiotic resistance, Coulter counting, etc.

Many of these applications account for temperature-dependent processes, and it is no surprise that a lot of effort has been made in recent years to develop methods for temperature measurements in microfluidic and nanofluidic devices. Most of them have already been mentioned in the previous section. Here, we will focus on the use of luminescence thermometry. Temperature measurement approaches which use luminescence can be categorized based on the material type used as the temperature indicator, type of temperature read-out scheme, and whether the temperature sensor is integrated into the device or whether is it added to the fluid. So far, organic dyes, organic–metal complexes, and quantum dots have been the sensing materials of choice in microfluidics, while emission intensity, luminescence intensity ratio, and luminescence lifetimes have been commonly used as temperature indicators. Regarding sensor integration, the soft lithography in polydimethylsiloxane (PDMS) is the most common fabrication process for microchips and probes and can be immobilized in PDMS by direct staining. Other integration methods include spin and blade coating, inkjet printing, recess coating, photopolymerization, etc. [24].

A typical example of the temperature mapping in a microfluidic system using a luminescence probe added to the fluid can be found in the work of Ross et al. [25]. They used the luminescence of rhodamine B to observe steady-state temperature as well as to monitor temperature transients in an acrylic microchannel and a fused-silica capillary (embedded in a sheet of PDMS). The temperature map of a multibranched microfluidic circuit isJ10 in Fig. 11.4. One can clearly see the difference in temperature in different branches of the circuit. The difference occurs since each branch carries a different fraction of the current. Measurements were carried out with a conventional fluorescence microscope equipped with a 20× objective, mercury arc lamp, set of filters (excitation, 500–550 nm; emission, >565 nm), and charge-coupled device (CCD) camera. With such experimental conditions, measurements could cover fluid temperatures from room temperature to 90°C, and provide, depending on the signal-averaging time, measurement precisions from 0.03 to 3.5°C. The achieved temporal and spatial resolutions were 33 ms and 1 μm, respectively.

However, two major problems occur with this type of measurement. First, as discussed in Chapter 4, determination of temperature from the intensity of a single emission band can be seriously compromised by fluctuations in the excitation power. Here, temperature maps were derived by comparison of emission intensity with a reference intensity determined at 22°C. If any change of experimental conditions occurs between the reference and practical measurements it will induce a measurement error. The second problem arises due to the adsorption of a dye onto the surfaces of plastics. It is well known that PDMS absorbs solvents and small molecules [26], such as rhodamine B, and therefore the dye cannot be rinsed from the surface. The dye experiences a different local environment in the solid material than in the fluid and presents altered luminescence behavior with its temperature not following the calibration curve obtained from dye molecules in a buffer. These problems can be eliminated by using ratiometric intensity read-out of temperature, and by incorporating sensing material into a device. For example, Schreiter et al. [27] performed temperature mapping in a microfluidic device using a fluorescent thin film containing *N*,*N*-bis(2,5-di-tertbutylphenyl)-3,4,9,10-perylenedi carboxamide (BTBP) dye. The film was incorporated into the device during the bonding stage. BTBP is an inexpensive,

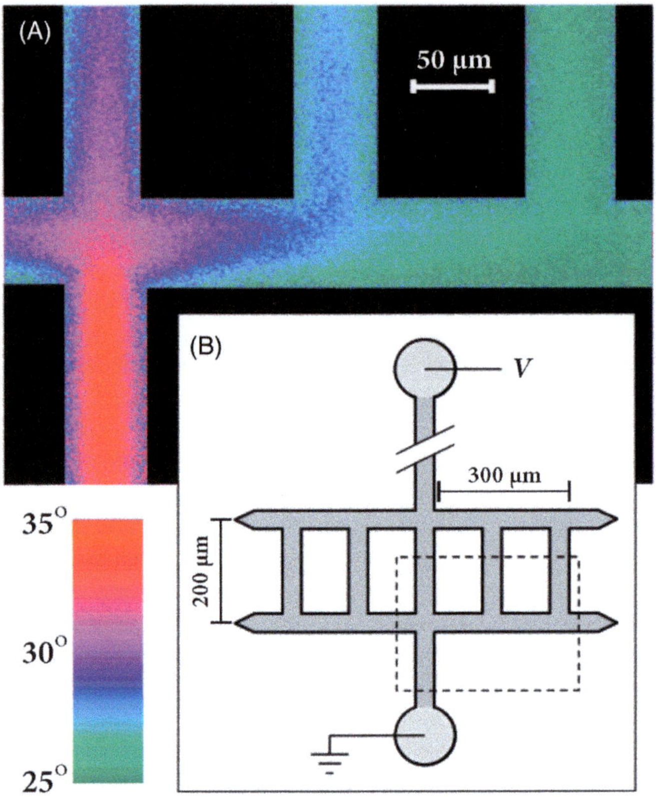

FIGURE 11.4 (A) The temperature map of a multibranched microfluidic circuit. (B) Schematics of a multibranched circuit. The area enclosed by the *dashed line* is the field of view for image (A). The measurement was made at an applied voltage of 1000 V. *(Adapted from Ross D, Gaitan M, Locascio LE. Temperature measurement in microfluidic systems using a temperature-dependent fluorescent dye. Anal Chem 2001;73(17):4117–23).*

commercially available dye, suitable for ratiometric intensity temperature sensing since it offers several emission bands with different temperature sensitivities. It is also temperature stable up to 270°C, resistant to quenching by oxygen, and can be easily introduced into various polymers (polystyrene, poly(methyl methacrylate), and polydimethylsiloxane) that are traditionally used in microfluidics. The performance of rhodamine B luminescence thermometry, realized through an in-channel measurement and using a thin film located slightly below the insulator-based dielectrophoretic devices (iDEP) channel, has been assessed by Nakano et al. [28]. They found that the thin film method provides closer agreement with numerical simulations than in-channel measurements, and that results do not depend on the iDEP channel material.

Benninger et al. [29] have shown that two- and three-dimensional temperature mapping in microfluidic systems can be realized with a temperature readout from the fluorescence lifetime of rhodamine B. As discussed in detail in Chapter 4, luminescence thermometry based on changes of excited state lifetimes with temperature is not affected by light scattering, reflection, excitation/detection efficiency, concentration, or inhomogeneous distribution of the sensing material, all of which may compromise ratiometric intensity methods [30, 31]. In addition, lifetime-based thermometry involves monitoring of one emission band, while ratiometric intensity measurements require measurements of two emission bands, which sometimes is not practical. To obtain two- and three-dimensional temperature maps of fluidic temperature distributions with a precision of ±1°C, Benninger et al. [29] employed 2-photon excitation by a Ti:Sapphire femtosecond laser (tuned to 840 nm) through a quasi-wide-field, multibeam, multiphoton microscope via an air objective (×10, N.A. 0.25), which resulted in a femtoliter-sized excitation volume that could be rapidly scanned over the complete field of view. Emission was detected by an ultrafast, wide-field detector based on a gated image intensifier. The precision of lifetime-based thermal imaging of microfluidic devices can be considerably improved by using probes whose fluorescent lifetimes show steep changes with temperature. For example, Graham et al. [32] demonstrated fluorescent lifetime imaging with a poly(DBD-AE-co-NIPAM) polymer probe and were able to map microfluidic temperature with a resolution of less than 0.1 K.

Reliable temperature measurements in microfluidic and nanofluidic devices may facilitate the development of analytical chemistry methods at microscales and nanoscale. For example, due to short optical path lengths in microfluidic devices, the precise direct measurement of absorbance by the attenuated light transmission technique is not feasible. However, absorbance can be indirectly measured via the photothermal effect, that is, by measuring the rise in temperature induced by optical absorption. After optical excitation, a molecule dissipates the excess energy by nonradiative relaxation processes, so it acts as a local heating source. Pfeiffer and Nagl [33] prepared luminescent temperature sensor spots (26 μm in diameter) inside microfluidic channels from the tris(1,10-phenanthroline)ruthenium(II) doped NOA 81 ultraviolet-curable thiolene prepolymer. Luminescence of these spots is sensitive to temperature changes in the 20–50°C range, with a temperature resolution better than 0.3 K. With an optical path of only 125 μm, the measurement system provided a minimal detectable absorbance of 3.2×10^{-3} AU, with the limit of detection for the model dye (food azo-dye Amaranth, E123) being 13 μM. Luminescent thermometry layers incorporated into microfluidic devices can also be used as components of a microfluidic anemometer. In this device, a heating element in a fluid-filled channel is used to generate a temperature gradient. When there is no fluid flow the temperature profile is symmetric around the heating element spot. The profile becomes asymmetric in

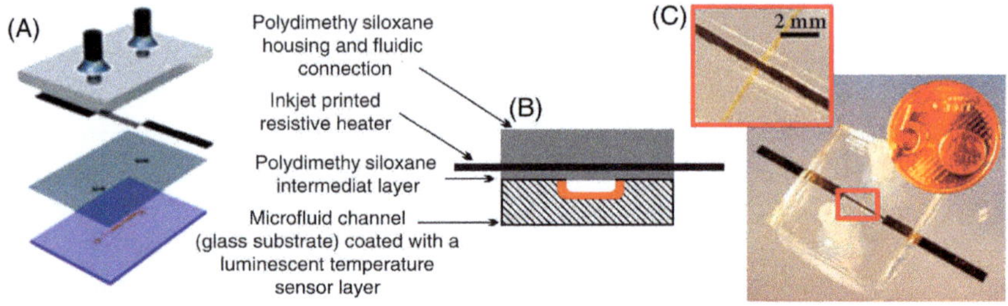

FIGURE 11.5 Microfluidic anemometer chip based on a measurement of the temperature profile from an integrated luminescence thermometry layer: schematic representation of the chip in an (A) exploded view, (B) sectional view, and (B) photograph of the flow sensor chip. *(Reprinted from Hoera C, Skadell MM, Pfeiffer SA, Pahl M, Shu Z, Beckert E, et al. A chip-integrated highly variable thermal flow rate sensor. Sensors Actuat B Chem 2016;225(Suppl. C):42–9, copyright (2016), with permission from Elsevier).*

the presence of a flow, and the location of the maximum temperature moves in the flow direction and is related to the flow rate. Hoera et al. [34] have developed a microfluidic flow rate sensor chip (Fig. 11.5) from a combination of a luminescent sensor layer (containing dichloro-tris(1,10-phenanthroline)-ruthenium(II)-hydrate ($Ru(phe)_3$) temperature sensitive fluorescent metal–organic complex) and an inkjet-printed heating element inside microfluidic channels. The device is able to track a wide range of flow rates, from very low (10 nL min^{-1}) to 0.5 ml min^{-1}.

11.4 LUMINESCENCE THERMAL IMAGING IN NANOELECTRONICS

As discussed in Chapter 9, electronic circuits and devices can be thermally imaged via luminescence. Once the surface of a nanodevice is covered with a thin layer of temperature-sensitive luminescence material, changes in temperature of the surface will be reflected through spatially-resolved changes in luminescence. For example, dried rhodamine B has been used to cover microwires and nanowires and its luminescence has been exploited to assess surface temperatures (between 25 and 70°C) with high spatial resolution [35] (Fig. 11.6). The spatial resolution is much lower than the resolution readily obtained with SThM, but, the achieved temporal resolution of 100 ms is considerably faster than in the case of SThM. Recently, Rodrigues et al. [36] demonstrated a method for surface functionalization by lanthanide doped (Tb^{3+} and Eu^{3+} complexes) polymer monolayers. Due to their small mass (monolayers), this type of layer causes a small perturbance on the thermal properties of a surface while providing excellent temperature-sensing performance.

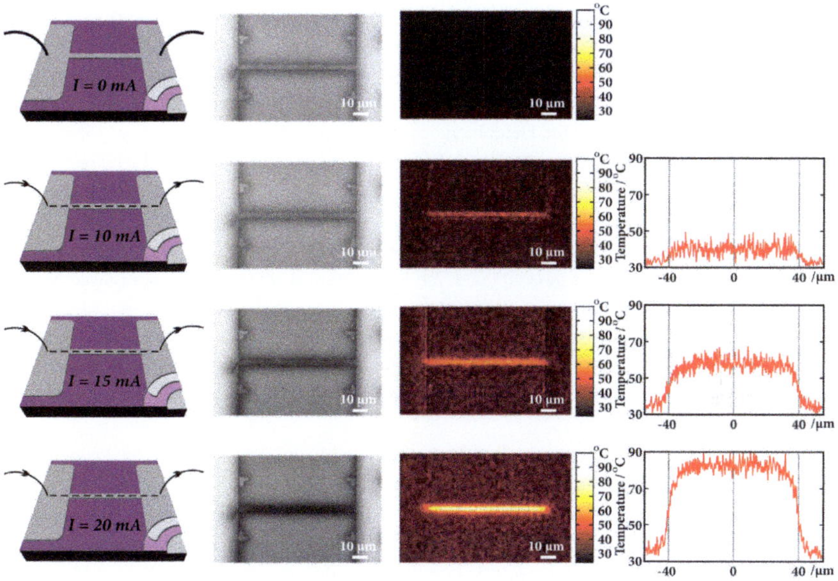

FIGURE 11.6 Resistive heating in a 40-nm-thick, 80-mm-long, and 2-mm-wide wire. When current is passed through the wire, the local resistive heating occurs due to geometrical confinement at the wire. The top row shows the case where no current is applied. The next three rows show the cases for gradually increasing currents of 10, 15, and 20 mA. The first column shows the applied current value; the second column shows the rhodamine B fluorescence images; the third column shows the temperature images after the conversion of the intensity changes; and the fourth column shows the temperature along the middle of the respective wire. *(Reprinted from Löw P, Kim B, Takama N, Bergaud C. High-spatial-resolution surface-temperature mapping using fluorescent thermometry. Small 2008;4(7):908–14, with permission from Wiley).*

REFERENCES

[1] Childs PRN. Nanoscale thermometry and temperature measurement. In: Carlos LD, Palacio F, editors. Thermometry at the nanoscale: techniques and selected applications. The Royal Society of Chemistry; 2016. p. 1–22.

[2] Hartmann M. Minimal length scales for the existence of local temperature. Contemp Phys 2006;47(2):89–102.

[3] Jaque D, Vetrone F. Luminescence nanothermometry. Nanoscale 2012;4(15):4301–26.

[4] Gao Y, Bando Y. Nanotechnology: carbon nanothermometer containing gallium. Nature 2002;415(6872):599.

[5] Li YB, Bando Y, Golberg D, Liu ZW. Ga-filled single-crystalline MgO nanotube: Wide-temperature range nanothermometer. Appl Phys Lett 2003;83(5):999–1001.

[6] Wang C-Y, Gong N-W, Chen L-J. High-sensitivity solid-state Pb(Core)/ZnO(shell) nanothermometers fabricated by a facile galvanic displacement method. Adv Mater 2008;20(24):4789–92.

[7] Li Y, Bando Y, Golberg D. Single-crystalline In_2O_3 nanotubes filled with In. Adv Mater 2003;15(7–8):581–5.

[8] Gong NW, Lu MY, Wang CY, Chen Y, Chen LJ. Au(Si)-filled β-Ga_2O_3 nanotubes as wide range high temperature nanothermometers. Appl Phys Lett 2008;92(7):73101.

[9] Wang C-Y, Chen S-H, Tsai P-H, Chiou C-H, Hsieh S-J. The fabrication and thermal properties of bismuth–aluminum oxide nanothermometers. Nanotechnology 2017;28(4):045705.
[10] Asheghi M, Yang Y. Micro- and nano-scale diagnostic techniques for thermometry and thermal imaging of microelectronic and data storage devices. In: Breuer KS, editor. Microscale diagnostic techniques. Berlin, Heidelberg: Springer Berlin Heidelberg; 2005. p. 155–96.
[11] Pfenning A, Hartmann F, Rebello Sousa Dias M, Castelano LK, Süßmeier C, Langer F, et al. Nanothermometer based on resonant tunneling diodes: from cryogenic to room temperatures. ACS Nano 2015;9(6):6271–7.
[12] Chu D, Wong W-K, Goodson KE, Pease RFW. Transient temperature measurements of resist heating using nanothermocouples. J Vac Sci Technol B Microelect Nanom Struct Process Meas Phenom 2003;21(6):2985–9.
[13] Pylkki RJ, Moyer PJ, West PE. Scanning near-field optical microscopy and scanning thermal microscopy. Jpn J Appl Phys 1994;33(6S):3785–90.
[14] Majumdar A, Carrejo JP, Lai J. Thermal imaging using the atomic force microscope. Appl Phys Lett 1993;62(20):2501–3.
[15] Nonnenmacher M, Wickramasinghe HK. Scanning probe microscopy of thermal conductivity and subsurface properties. Appl Phys Lett 1992;61(2):168–70.
[16] Nakabeppu O, Chandrachood M, Wu Y, Lai J, Majumdar A. Scanning thermal imaging microscopy using composite cantilever probes. Appl Phys Lett 1995;66(6):694–6.
[17] Saïdi E, Samson B, Aigouy L, Volz S, Löw P, Bergaud C, et al. Scanning thermal imaging by near-field fluorescence spectroscopy. Nanotechnology 2009;20(11):115703.
[18] Majumdar A. Scanning thermal microscopy. Annu Rev Mater Sci 1999;29(1):505–85.
[19] Pereira MJ, Amaral JS, Silva NJO, Amaral VS. Nano-localized thermal analysis and mapping of surface and sub-surface thermal properties using scanning thermal microscopy (SThM). Microsc Microanal 2016;22(6):1270–80.
[20] Sadat S, Tan A, Chua YJ, Reddy P. Nanoscale thermometry using point contact thermocouples. Nano Lett 2010;10(7):2613–7.
[21] Crossno J, Liu X, Ohki TA, Kim P, Fong KC. Development of high frequency and wide bandwidth Johnson noise thermometry. Appl Phys Lett 2015;106(2):23121.
[22] Christofferson J, Maize K, Ezzahri Y, Shabani J, Wang X, Shakouri A. Microscale and nanoscale thermal characterization techniques. J Elect Packag 2008;130(4):41101–6.
[23] Brites CDS, Lima PP, Silva NJO, Millan A, Amaral VS, Palacio F, et al. Thermometry at the nanoscale. Nanoscale 2012;4(16):4799–829.
[24] Pfeiffer SA, Nagl S. Microfluidic platforms employing integrated fluorescent or luminescent chemical sensors: a review of methods, scope and applications. Methods Appl Fluoresc 2015;3(3):034003.
[25] Ross D, Gaitan M, Locascio LE. Temperature measurement in microfluidic systems using a temperature-dependent fluorescent dye. Anal Chem 2001;73(17):4117–23.
[26] Mukhopadhyay R. When PDMS isn't the best. Anal Chem 2007;79(9):3248–53.
[27] Schreiter KM, Glawdel T, Forrest JA, Ren CL. Robust thin-film fluorescence thermometry for prolonged measurements in microfluidic devices. RSC Adv 2013;3(38):17236–43.
[28] Nakano A, Luo J, Ros A. Temporal and spatial temperature measurement in insulator-based dielectrophoretic devices. Anal Chem 2014;86(13):6516–24.
[29] Benninger RKP, Koç Y, Hofmann O, Requejo-Isidro J, Neil MAA, French PMW, et al. Quantitative 3D mapping of fluidic temperatures within microchannel networks using fluorescence lifetime imaging. Anal Chem 2006;78(7):2272–8.
[30] Lojpur V, Antić Ž, Dramićanin MD. Temperature sensing from the emission rise times of Eu^{3+} in SrY_2O_4. Phys Chem Chem Phys 2014;16(46):25636–41.

[31] Dramićanin MD. Sensing temperature via downshifting emissions of lanthanide-doped metal oxides and salts. A review. Methods Appl Fluoresc 2016;4(4):42001.
[32] Graham EM, Iwai K, Uchiyama S, Prasanna de Silva A, Magennis SW, Jones AC. Quantitative mapping of aqueous microfluidic temperature with sub-degree resolution using fluorescence lifetime imaging microscopy. Lab Chip 2010;10(10):1267–73.
[33] Pfeiffer SA, Nagl S. On-chip photothermal analyte detection using integrated luminescent temperature sensors. Anal Chem 2017;89(17):9400–6.
[34] Hoera C, Skadell MM, Pfeiffer SA, Pahl M, Shu Z, Beckert E, et al. A chip-integrated highly variable thermal flow rate sensor. Sensors Actuat B Chem 2016;225(Suppl C):42–9.
[35] Löw P, Kim B, Takama N, Bergaud C. High-spatial-resolution surface-temperature mapping using fluorescent thermometry. Small 2008;4(7):908–14.
[36] Rodrigues M, Piñol R, Antorrena G, Brites CDS, Silva NJO, Murillo JL, et al. Implementing thermometry on silicon surfaces functionalized by lanthanide-doped self-assembled polymer monolayers. Adv Funct Mater 2016;26(2):200–9.

Chapter 12

Achieving Multifunctionality by Combining Thermometry With Other Luminescence Applications

12.1 INTRODUCTION

To date, luminescence has found numerous practical uses, from traditional ones such as lightning, displays, lasers, and chemical sensors, to contemporary ones that are developing at accelerating pace mostly in life sciences, nanosciences, food processing and safety, and environmental monitoring. Such a huge variety of applications offers more than sufficient elements for the construction of luminescent multifunctional and smart materials and devices, which, in turn, may facilitate the development of critical and game-changing technologies. This Chapter aims to familiarize readers with several applications of luminescence which can be combined with luminescence thermometry to provide materials and devices with polyvalent functions. Combined functionality with thermometry has already been demonstrated in some cases, in others, it is reasonable to anticipate it, knowing that these applications utilize similar materials to those in luminescence temperature sensors.

At the beginning of this chapter, the concept of multifunctional and smart materials is presented along with their potential role in the economy and technology of the near future. Brief overviews of physical, chemical, and structural sensing techniques, based on the use of luminescence, will then be outlined. The selection of applications is made based on their close relationship with luminescence temperature sensing, and includes sensing of pressure, electric and magnetic fields, evaluation of structural and compositional changes of materials via luminescence, and chemical sensing. The chapter concludes with examples of multifunctional bioimaging and thermal imaging.

12.2 MULTIFUNCTIONAL AND SMART MATERIALS

The approach of designing, producing, and using materials has considerably evolved through history. For a long period of time, people mainly dealt with so-called structural materials. These materials have only one function, to provide

Luminescence Thermometry. http://dx.doi.org/10.1016/B978-0-08-102029-6.00012-9

mechanical strength (i.e., to bear loads). Over time, when new functions of materials were discovered, such as conduction of electricity or emission of light, the era of the functional material had emerged. The last few decades witnessed a shift in research toward multifunctional (polyvalent) materials due to the expectation and necessity that materials and structures should be able to perform several functions. By using such materials and structures, it is now possible to produce more services and goods while using fewer resources and creating less waste and pollution, which is the current trend in global economy and technology markets. Multifunctional materials already deliver direct and positive effects on the environment, our quality of life, and economic growth. They impact many areas by lowering the weight of products, reducing power consumption, reducing product complexity and their size and cost, minimizing environmental effects, and increasing safety and versatility. NASA (National Aeronautics and Space Administration) in its Roadmap Report has recognized the multifunctional concept of materials and structures as one of the key elements to technologies that will enable future deep-space missions, advanced aeronautic capabilities, and long-term space travel [1]. They recognize that roadmap technologies will not only enable future NASA missions, but will also deliver spin-offs that will benefit various sectors of the economy, increase global competitiveness, and improve security and the quality of life worldwide. In recent years, the European Union has financed several large research projects related to the development and use of multifunctional and smart materials. These include projects related to the development of adaptive and smart materials and structures for more efficient vessels, and a completely new generation of smart materials that can sense, touch, and respond to luminescence.

Today, multifunctional design incorporates materials, structures and material systems, both naturally existing and specially engineered, that have the capacity to perform multiple functions through particular combinations of structural properties and at least one added functional capability. In fact, nature has recognized that superior constructions require building blocks having polyvalent functions. Thus, life as it developed and adapted through evolution utilized most of it materials as multifunctional. For example, the human skin provides the basic structure and protection for our internal organs, but also contains sensory receptors and other functional moieties. Bones provide structure and support for the body, facilitate mobility, support and protect the various organs of the body, produce red and white blood cells, store minerals, etc. Many other examples of multifunctionality may be found by just observing nature.

By looking further into nature's design of materials and structures, we can observe that many of them may be considered as a “smart” since they can change their properties in a specific manner in response to specific stimuli. Similarly, smart materials, the new generation of materials surpassing functional and multifunctional materials, possess adaptive capabilities to external stimuli with inherent intelligence [2]. In this sense, the term “smartness” explains the self-adaptability, self-sensing, memory and multiple functionalities of materials

TABLE 12.1 Examples of Smart Materials and Their Respective Stimuli and Responses

Smart material type	Stimuli (input)	Response (output)
Thermochromic	Temperature	Color change
Electrochromic	Electric field	Color change
Photoochromic	Radiation	Color change
Thermoelectric	Temperature	Potential difference
Shape memory alloys	Temperature	Shape deformation
Piezoelectric	Mechanical stress	Potential difference
Electrostrictive	Potential difference	Shape deformation
Magnetostrictive	Magnetic field	Shape deformation

or structures. The classical example of smart materials is the piezoelectric element which produces a voltage when mechanical stress is applied to it. Another example includes chromic materials (thermochromic, electrochromic, and photochromic) which change their color with a change in temperature, electric field, or radiation, respectively. Generally speaking, smart materials can be divided into two classes, active and passive. Active smart materials can transduce energy and perform actuation, while passive materials lack these capabilities. Piezoelectrics are one example of active smart materials while chromic materials are passive. Table 12.1 gives several examples of smart materials. It is important to note that the multifunctionality of materials and their "smartness" is possible to achieve at spatial scales from nano through micro to macro, as well as on various temporal and compositional levels.

A new generation of smart materials that will be capable of continuous monitoring of physical (temperature, pressure), chemical, and structural (impacts, cracks, etc.) parameters is under development. By gathering data about their surroundings, these sentient materials will make conventional sensor devices obsolete in the near future. It is foreseen that smart materials will be able to capture and relay data to the cloud, to reconfigure as they react to environmental changes, and, therefore, will impart startling changes to the way things are designed and used.

Luminescent materials are inherently multifunctional and smart since luminescence is sensitive to changes in a materials' environment. Luminescence is used for a great number of diverse functions, and host materials which accommodate luminescence centers may have their own functional abilities. Thus, it is reasonable to expect that with many well-known luminescent materials, multifunctionality and/or smartness may be achieved by the addition of luminescent thermometric functionality. Alternatively, it is possible to create an entirely new multifunctional material with a thermometric function by designing a structure with adequate luminescent properties [3]. Luminescence is a powerful tool for investigating the structure and dynamics of matter or living systems at a molecular or supramolecular level. Significant benefits may be expected to research from improved capabilities in temperature measurements and temperature mapping. Similarly, benefits may be gained from probes capable of physical, chemical, or structural sensing along with temperature.

12.3 LUMINESCENCE SENSING OF PRESSURE

Pressure is a thermodynamic parameter that has to be determined for many practical needs, in both research and in many industrial applications. For example, pressure measurements are crucial for practical aerodynamic testing or fluid mechanics experiments. Regarding the use of luminescence for pressure measurements, the quenching of luminescence of different materials in the presence of oxygen has been the method of choice since Kautsky and Hirsch discovered this phenomenon in 1935 [4]. Pressure-sensitive paints (PSP) and thin pressure-sensitive polymer coatings (PSPC) contain luminescent species incorporated into oxygen permeable medium, see Fig. 12.1, whose luminescence depends on the quantity of oxygen which diffuses into the paint or film. Since the partial pressure of oxygen depends on the air pressure, changes of luminescence can be directly related to the air pressure. PSPs and PSPCs can be used to cover large objects, and the pressure distribution on objects can be evaluated by observing

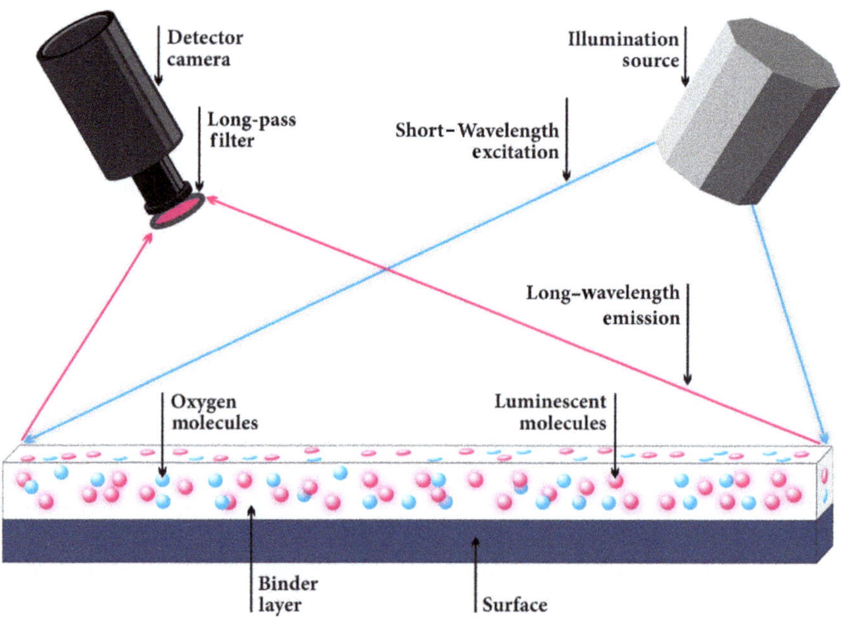

FIGURE 12.1 **Pressure measurements via oxygen quenching of luminescence of pressure sensitive paints and thin polymer films.**

pressure-dependent emission changes at long wavelengths after surface illumination by short-wavelength radiation, see Fig. 12.1. Both spectroscopic and lifetime read-outs of pressure are possible since the mechanism of luminescence quenching by oxygen is collisional. Enthusiasm for the use of this optical technique in wind tunnels is stimulated by the substantial savings it offers in terms the instrumentation cost and model construction time [5,6].

So far, platinum porphyrins and ruthenium complexes are mainly used as luminescence pressure probes for aeronautical testing. They are embedded in binders or coatings which can be easily be applied to, and removed from, the surface of interest, and which are permeable to oxygen and possess adequate hardness, optical properties, and adhesion properties hardness and resistance to abrasion.

Since luminescence of metal–organic complexes is also influenced by temperature (see Chapter 8) paints and coatings (see Chapter 9) with two types of luminescent probes can be prepared which can serve as a multifunctional temperature–pressure responsive layer/coating. Luminescence of one probe should be strongly sensitive to temperature and luminescence of the other to pressure. In addition, this approach enables pressure measurements which are compensated to changes in temperature [7]. The additional luminescence probe in the paint or film may be needed for the compensation of fluctuations in excitation. This probe should be insensitive to pressure and its emission should be clearly distinguishable from that of the pressure-sensitive probe [5]. In cases of low speed air flow, such as those used for testing cars, pressure variations are rather small and, thus, PSPs with higher sensitivities are needed. Also, compensation for the excitation fluctuations is more critical. To overcome these problems, Engler et al. [8] used PSP that contained pyrene, as the pressure-sensitive dye and a gadolinium oxysulfide as the reference dye, along with a specialized procedure and devoted software to obtain measurements with a maximal error of ±1 mbar.

Recently, a dual-fluorescence sensor for tracing of oxygen levels and temperature, with unprecedented range and sensitivity, has been presented [9]. The sensor has two components: ruthenium tris-1,10-phenanthroline incorporated in poly(acrylonitrile), whose luminescence is strongly temperature-dependent but insensitive to oxygen, and fullerene C_{70} as the oxygen-sensitive probe. Temperature and oxygen concentration are derived from fluorescence lifetimes in the 0–120°C temperature range, with detection limits for oxygen in the parts-per-billion range.

Measurements of high pressures in diamond anvil cells are quite important in research. Historically, values of pressure have been derived from the shift of ruby R emission lines [10]. Over time, some other probes (so-called "luminescence manometers") have been introduced [11], such as Cr^{3+} doped $YAlO_3$, Eu^{3+} doped LaOCl and LaOBr (shift of emission from the $^5D_0 \rightarrow {}^7F_0$ transition), Eu^{3+} doped $Y_3Al_5O_{12}$ (shift of emission from the $^5D_0 \rightarrow {}^7F_0$ transition), etc. As discussed in Chapter 6, emissions of transition metal and lanthanide doped materials can be effectively used for temperature measurements. By using the dual-sensing functionality of these materials, it is possible to simultaneously measure pressure and temperature in diamond anvil cells.

12.4 LUMINESCENCE SENSING OF AN ELECTRIC FIELD

It is well known that an external electric field acting upon material can induce and influence luminescence. For example, injection luminescence may occur due to recombination of the electrons with centers, or positive holes, when electric current passes through a material. Optical properties of atoms and molecules change when they interact with external electric fields. Effects of this interaction can be seen in the absorption and emission spectra as changes in intensity, position, width, and polarization of absorption/emission lines. This phenomenon is termed the Stark effect, after German physicist Johannes Stark who observed in 1913 the splitting of the Balmer transitions in atomic hydrogen when exposed to an external electric field. The phenomenon is also known as electrochromism, electroabsorption, and electrooptic absorption, and it is much utilized in probing of the electronic structure of assorted systems, from isolated molecules in gas phase, band structures of semiconductors and crystalline materials, to biological systems [12].

Luminescence of almost all materials is cross-sensitive to temperature and external electric field. On one hand, this presents an obstacle for sensing. On the other hand, it provides an opportunity to use luminescence temperature probes to sense electric fields and vice versa. Even though humans do not perceive electric fields (unlike some species of fish), it does not mean that sensing electric fields is unimportant to human life. On the contrary, it is required in atmospheric science and geology and in many engineering fields including emerging ones such as robot proximity sensing and vision. Sensing of electric fields at the nanoscale is of exceptional interest to estimate biomembrane potential, the surface potential of protein molecules or nanoparticles, etc. Electrochromic dyes, that is, dyes which change color in response to electric potential changes, have been of huge aid to neuroscientists interested in mapping patterns of electrical activity in large neuronal populations or interested in the activity of myocytes. The following paragraphs present several examples of electric field influence on the luminescence of different materials. These materials belong to groups of materials which are also suitable for luminescence temperature sensing, and which are described in the Chapters 6–8 of this book.

In molecules, the magnitude of the Stark effect on optical transitions depends on the value of the electric dipole moment and the molecular polarizability [13]. The electronic coupling between the initial and final states of transition may also be affected, depending on the overlap of the wavefunctions of the donor and acceptor. Consequently, electron-transfer processes are either enhanced or quenched in the presence of external electric fields. In some π-conjugated polymers, electric field–induced dissociation of the exciton state is observed [14].

Intensity of fluorescence of zinc porphyrin in zinc porphyrin–C_{60} dyads in poly(methyl methacrylate) (PMMA) film (ZnP–P23–C_{60}, see Fig. 12.2A) is increased by applying an electric field because of the field causing deceleration of the electron transfer rate from the singlet excited state of zinc porphyrin to C_{60} (see Fig. 12.2B). Free-base porphyrin–C_{60} dyads (H_2P–P23–C_{60}, see Fig. 12.2A), on the other hand, show the opposite effect (see Fig. 12.2C) [15].

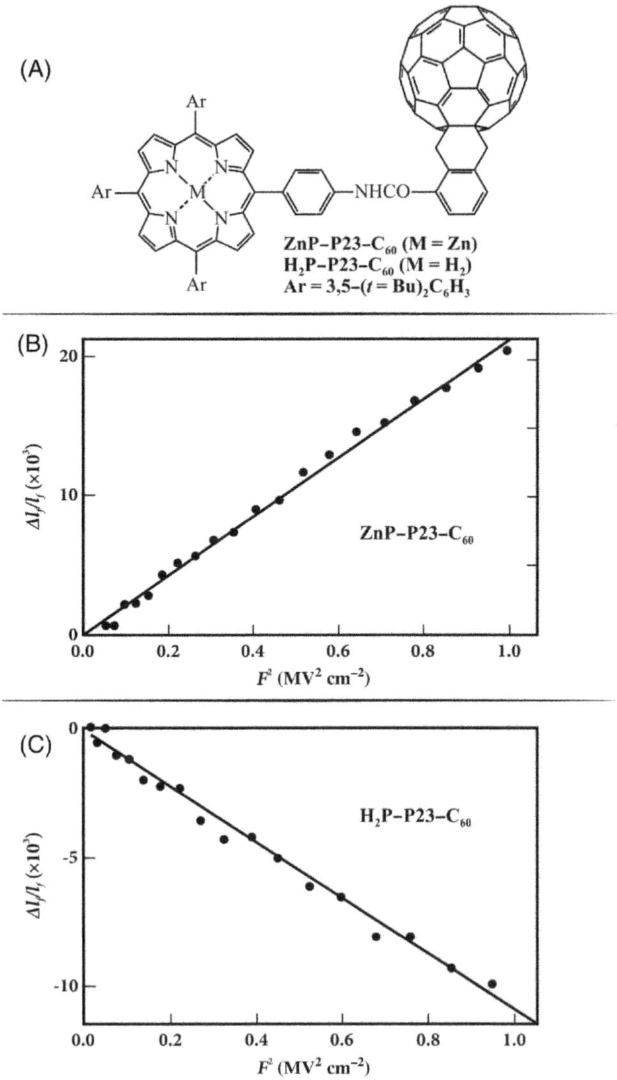

FIGURE 12.2 (A) Molecular structure of ZnP–P23–C_{60} and H_2P–P23–C_{60}. (B) Plots of $\Delta I_f/I_f$ of porphyrin for ZnP–P23–C_{60}; (C) plot for H2P–P23–C_{60}; as a function of the square of the applied field strength, F. *(Reprinted from Ohta N, Mikami S, Iwaki Y, Tsushima M, Imahori H, Tamaki K, et al. Acceleration and deceleration of photoinduced electron transfer rates by an electric field in porphyrin-fullerene dyads. Chem Phys Lett 2003;368(1):230–5, copyright (2003), with permission from Elsevier).*

The magnitude of the fluorescence intensity change (ΔI_f) relative to the fluorescence intensity (I_f) is proportional to the square of the applied field strength in both compounds (see Fig. 12.2B and C).

Takehara et al. [13] have shown that fluorescence from an exciplex formed between 1,1-diphenylpropene (DPP) and chiral 1,4-naphthalenedicarboxylate (NDC) in PMMA polymer films is enhanced by an external electric field. At the same time, the fluorescence of NDC codoped with DPP was quenched. Nakabayashi et al. [16] investigated the effects of an electric field on the emission of p-HBI (p-hydroxybenzylidene-imidazolidinone) in a PMMA film. This chromophore may be considered as a model chromophore of the wild-type green fluorescent protein whose emission is known to be temperature sensitive. The magnitude of the field-induced increase in fluorescence intensity was found to be proportional to the square of applied field strength. Emission decay measurements revealed that the electric field–induced reduction of the nonradiative transition was responsible for the increase in the fluorescence.

The electric field considerably influences the emission properties of semiconducting quantum dots. Emission of CdSe quantum dots is quenched by electric fields, and the quenching is stronger in larger particles. The effect may be ascribed to the field-induced reduction, in a population, of the emitting state, as revealed by emission decay measurements [17]. Gurinovich et al. [18] analyzed the influence of an electric field on photoluminescence of two types of CdSe nanoparticles in PMMA films: quantum dots (4 nm in diameter) and nanorods (25 nm long and 4 nm width). Emission spectra of both types of nanostructures are altered by an external electric field in terms of reduced intensity and spectral shift of the band maxima, see Fig. 12.3A and B. Spectral changes are more distinct in CdSe quantum dots than in nanorods. In quantum dots, the decrease of emission intensity and redshift of the band maximum are proportional to the squared electric field strength, while in nanorods these dependencies are linear, see Fig. 12.3C and D. The authors [18] argued that quenching of the photoluminescence may be related to a reduction in the overlap of wavefunctions of electrons and holes upon the application of an external electric field, and, thus, to the decrease in the probability of radiative recombination of the electron–hole pair.

A similar observation has been made with benzyl mercaptan–capped CdS quantum dots embedded in a PMMA film [19] and CdTe quantum dots embedded in poly(vinyl alcohol) (PVA) film [20]. Under an external electric field the maximum of the emission band of CdS quantum dots was redshifted, and emission intensity was quenched. The electric field–induced change in emission decay suggested that the emission intensity decrease was a result of the occurrence of an electric field-assisted dissociation into electron and hole at the photoexcited state having a charge-separated character [19]. In the case of CdTe quantum dots, the emission spectrum of the small size CdTe dots (2.2 nm) shows a blueshift in the presence of an electric field, while the large size dots (4.1 nm) exhibit a redshift in emission while exposed to an electric field [20]. For all particle sizes, emission intensity decreases proportionally to the square of the electric field strength.

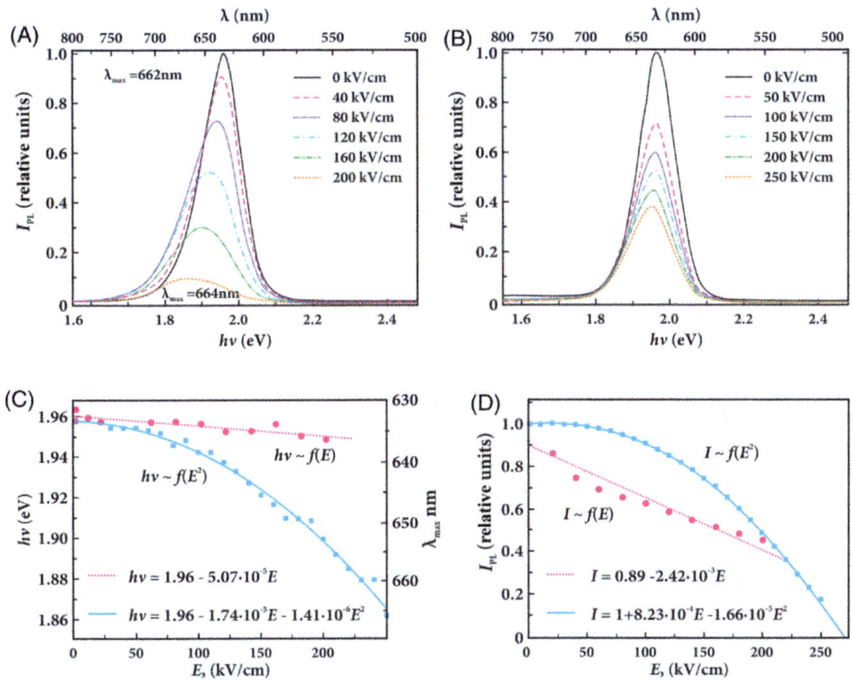

FIGURE 12.3 Electric field influence on emission (excited at 488 nm) of two types of CdSe nanoparticles in PMMA films: (A) emission spectra of 4 nm quantum dots; (B) emission spectra of 25 nm long and 4 nm wide nanorods; (C) spectral position of the maxima of emission band under the electric field of different strength quantum dots (*squares and solid line in gray* (*magneta in the web version*)) and nanorods (*circles and dotted line in dark gray* (*cyan in the web version*)); (D) relative change of the emission intensity with increasing electric field of quantum dots (*squares and solid line in gray* (*magneta in the web version*)) and nanorods (*circles and dotted line in dark gray* (*cyan in the web version*)). *(Adapted from Gurinovich LI, Lyutich AA, Stupak AP, Artem'ev MV, Gaponenko SV. Effect of an electric field on photoluminescence of cadmium selenide nanocrystals. J Appl Spectrosc 2010;77(1):120–5, with permission from Springer).*

Variations of the crystal field around lanthanide luminescence centers, caused either by temperature or an electric field, enable the use of lanthanide doped ferroelectric materials for multifunctional sensing. Hao et al. [21] have shown that upconversion luminescence of epitaxial $BaTiO_3{:}Yb^{3+}/Er^{3+}$ film is greatly enhanced by an external electric field. In this material, the intensity of green emission bands centered at 523 and 548 nm ($^2H_{11/2}, ^4S_{3/2} \rightarrow {^4I_{15/2}}$ electronic transitions of Er^{3+}) are strongly enhanced when the material is in an electric field, while the intensity of the red emission band centered at 656 nm ($^4F_{9/2} \rightarrow {^4I_{15/2}}$ electronic transitions of Er^{3+}) practically remained unaltered, see Fig. 12.4A and B. The origin of the emission intensity enhancement lies in the local symmetry change around Er^{3+} caused by an external electric field. Er^{3+} ions substitute Ti^{4+} in $BaTiO_3$ [22] and, thus, reside in noncentrosymmetric sites in a crystal lattice. An electric field applied along the direction of spontaneous polarization produces the elongation of the lattice along

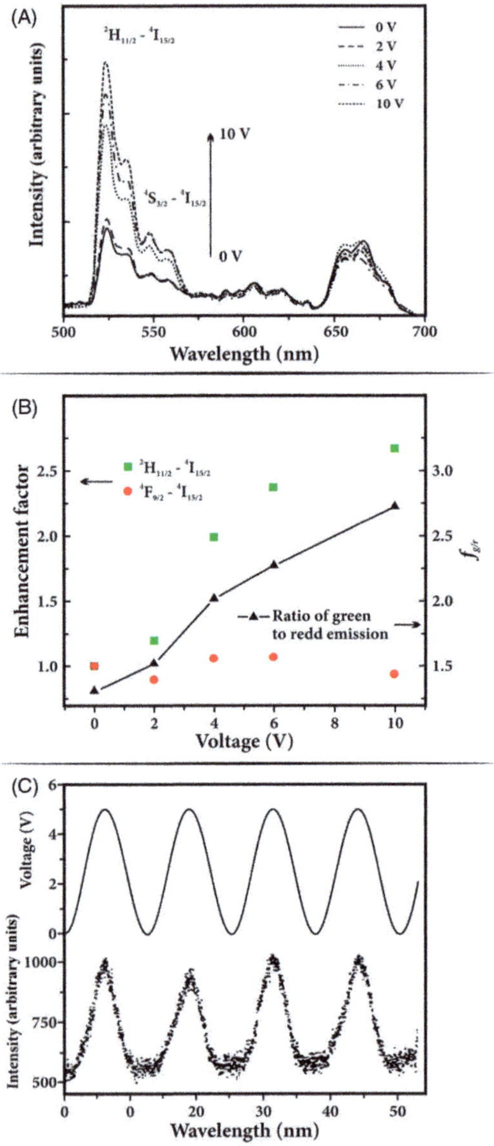

FIGURE 12.4 (A) The upconversion emission spectra of the $BaTiO_3$:Yb^{3+}/Er^{3+} film under DC bias voltage ranging from 0 to 10 V. (B) The enhancement factors (left axis) for the green (dark gray (green in the web version) squares) and the red (darker gray (red in the web version) circles) upconversion emissions as a function of the applied DC voltage; the green to red emission intensity ratio's (fg/r) (triangles, right axis) dependence on the bias. (C) Sinusoidal AC electric voltage applied to the $BaTiO_3$:Yb^{3+}/Er^{3+} film (top) and the intensity of the green upconversion emission measured at 532 nm as a function of time (bottom). The upconversion emission was excited by a 980 nm diode laser. *(Adapted from Hao J, Zhang Y, Wei X. Electric-induced enhancement and modulation of upconversion photoluminescence in epitaxial $BaTiO_3$:Yb/Er thin films. Angew Chem Int Ed 2011;50(30):6876–80, with permission from Wiley).*

the *c*-axis and the shrinkage of the lattice along the *a*-axis (at rates of d_{33}= 149 ± 54 pm $V^{-1}$and d_{31} = −82 ± 61 pm V^{-1}, respectively) [23]. In this way, the crystal field asymmetry around Er^{3+} ions becomes more pronounced, increasing the electric dipole transition probabilities of the dopant ions. Knowing that nonradiative transition probabilities are not influenced by changes in the site symmetry, this displacement of Ti^{4+}/Er^{3+} and O_2 ions by the external electric field is responsible for the enhancement of upconversion emission through the increase of the radiative transition probabilities. One should note that the transitions from $^2H_{11/2}$ and $^4S_{3/2}$ levels are more sensitive on symmetry around Er^{3+} ions than a transition from the $^4F_{9/2}$ level. Consequently, green emissions are strongly enhanced by external electric fields and red emission is not. Therefore, the ratio of the green to red emission intensity, see Fig. 12.4B, can be used for the evaluation of the electric field strength. Moreover, the ratio of intensities of green emissions at 523 and 548 nm can be effectively used for temperature measurements [24], since these emissions come from thermally coupled $^2H_{11/2}$ and $^4S_{3/2}$ levels, as described in detail in Chapter 6. Hao et al. [21] also demonstrated the variation of the green upconversion emission intensity when $BaTiO_3{:}Yb^{3+}/Er^{3+}$ film was put in an AC electric field, see Fig. 12.4C. It is clear that the optical signal follows the electrical one.

12.5 LUMINESCENCE SENSING OF A MAGNETIC FIELD

Nanoparticles that exhibit both luminescence and superparamagnetic behavior (so-called bifunctional magnetic fluorescent nanostructures) are of exceptional interest in biomedicine for use in magnetically guided drug delivery, cell sorting, protein immobilization, hyperthermia, magnetic resonance imaging (MRI), etc. [25]. The value of these nanoparticles comes from the ability to manipulate them (i.e., to guide them to desired locations) by an external magnetic field, to heat them by an external magnetic field, to observe them by MRI and luminescence imaging, and, finally, to use their luminescence for temperature sensing. For example, the exclusion of all tumor cells with minimal damage to the adjacent healthy tissues may be facilitated by a preoperative MRI diagnosis of tumor cells and by using luminescence as an intraoperative optical probe during surgery [26]. The emitting component in bifunctional magnetic luminescent nanostructures can be semiconductor quantum dots [27–29], rare earth–based phosphors [30], organic dyes [31,32], conjugated polymers [33], and carbon quantum dots [34]. All of these, as discussed in Chapters 6–8 of this book, can also be used for the luminescence temperature sensing. For MRI functionality, the most frequently used component is superparamagnetic iron oxide (magnetite Fe_3O_4 and maghemite γ-Fe_2O_3) and, to a lesser extent, different types of particles containing elements with unpaired electron spins such as gadolinium and manganese. Since they are biocompatible and nontoxic to humans, magnetite and maghemite nanocrystals are favored for use in biomedicine and, so far, many of the iron oxide–based contrast agents have been approved by the Food and Drug Administration (FDA). The spin–spin relaxation effects of iron oxide nanoparticles cause the shortening of both *T*1 and *T*2 relaxation times of protons located in their vicinity. Still, they are mainly used as

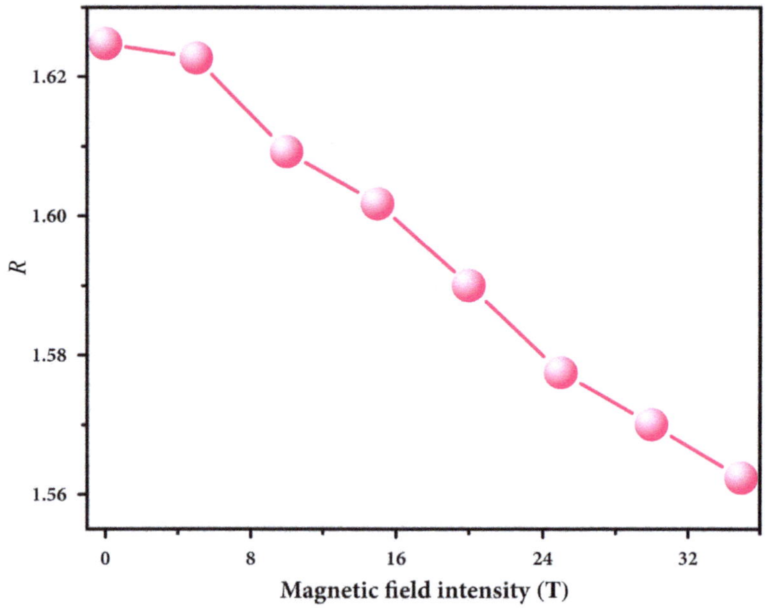

FIGURE 12.5 The dependence of index R and the integrated emission intensity on the magnetic field of the transitions of $^5D_0 \rightarrow {}^7F_1$ and $^5D_0 \rightarrow {}^7F_2$ intensity at 80K. *(Adapted from Chen P, Zhang J, Xu B, Sang X, Chen W, Liu X, et al. Lanthanide doped nanoparticles as remote sensors for magnetic fields. Nanoscale 2014;6(19):11002–6, with permission from The Royal Society of Chemistry).*

T2 contrast agents for generation of negative enhanced *T2*– and *T2**–weighted images [35]. Gd^{3+}-containing contrast agents (such as Gd^{3+} complexes, diamagnetic matrices doped with Gd cations, or Gd compounds such as Gd_2O_3 or $GdVO_4$) can reduce the *T1* relaxation time and are used as positive contrast agents.

Notable alternations in emission spectra of luminescent materials appear when they are exposed to magnetic fields since it splits degenerate quantized energy levels of electronic orbitals and shifts their positions due to the Zeeman effect (the effect named after a Dutch physicist Pieter Zeeman) [36]. Such behavior provides a means to use lanthanide doped nanoparticles as potential remote magnetic sensors in the nanoscale environment [37]. For example, in Eu^{3+} doped $NaYF_4$, a magnetic field induces the change in the positions of 4f levels. Also, intensities of emission lines are quenched with an increase in the magnetic field intensity [37] due to the enhanced cross-relaxation rates which differ among different levels (stronger for the $^5D_1$ level than for the $^5D_0$ level). It has been suggested [37] that a magnetic field also induces improvement in symmetry of the local environment around Eu^{3+} ions, changing in this way the so-called symmetry ratio index *R* which is defined as the ratio of integrated intensities of the magnetic dipole transition ($^5D_0 \rightarrow {}^7F_1$) and the electric dipole transition ($^5D_0 \rightarrow {}^7F_2$), see Fig. 12.5. Similar magnetic field effects on luminescence have been observed with Eu^{3+} doped hexagonal $NaGdF_4$ nanoparticles [38]. The tunable multicolor luminescence and paramagnetic properties are simultaneously realized in single-phase $NaGdF_4:Dy^{3+},Eu^{3+}$ nanofibers [39]. Bearing in mind that Dy^{3+} and Eu^{3+} emissions can be used for sensing temperature, this material can be regarded as an excellent platform for multifunctional physical sensing.

In Er^{3+} doped YVO_4 and $GdVO_4$ crystals, a strong increase in the magnitude of both green and mid-infrared emissions under a magnetic field of 4 T has been observed [40,41]. Accurate optical calibration of pulsed high magnetic fields is thus possible utilizing the magnetic field dependent photoluminescence of these materials. On the other hand, Singh et al. [42] noticed the eightfold decrease of green $Gd_2O_3:Er^{3+}/Yb^{3+}$ upconversion emission intensity with increase in the intensity of magnetic field from 0 to 1 T. Similar magnetic field–induced quenching of emission has been noted by Tikhomirov et al. [43] with Er^{3+} doped nano-glass-ceramics.

A magnetic field may also influence fluorescence of organic compounds. For example, an increase in exciplex luminescence was observed for the N-ethylcarbazole–dimethylterephthalate and N-ethylcarbazole–1,4-dicyanobenzene system in the presence of an external magnetic field [44].

12.6 LUMINESCENCE SENSING OF STRUCTURAL AND COMPOSITIONAL CHANGES IN MATERIALS

When discussing different applications of luminescence that can be combined with thermometry, one should highlight the use of luminescence in monitoring compositional and structural changes in materials. For example, luminescence is much used to detect the presence of certain gasses by the reaction of designated probes with CO_2, SO_2, and reactive oxygen and nitrogen species, as well as many other molecules [45,46]. Parallel factor analysis (PARAFAC) of luminescence has been demonstrated as a tool for the determination of the composition of rare earth–based phosphors in mixtures of powders, with a detection limit of 0.1% [47]. Sensitive detection of water content, to 0.008 vol% (80 ppm), in an organic solvent has been demonstrated using upconversion emission of Yb/Er codoped NaYF4 nanoparticles [47]. Luminescence is used to reveal structures and membrane interactions of proteins [48] and changes of cellular membranes [49]. Single-molecule spectroscopy has been used to understand the nature of the glass transition of poly(methacrylate) in which rhodamine 6G dye is dispersed [50].

In recent application of nanophosphor luminescence, Labrador-Páez et al. [51] have shown the crossover between different water properties/structures that occurs at about 45 °C. Two water colloids of Nd^{3+} doped and Yb^{3+} doped $GdVO_4$ ultrasmall nanoparticles (average diameter of 2.9 ± 0.5 nm) were mixed together in one nanofluid, excited by 808 nm laser radiation into an Nd^{3+} absorption line (donor nanoparticle), while monitoring emission of Yb^{3+} at 980 nm (acceptor nanoparticle) at different temperatures, see Fig. 12.6A. Since no Yb^{3+}

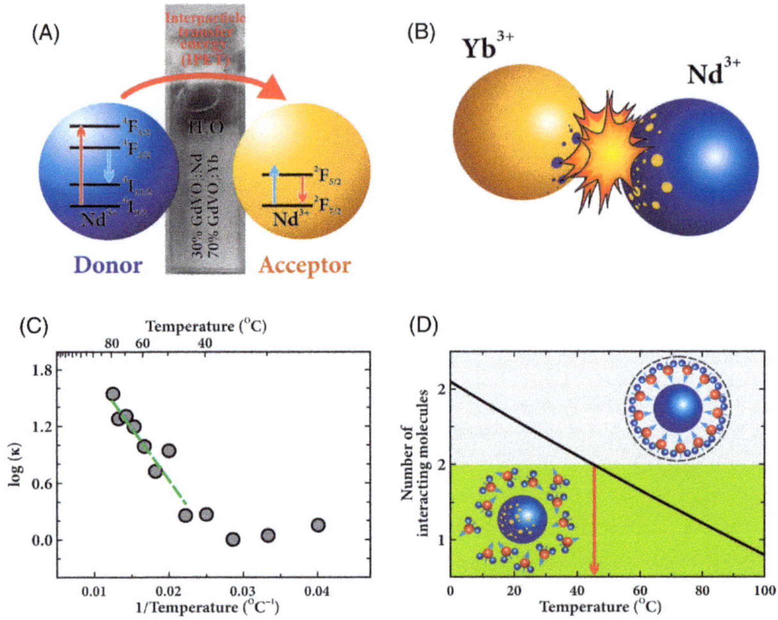

FIGURE 12.6 (A) An optical image of the cuvette containing the mix of Nd^{3+} and Yb^{3+} doped $GdVO_4$ ultrasmall water colloid nanoparticles with schematic representation of the energy levels of Nd^{3+} and Yb^{3+} ions responsible for the excitation and emission via interparticle transfer energy (IPET). (B) Schematic representation of a collision-assisted interparticle ion exchange process. (C) Logarithm of the ion transfer probability as a function of the inverse of temperature. A 1/T linear dependence is found for temperatures above 45°C (dashed line). (D) Number of interacting molecules in water as a function of temperature. Insets: Schematic diagram of water molecules: (top) forming a cage around particles which inhibit ion exchange in collisions; (bottom) being unable to interact among themselves and so allowing for ion exchange during interparticle collisions. The dashed arrow points out the crossover temperature. *(Reprinted from Labrador-Páez L, Jovanović DJ, Marqués MI, Smits K, Dolić SD, Jaque F, et al. Unveiling molecular changes in water by small luminescent nanoparticles. Small 2017;13(30):1700968 7 pp.), with permission from Wiley).*

emission was detected when only Yb^{3+} doped nanoparticles were excited by 808 nm radiation, Yb^{3+} emission in the presence of Nd^{3+} doped particles in the mixed nanofluid showed that interparticle energy transfer (IPET) occurs when particles collide in a solution. The IPET efficiency increased upon a rise in nanofluid temperature because of the enlarged particle collision probability due to the faster Brownian motion of particles. However, when nanofluid was heated above 45 °C, the variation of Yb^{3+} emission with temperature becomes irreversible suggesting that exchange of ions occurs between particles during particle collisions at high temperatures, see Fig. 12.6B. The probability of ion exchange is quite small at low temperatures but becomes significant at temperatures above 45 °C, at which it increases with temperature, see Fig. 12.6C. Thus, it is reasonable to assume that water properties change at about 45°C, see Fig. 12.6D, as it has been already reported in several studies [52–54]. At low temperatures, the hydration shells possess a hydrophobically improved water structure with larger tetrahedral order and fewer weak hydrogen bonds than neighboring bulk water, forming a "cage" around particles, see right inset in Fig. 12.6D. This structure disappears with rising temperature and is a more disordered structure with weaker hydrogen bonds than bulk water, see left inset in Fig. 12.6D [52]. These observations suggest that the solvent plays a decisive role in the interparticle dynamics of nanofluids, and that luminescence of colloidal small luminescent nanoparticles can be used as a valuable tool in the study of molecular interactions of not just water but also a variety of other liquids.

12.7 LUMINESCENCE SENSING OF CHEMICAL AND BIOCHEMICAL ANALYTES

The field of chemical and biochemical sensing is immense, active, and multidisciplinary, and, in many applications, its research is critically needed for the improvement of existing, and the development of novel, technologies. Optical chemical sensors have the ability to continuously record chemical species. Among them, luminescence sensors have shown significant value because of the versatility of features that can be utilized for obtaining analytical information [55], such as luminescence intensity, lifetime, polarization, etc. Luminescence methods are enormously sensitive and even allow the detection of single molecules [56]. Generally, with luminescence read-outs, there is no loss of analytes and no need for use of the reference material. They are feasible for remote monitoring since light can be easily detected over distances. Also, luminescence read-outs can be easily implemented or combined with luminescence imaging techniques. Consequently, it is possible to visualize many parameters or species which exhibit no intrinsic color or luminescence, for example, oxygen, Ph value, CO_2, H_2O_2, or Ca^{2+}, to name just a few, which is important in diverse areas, such as medicine, biotechnology, aerodynamics, chemical engineering, and marine research [57]. For these reasons, the largest single group of chemical sensors comprises, at present, sensors based on the use of fluorescent probes or labels [55]. The majority of materials used in luminescence chemical sensors also show appreciable luminescence dependence on temperature, and, thus, are suitable for multifunctional chemical–temperature sensing. The field of luminescence chemical sensing is huge and complex, and, thus, it would need a separate book (or more than one) to cover all the relevant information. This is, however, beyond the scope of this book and those interested in this subject should refer to the many excellent books and papers, such as, for example, those listed in Refs. [55–65].

It is important to note that luminescence sensor devices can be miniaturized and portable, providing in this way a means for a noninvasive monitoring of clinically relevant species and physiological parameters and a storage of data and/or their transfer to a physician [65]. Also, regarding multifunctionality, luminescence facilitates multiple sensing of chemical species using probes whose signals can

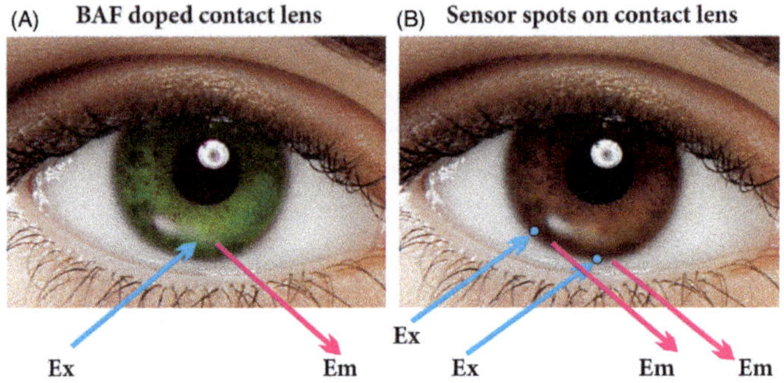

FIGURE 12.7 Fluorescence method for noninvasive continuous tear glucose monitoring. (A) Plastic contact lenses doped with the boronic acid-containing fluorophores (BAFs). (B) Spots on the surface of the lens containing additional sensor material for the monitoring of other analytes in addition to glucose, such as chloride or oxygen. Sensor spot regions may also allow for ratiometric-, lifetime-, or polarization-based fluorescence glucose sensing. *(Adapted from Badugu R, Lakowicz JR, Geddes CD. A glucose sensing contact lens: a non-invasive technique for continuous physiological glucose monitoring. J Fluoresc 2003;13(5):371–4, with permission from Springer).*

be differentiated by spectral and/or temporal resolution [64]. Such multiple sensors (dual, triple, etc.) are of substantial benefit to noninvasive and fast medical diagnostics. For example, Badugu et al. [66] demonstrated a potential method for noninvasive tear glucose monitoring befitting continuous monitoring in the range 50–1000 μM. Note that tear glucose levels are corelated to blood glucose levels (which are ≈ 10 fold higher). The sensor consists of commonly used plastic contact lenses doped with boronic acid contains fluorophores (BAFs) or containing spots of BAFs, which are well-known for sensing monosaccharides. BAFs are based on a quaternized form of the quinolinium nucleus, which react well to glucose within the contact lens, and which also show high monosaccharide affinity due to the stabilization of the anionic glucose bound form (the boronate ester form) by the cationic quaternary nitrogen heterocyclic center. Badugu et al. [66] envisaged that multiple modes of glucose sensing could be undertaken in contact lenses (ratiometric, lifetime, and polarization), and that sensor spots of different types can be used to monitor other analytes in addition to glucose, such as chloride or oxygen, see Fig. 12.7.

12.8 LUMINESCENCE BIOMEDICAL AND THERMAL IMAGING

Development of new multifunctional nanostructures capable of simultaneous imaging, heating, and local thermal sensing is expected to provide a means for overcoming the limitations of existing biomedical detection systems and the improvement of self-monitored thermal therapies. Development of two-dimensional thermal–imaging technologies would not only strongly impact real-time clinical imaging in humans and theranostics, but also other fields where there is a critical need for knowledge of submicron thermal gradients, such as nanoelectronics and micromechanics. Luminescence nanoparticles having both strong excitation and emission bands in the so-called biological windows (near-infrared spectral region) are almost ideal candidates for two-dimensional thermal–bioimaging probes. This is because small nanoparticles can flow through the bloodstream to be distributed along the organism after intravenous injection [67], and because many of them present efficient emission with an appropriate temperature dependence in the physiological temperature range (10–50°C).

Recently, Carrasco et al. [68] demonstrated Nd^{3+} ion doped LaF_3 nanoparticles capable of in vivo photothermal heating, fluorescent tumor localization, and intratumoral temperature sensing, see Fig. 12.8A. Luminescence imaging functionality comes from the Nd^{3+} emission in the near-infrared after excitation with a 808 nm laser beam. Due to heavy doping (5.6 at% of Nd), nonradiative transitions are enhanced and, consequently, nanoparticles generate heat in addition to luminescence (see Fig. 12.8, image on the right). The temperature-sensing functionality is based on the ratio between emitted intensities at 865 and 885 nm, which correspond to electronic transitions between two Stark sublevels of the $^4F_{3/2}$ excited level and $^4I_{9/2}$ ground level of Nd^{3+} ions (see Chapter 6).

The performance of nanoparticles has been tested in athymic nude mice in which human breast cancer cells (MDA-MB-231 line) have been introduced in both flanks of each mouse to develop tumors. When tumor size exceeded 9 mm^3, a solution of Nd^{3+}:LaF_3 nanoparticles dispersed in a phosphate buffered saline (PBS) was directly injected into one tumor while pure PBS solution was injected into another. Both tumors were then irradiated with an 808 nm laser. Optical, luminescent, and thermal images of the mouse are shown in Fig. 12.8B. It can be clearly seen that under 808 nm irradiation Nd^{3+}:LaF_3 nanoparticles enable both luminescence and thermal imaging in a living animal. What is more, under 808 nm irradiation they developed sufficient heat to be used as a photothermal agent, see Fig. 12.8C, and, finally, to ablate the tumor. Fig. 12.8C also shows the difference in temperature measurements performed by luminescence thermometry (the intratumoral temperature) and by an infrared thermal camera (the surface/skin temperature). It was found that the intratumoral temperature was about 20% higher than the surface temperature, which can be interpreted by considering the heat diffusion through tissue from the tumor's location to the surface and heat losses at tissue–air interfaces. Thus, it is important to emphasize that luminescence thermometry has provided local temperature at the site of a tumor, something which is important for photothermal tumor treatments since proper temperature monitoring is key to achieving optimal therapeutic results.

The need for precise temperature monitoring of photothermal cancer treatments has boosted research in developing luminescence nanoprobes capable of both heating and temperature sensing, the so-called nanoheater/nanosenor platforms. As a result, in recent times, temperature-controlled photothermal therapy has also been demonstrated with PbS/CdS/ZnS quantum dots [69], and with the hybrid structure composed of a core of Er^{3+}/Yb^{3+} codoped $NaLuF_4$ nanocrystal and a passive shell of undoped $NaLuF_4$ covered with an external carbon shell [70].

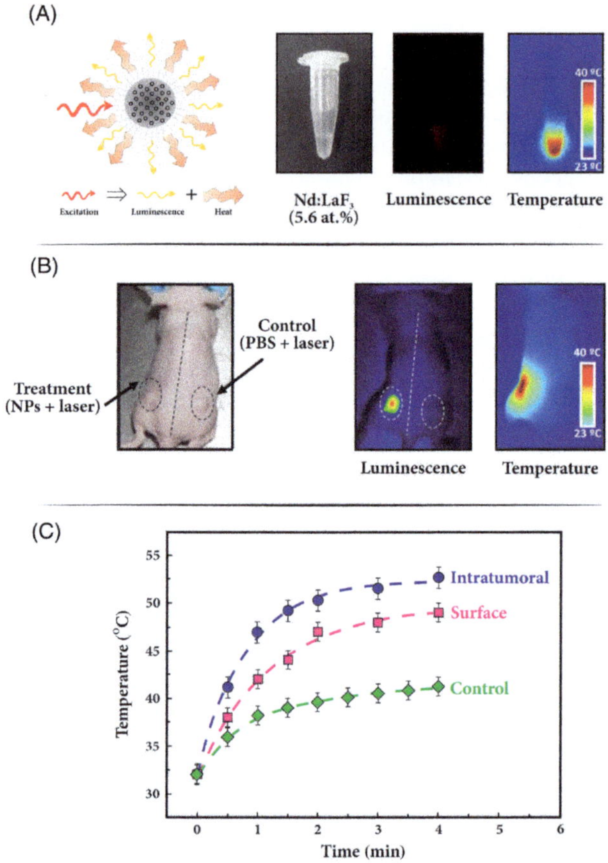

FIGURE 12.8 (A) Heavily doped Nd^{3+}:LaF_3 nanoparticles: schematic of the excitation–emission heat generation process; the optical image of the tube containing an aqueous solution of nanoparticles; and luminescence and thermal image of the same solution under 808 nm optical excitation. (B) Optical image of a mouse with two tumors. A solution of "dense" Nd:LaF_3 nanoparticles was injected only into the left-side tumor whereas the right-side tumor was used as a control. (C) and (D) Infrared fluorescence and thermal images of the same mouse under 808 nm (4 W cm^{-2}) laser irradiation, respectively. The images show the fluorescent and heating signals differentially emitted by the treated tumor. (E) Time evolution of the temperature at the tumor surface as obtained from the analysis of infrared thermal images (dark gray (red in the web version) squares), and the time evolution of intratumoral temperature obtained from the analysis of subtissue fluorescence (black (blue in the web version) circles)); the control is also shown (darker gray (green in the web version) diamonds). *(Reprinted from Carrasco E, del Rosal B, Sanz-Rodríguez F, de la Fuente ÁJ, Gonzalez PH, Rocha U, et al. Intratumoral thermal reading during photo-thermal therapy by multifunctional fluorescent nanoparticles. Adv Funct Mater 2015;25(4):615–26 with permission from Wiley).*

REFERENCES

[1] Piascik B, Vickers J, Lowry D, Scotti S, Stewart J, Calomino A. Materials, structures, mechanical systems, and manufacturing roadmap-technology area 12. Washington; 2012.
[2] Kamila S. Classification and applications of smart materials: an overview. Am J Appl Sci 2013;10(8):876–80.
[3] Dramićanin MD. Sensing temperature via downshifting emissions of lanthanide-doped metal oxides and salts. A review. Methods Appl Fluoresc 2016;4(4):42001. 23 pp..
[4] Kautsky H, Hirsch A. Nachweis geringster Sauerstoffmengen durch Phosphoreszenztilgung. Z Anorg Allg Chem 1935;222(2):126–34.
[5] Le Sant Y, Mérienne M-C. Surface pressure measurements by using pressure-sensitive paints. Aerosp Sci Technol 2005;9(4):285–99.
[6] McLachlan BG, Bell JH. Pressure-sensitive paint in aerodynamic testing. Exp Therm Fluid Sci 1995;10(4):470–85.
[7] Köse ME, Carroll BF, Schanze KS. Preparation and spectroscopic properties of multiluminophore luminescent oxygen and temperature sensor films. Langmuir 2005;21(20):9121–9.
[8] Engler RH, Mérienne M-C, Klein C, Le Sant Y. Application of PSP in low speed flows. Aerosp Sci Technol 2002;6(5):313–22.
[9] Baleizão C, Nagl S, Schäferling M, Berberan-Santos MN, Wolfbeis OS. Dual fluorescence sensor for trace oxygen and temperature with unmatched range and sensitivity. Anal Chem 2008;80(16):6449–57.
[10] Forman RA, Piermarini GJ, Barnett JD, Block S. Pressure measurement made by the utilization of ruby sharp-line luminescence. Science 1972;176(4032):284–5.
[11] Shen YR, Gregorian T, Holzapfel WB. Progress in pressure measurements with luminescence sensors. High Press Res 1991;7(1–6):73–5.
[12] Bublitz GU, Boxer SG. Stark spectroscopy: applications in chemistry, biology, and materials science. Annu Rev Phys Chem 1997;48(1):213–42.
[13] Takehara Y, Ohta N, Shiraishi S, Asaoka S, Wada T, Inoue Y. External electric field effects on exciplex formation of 1,1-diphenylpropene with chiral 1,4-naphthalenedicarboxylate in PMMA polymer films. J Photochem Photobiol A Chem 2001;145(1):53–60.
[14] Ohta N. Electric-field effects on photoinduced dynamics and function. Pure Appl Chem 2013;85(7):1427–35.
[15] Ohta N, Mikami S, Iwaki Y, Tsushima M, Imahori H, Tamaki K, et al. Acceleration and deceleration of photoinduced electron transfer rates by an electric field in porphyrin-fullerene dyads. Chem Phys Lett 2003;368(1):230–5.
[16] Nakabayashi T, Hino K, Ohta Y, Ito S, Nakano H, Ohta N. Electric-field-induced changes in absorption and fluorescence of the green fluorescent protein chromophore in a PMMA film. J Phys Chem B 2011;115(26):8622–6.
[17] Nakabayashi T, Ohshima R, Ohta N. Electric field effects on photoluminescence of CdSe nanoparticles in a PMMA film. Crystals 2014;4:152–67.
[18] Gurinovich LI, Lyutich AA, Stupak AP, Artem'ev MV, Gaponenko SV. Effect of an electric field on photoluminescence of cadmium selenide nanocrystals. J Appl Spectrosc 2010;77(1):120–5.
[19] Mehata MS, Majumder M, Mallik B, Ohta N. External electric field effects on optical property and excitation dynamics of capped CdS quantum dots embedded in a polymer film. J Phys Chem C 2010;114(37):15594–601.
[20] Ohshima R, Nakabayashi T, Kobayashi Y, Tamai N, Ohta N. External electric field effects on state energy and photoexcitation dynamics of water-soluble CdTe nanoparticles. J Phys Chem C 2011;115(31):15274–81.

[21] Hao J, Zhang Y, Wei X. Electric-induced enhancement and modulation of upconversion photoluminescence in epitaxial $BaTiO_3$:Yb/Er thin films. Angew Chem Int Ed 2011;50(30):6876–80.

[22] Zhang Y, Hao J, Mak CL, Wei X. Effects of site substitutions and concentration on upconversion luminescence of Er^{3+}-doped perovskite titanate. Opt Express 2011;19(3):1824–9.

[23] Tazaki R, Fu D, Itoh M, Daimon M, Koshihara S. Lattice distortion under an electric field in $BaTiO_3$ piezoelectric single crystal. J Phys Condens Matter 2009;21(21):215903.

[24] Gavrilović TV, Jovanović DJ, Lojpur V, Dramićanin MD. Multifunctional Eu^{3+}- and Er^{3+}/Yb^{3+}-doped $GdVO_4$ nanoparticles synthesized by reverse micelle method. Sci Rep 2014;4(4209). 9 pp..

[25] Demchenko AP. Fluorescent nanocomposites. In: Demchenko AP, editor. Introduction to fluorescence sensing. 2nd ed. Cham: Springer International Publishing; 2015. p. 263–99.

[26] Kircher MF, Mahmood U, King RS, Weissleder R, Josephson L. A multimodal nanoparticle for preoperative magnetic resonance imaging and intraoperative optical brain tumor delineation. Cancer Res 2003;63(23):8122–5.

[27] Xie H-Y, Zuo C, Liu Y, Zhang Z-L, Pang D-W, Li X-L, et al. Cell-targeting multifunctional nanospheres with both fluorescence and magnetism. Small 2005;1(5):506–9.

[28] Wang G-P, Song E-Q, Xie H-Y, Zhang Z-L, Tian Z-Q, Zuo C, et al. Biofunctionalization of fluorescent-magnetic-bifunctional nanospheres and their applications. Chem Commun 2005;34:4276–8.

[29] Sathe TR, Agrawal A, Nie S. Mesoporous silica beads embedded with semiconductor quantum dots and iron oxide nanocrystals: dual-function microcarriers for optical encoding and magnetic separation. Anal Chem 2006;78(16):5627–32.

[30] Nunez NO, Rivera S, Alcantara D, de la Fuente JM, Garcia-Sevillano J, Ocana M. Surface modified Eu:$GdVO_4$ nanocrystals for optical and MRI imaging. Dalt Trans 2013;42(30):10725–34.

[31] Lu C-W, Hung Y, Hsiao J-K, Yao M, Chung T-H, Lin Y-S, et al. Bifunctional magnetic silica nanoparticles for highly efficient human stem cell labeling. Nano Lett 2007;7(1):149–54.

[32] Lee J-H, Jun Y, Yeon S-I, Shin J-S, Cheon J. Dual-mode nanoparticle probes for high-performance magnetic resonance and fluorescence imaging of neuroblastoma. Angew Chem Int Ed 2006;45(48):8160–2.

[33] Tan H, Wang M, Yang C-T, Pant S, Bhakoo KK, Wong SY, et al. Silica nanocapsules of fluorescent conjugated polymers and superparamagnetic nanocrystals for dual-mode cellular imaging. Chem Eur J 2011;17(24):6696–706.

[34] Mohapatra S, Sahu S, Nayak S, Ghosh SK. Design of Fe_3O_4@SiO_2@carbon quantum dot based nanostructure for fluorescence sensing, magnetic separation, and live cell imaging of fluoride ion. Langmuir 2015;31(29):8111–20.

[35] Qiao R, Yang C, Gao M. Superparamagnetic iron oxide nanoparticles: from preparations to in vivo MRI applications. J Mater Chem 2009;19(35):6274–93.

[36] Judd BR, Lindgren I. Theory of Zeeman Effect in the ground multiplets of rare-earth atoms. Phys Rev 1961;122(6):1802–12.

[37] Chen P, Zhang J, Xu B, Sang X, Chen W, Liu X, et al. Lanthanide doped nanoparticles as remote sensors for magnetic fields. Nanoscale 2014;6(19):11002–6.

[38] Chen P, Jia H, Zhang J, Han J, Liu X, Qiu J. Magnetic tuning of optical hysteresis behavior in lanthanide-doped nanoparticles. J Phys Chem C 2015;119(10):5583–8.

[39] Li D, Ma Q, Xi X, Dong X, Yu W, Wang J, et al. Dy^{3+} and Eu^{3+} Co-doped $NaGdF_4$ nanofibers endowed with bifunctionality of tunable multicolor luminescence and paramagnetic properties. Chem Eng J 2017;309:230–9.

[40] Ma Z-W, Zhang J-P, Wang X, Yu Y, Han J-B, Du G-H, et al. Magnetic field induced great photoluminescence enhancement in an Er^{3+}:YVO_4 single crystal used for high magnetic field calibration. Opt Lett 2013;38(19):3754–7.
[41] Zhang J, Zhong Z, Wang X, Ma Z, Wang S, Han Y, et al. Large photoluminescence enhancement of Er^{3+}: $GdVO_4$ crystal in both green and middle infrared regions under moderate low magnetic fields. Opt Mater Express 2016;6(11):3446–54.
[42] Singh SK, Kumar K, Srivastava MK, Rai DK, Rai SB. Magnetic-field-induced optical bistability in multifunctional Gd_2O_3:Er^{3+}/Yb^{3+} upconversion nanophosphor. Opt Lett 2010;35(10):1575–7.
[43] Tikhomirov VK, Chibotaru LF, Saurel D, Gredin P, Mortier M, Moshchalkov VV. Er^{3+}-doped nanoparticles for optical detection of magnetic field. Nano Lett 2009;9(2):721–4.
[44] Choudhury SD, Basu S. Magnetic field effect on N-ethylcarbazole–dimethylterephthalate and N-ethylcarbazole–1,4-dicyanobenzene: a comparitive study focusing on steric effect. Chem Phys Lett 2005;408(4):274–8.
[45] Zhou X, Lee S, Xu Z, Yoon J. Recent progress on the development of chemosensors for gases. Chem Rev 2015;115(15):7944–8000.
[46] Chen X, Tian X, Shin I, Yoon J. Fluorescent and luminescent probes for detection of reactive oxygen and nitrogen species. Chem Soc Rev 2011;40(9):4783–804.
[47] Lenhardt L, Dramićanin MD, PARAFA C. A tool for the analysis of phosphor mixture luminescence. J Lumin 2016;170:136–40.
[48] Munishkina LA, Fink AL. Fluorescence as a method to reveal structures and membrane-interactions of amyloidogenic proteins. Biochim Biophys Acta Biomembr 2007;1768(8):1862–85.
[49] Demchenko AP. The change of cellular membranes on apoptosis: fluorescence detection. Exp Oncol 2012;24(3):263–8.
[50] Deschenes LA, Vanden Bout DA. Single-molecule studies of heterogeneous dynamics in polymer melts near the glass transition. Science 2001;292(5515):255–8.
[51] Labrador-Páez L, Jovanović DJ, Marqués MI, Smits K, Dolić SD, Jaque F, et al. Unveiling molecular changes in water by small luminescent nanoparticles. Small 2017;13(30):1700968. 7 pp..
[52] Davis JG, Gierszal KP, Wang P, Ben-Amotz D. Water structural transformation at molecular hydrophobic interfaces. Nature 2012;491(7425):582–5.
[53] Mallamace F, Corsaro C, Stanley HE. A singular thermodynamically consistent temperature at the origin of the anomalous behavior of liquid water. Sci Rep 2012;2(993):5.
[54] Maestro LM, Marqués MI, Camarillo E, Jaque D, Solé JG, Gonzalo JA, et al. On the existence of two states in liquid water: impact on biological and nanoscopic systems. Int J Nanotechnol 2016;13(8/9):667. 11 pp.
[55] Wolfbeis OS. Materials for fluorescence-based optical chemical sensors. J Mater Chem 2005;15(27–28):2657–69.
[56] Basabe-Desmonts L, Reinhoudt DN, Crego-Calama M. Design of fluorescent materials for chemical sensing. Chem Soc Rev 2007;36(6):993–1017.
[57] Schäferling M. The Art of Fluorescence Imaging with Chemical Sensors. Angew Chem Int Ed 2012;51(15):3532–54.
[58] Demchenko AP. Introduction to Fluorescence Sensing. Cham: Springer International Publishing; 2015.
[59] Valeur B, Berberan-Santos MN. Molecular fluorescence: principles and applications. 2nd ed. Weinheim: Wiley-VCH; 2013. 569.
[60] Wang X-D, Wolfbeis OS. Fiber-optic chemical sensors and biosensors (2008–2012). Anal Chem 2013;85(2):487–508.

[61] Wang X, Wolfbeis OS. Fiber-optic chemical sensors and biosensors (2013–2015). Anal Chem 2016;88(1):203–27.

[62] Wang S, Li N, Pan W, Tang B. Advances in functional fluorescent and luminescent probes for imaging intracellular small-molecule reactive species. TrAC – Trends Anal Chem 2012;39:3–37.

[63] Achatz DE, Ali R, Wolfbeis OS. Luminescent chemical sensing, biosensing, and screening using upconverting nanoparticles. In: Prodi L, Montalti M, Zaccheroni N, editors. Luminescence applied in sensor science. Berlin, Luminescence applied in sensor science: Springer Berlin Heidelberg; 2011. p. 29–50.

[64] Stich MIJ, Fischer LH, Wolfbeis OS. Multiple fluorescent chemical sensing and imaging. Chem Soc Rev 2010;39(8):3102–14.

[65] Lakowicz JR. Principles of fluorescence spectroscopy. In: Lakowicz JR, editor. Fluorescence sensing. Boston, MA: Springer US; 2006. p. 623–73.

[66] Badugu R, Lakowicz JR, Geddes CD. A glucose sensing contact lens: a non-invasive technique for continuous physiological glucose monitoring. J Fluoresc 2003;13(5):371–4.

[67] Alexis F, Pridgen E, Molnar LK, Farokhzad OC. Factors affecting the clearance and biodistribution of polymeric nanoparticles. Mol Pharm 2008;5(4):505–15.

[68] Carrasco E, del Rosal B, Sanz-Rodríguez F, de la Fuente ÁJ, Gonzalez PH, Rocha U, et al. Intratumoral thermal reading during photo-thermal therapy by multifunctional fluorescent nanoparticles. Adv Funct Mater 2015;25(4):615–26.

[69] del Rosal B, Carrasco E, Ren F, Benayas A, Vetrone F, Sanz-Rodríguez F, et al. Infrared-emitting QDs for thermal therapy with real-time subcutaneous temperature feedback. Adv Funct Mater 2016;26(33):6060–8.

[70] Zhu X, Feng W, Chang J, Tan Y-W, Li J, Chen M, et al. Temperature-feedback upconversion nanocomposite for accurate photothermal therapy at facile temperature. Nat Commun 2016;7(10437):10 pp..

Index

A

Absolute primary thermometer, 16
Absolute sensitivity, 30
Absorption filters, 59
Absorptive polarizers, 59
Accuracy of measurement, 29
Acoustic gas thermometry (AGT), 13, 17
Air-thermoscope, 4
Anisotropy, 78
Autofluorescence, 240

B

Ball milling, 132
Band pass filters, 59
Bimetallic thermometers, 13, 24
Biomedical applications, luminescence thermometry
  biological tissues
    cancer diagnostics, 235
    food chemistry, 235
    molecular composition, 237
    penetration depth of radiation, 237
    refractive index, 236
    wavelength dependence, 238
  intracellular and intercellular measurements, 240
  in-tumor temperature, 245
  ischemia diagnosis, 246
  tissue fluorescence, 238
Biosystem temperature, 11
N,N-bis(2,5-di-tertbutylphenyl)-3,4,9,10-perylenedi carboxamide (BTBP) dye, 257
Blackbody radiation, 138, 139
Boltzmann constant, 16
Boltzmann-type ratiometric temperature reading, 70, 71
Boronic acid–containing fluorophores (BAFs), 279
Boyle's law, 5
Bunsen prism monochromator, 53, 55

C

Carbon nanothermometer, 252
Carnot's theorem, 14
Cathodoluminescence, 38
Celsius scale, 8, 15
Centigrades, 8
Ceramic method, 129
Charge-coupled device (CCD) cameras, 58
Charge transfer (CT) band, 65
Chemiluminescence, 37
Chlorophyll, 239
Collagen, 240
Combustion methods, 130
Conference on Weights and Measures (CGPM), 16
Configuration coordinate diagram, 42
Constant-volume gas thermometry, 18, 26
Cross-luminescence, 39
Cut off filters, 59
Czerny–Turner grating monochromator, 53, 55
Czochralski process, 129

D

Dichroic polarizers, 59
Dieke diagram, 120, 121
Dielectric constant gas thermometry, 19
Digital low-pass filtering, 92
Doppler-broadening thermometry, 19
Downshifting phosphors, 124

E

Elastin, 240
Electrical machines, 216
Electrical nanothermometry methods, 254, 255
Electroluminescence, 38
Electron-beam physical vapor deposition (EB-PVD), 225
Electron density of states (DOS), 160
Electron–hole pairs (EHP), 161
Electronic circuits, 216
Emission barycenter, 102
Environmental barrier coatings (EBCs), 224
Exciplex-type probes, 203
Excitation-emission matrix (EEM), 46
Exciton Bohr radius, 160, 162
Extrinsic luminescence, 39

F

Fabry–Perot filters, 59
Fahrenheit scale, 7
Flame fusion process, 129
Fluorescein dye
  chemical formula and structure, 191
  excitation and emission, 191
  optical properties, 192
Fluorescein isothiocyanate (FITC), 190
Fluorescence polarimeter, 50
Forward smoothing, 92
Fractoluminescence, 38
Frequency domain measurements of emission decay times, 73

G

Gay-Lussac's law, 6
Generic luminescence thermometry system, 78
Glan–Thompson polarizer, 59
Glasses and glass-ceramics, 134
Grating monochromators, 53

H

Hydrothermal method, 131

I

Infrared thermography, 221
Infrared thermometers, 27
Inokuti–Hirayama model, 108
Insulator-based dielectrophoretic devices (iDEP) channel, 257
Interference filters, 59
Interference wedge, 59
Internal conversion, 36
International Temperature Scale (ITS-90), 20
Interparticle energy transfer (IPET), 277
Intersystem crossing (ISC), 36
Intrinsic luminescence, 39

J

Jablonski diagram, 33, 36
Johnson noise thermometry, 19
J-selection rules, 122
Judd–Offelt theory, 121

K

Kelvin scale, 8
  definition, 13
  numerical value, 15

L

Lanthanide and transition metal ion doped materials
  crystal growth, 127
  crystalline environment, 114
  Dieke diagram, 120
  divalent lanthanide ions, 123
  downshifting phosphors, 124
  electronic configurations, 114, 115, 117
  energy level structure, 114
  4f–5d transitions, 123
  glasses and glass-ceramics, 134
  J-mixing, 122
  lanthanum and lutetium, 119
  nephelauxetic effect, 117
  octahedral symmetry, 114, 116
  powders preparation
    liquids–solution based process, 130
    low-temperature, self-propagating synthesis, 130
    molten salt method, 130
    solid-state method, 129
  quantum-cutting phosphors, 125, 126
  scintillating phosphors, 126
  spectroscopic properties, 114
  Tanabe–Sugano diagram, 116, 148
  temperature sensing. *See* Luminescence thermometry
  thin films and coatings preparation, 132
  trivalent europium ion centers, energy levels, 121, 122
  trivalent lanthanide ions, 120
  upconversion processes, 124, 125
Laser ablation method, 132
Liquid-in-glass thermometer, 24
Little Florentine thermometer, 4, 5
Localized luminescence centers, 40
Luminescence
  cathodoluminescence, 38
  chemiluminescence, 37
  electroluminescence, 38
  electromagnetic spectrum
    absorption process, 34
    frequencies, 33
    interaction with matter, 33, 34
    reflection, 34
    refraction, 34
    wavelength, 33
  history, 35
  mechanoluminescence, 38
  photoluminescence. *See* Photoluminescence
  quenching, 36
  in solids, 39

sonoluminescence, 39
temperature quenching
brightness, 41
collisional quenching, 41
configuration coordinate diagram, 42
lifetime of excited state, 40
Mott–Seitz theory, 42, 44
nonradiative transition rate, 41
quantum yield, 41
radiative lifetime, 41
temporal evolution, 40
temperature read-out from
anisotropy, 78
decay time, 73
emission and excitation bands, 64
generic luminescence thermometry system, 78
ratiometric temperature sensing, 69
rise time, 74
single emission band intensity, 68
static *vs.* time-resolved method, 80
time-integrated methods, 63
time-resolved methods, 63
types of, 64
thermoluminescence, 38
Luminescence intensity ratio temperature (LIRT), 90
Luminescence sensing
biomedical and thermal imaging, 280
chemical and biochemical analytes, 279
electric field, 270
magnetic field, 275
multifunctional and smart materials, 265
pressure, 268
structural and compositional changes, 277
Luminescence thermometry, 10
baseline offset
approximation, 87
black-body radiation, 86
dark current, 86
data preprocessing methods, 87
instrumental approach, 86, 87
luminescence intensity ratio temperature, 90
manual baseline removal produces, 89
stray light, 86
three-point correction, 89
biological tissues
cancer diagnostics, 235
food chemistry, 235
molecular composition, 237
penetration depth of radiation, 237
refractive index, 236
wavelength dependence, 238
digital signal processing, 86
dyes and organic pigments
conjugated molecules, 193
electronic transitions, 190
energy levels of molecular orbitals, 191, 193
fluorescein, 190
hybrid probes, 196
Pyronin B, 194
rhodamine, 192, 194
xanthene dyes, 194, 196
engineering
electrical machines, 216
electronic circuits, 216
environmental barrier coatings, 224
fluid flows, 222
heat flux, 215
phosphor coatings, 227
temperature-sensitive paints, 227
thermal barrier coating, 224
exciplex-type probes, 203
functional polymer dyes, 198
intracellular and intercellular temperature measurements, 240
in-tumor temperature, 245
ischemia diagnosis, 246
lanthanide ion doped materials
blackbody radiation, 138, 139
downshifting emission, 140
LIRT, 135
low-temperature dependence, 137, 138
phenomenological error parameter, 136
relative sensitivity, 136, 137
thermally coupled energy levels, 136
upconversion emission, 135, 145
luminescence spectra quantification, 98
measurement data, 85
metal–organic complexes, 204
metal–organic frameworks, 207
noise
digital low-pass filtering, 92
ensemble average method, 91
forward smoothing, 92
high-frequency fluctuation, 91
photon noise, 91
Savitzky–Golay smooting, 93, 95
smooth width, 92
triangular smoothing, 94
numerical resolution enhancement, 85
performance, 109

Luminescence thermometry (*cont.*)
quantum dots. *See* Semiconductor quantum dots
sensitivity, 109
spectral data differentiation, 95
time-resolved measurements, 104
tissue fluorescence, 238
transition ion doped materials, 147

M

Magnetic thermometry, 20
Manometric thermometers, 26
Mechanical milling, 132
Mechanoluminescence, 38
Microfluidics, 256
Mirror image rule, 45
Molten salt method, 130
Mott–Seitz theory, 42, 44
Multidimensional luminescence measurements, 46
Multiexponential decay law, 106

N

Nanoelectronics, 260
Nanofluidics, 256
Nanothermomometry
carbon nanothermometer, 252
contact thermocouple set-up, 255
current voltage characteristics, 252
electrical nanothermometry methods, 254, 255
hot-spots, 251
liquid and solid-in-glass thermometers, 252
microfluidics and nanofluidics, 256
nanoelectronics, 260
physical processes, 251
resonant tunneling diode, 252, 254
scanning thermal microscopy, 254
spatial resolution, 254
temperature, spatial, and temporal resolutions, 256
Nephelauxetic effect, 117
Neutral-density filters, 59
Neutral temperature, 2
Nicotinamide adenine dinucleotide (phosphate) (NAD(P)H), 239

O

Optical filters, 59
Optical parametric oscillators (OPOs), 53

P

Parallel factor analysis (PARAFAC) of luminescence, 277
Particle image velocimetry (PIV), 222, 223
Peak area, 100
Peak sharpening, 102, 103
Peak width, 100
Pechini method, 130
Penetration depth of radiation, 237
Phase-modulated technique, 104
Phenomenological error parameter, 136
Phenylalanine, 238
Phosphor coatings, 227
Phosphorescence (PHOS), 33, 36
Photoluminescence, 33
anisotropy measurements, 50
delayed luminescence, 37
emission spectrum, 45
excitation-emission matrix, 46
excitation spectrum, 45
excited state lifetime measurements, 46, 49
fluorescence polarimeter, 50
frequency domain measurements, 46
generic set-up, 51
instrumentation
charge-coupled device cameras, 58
gratings, 56
lasers, 53
light sources, 51, 52
monochromators, 53
optical filters, 59
photomultiplier tube, 56, 57
PIN photodiode, 58
polarizer, 59
solid state semiconductor detectors, 58
internal conversion, 36
lanthanide doped materials, 45
mirror image rule, 45
multidimensional luminescence measurements, 46
observed data, 45
phosphorescence, 33, 36
process involved, 33
quantum yield, 48
steady-state measurements, 44
synchronous luminescence spectrum, 46
3D TSLS, 46, 48
vibrational relaxation, 36
Photomultiplier tube (PMT), 56, 57
Photon noise, 91
Piezoelectrics, 266
Piezoluminescence, 38

Polarizer, 59
Polydiacetylene (PDA) supramolecules, 220
Porphyrins, 239
Precipitation method, 131
Pressure-sensitive paints (PSP), 268
Pressure-sensitive polymer coatings (PSPC), 268
Provisional Low Temperature Scale (PLTS-2000), 20
Pyridoxine, 239
Pyrometers, 27
Pyrometry, 11

Q

Quantum-cutting phosphors, 124–126
Quantum dots (QDs) emission bands, 66. *See also* Semiconductor quantum dots

R

Radiation thermometers, 27
Rankine scale, 8
Ratiometric temperature sensing, 69
Relative primary thermometer, 16
Relative sensitivity, 30
Repeatability measurement condition, 29
Reproducibility measurement condition, 29
Resistance thermometers, 25
Resonant tunneling diode (RTD), 254
Reverse micelles, 131
Rhodamine B dye
  absorption and emission spectra, 195
  fluorescence intensity, 196, 197
  fluorescence quenching, 194
  quantum yield, 194
  structural formula, 195
  temperature dependence of quantum yield, 194, 195
Riboflavin, 239

S

Savitzky–Golay smooting, 93, 95
Scanning thermal microscope (SThM), 216, 217
Scintillating phosphors, 126
Self-referencing read-out methods, 74, 80
Semiconductor quantum dots
  low-dimensional materials, 160
  nanostructures, 160
  optical properties, 159, 164
  principal applications, 159
  qantum confinement
    charge carriers, 164
    electron density of states, 160
    electron–hole pairs, 161
    energy dependence, 160
    exciton Bohr radius, 160
    intermediate-confinement regime, 161
    luminescence changes, 160
    weak-confinement regime, 161
  quantization effect, 159
  surface chemistry o, 164
  temperature sensing
    Ag/Ag_2S quantum dot, 177
    CdSe/CdS/ZnS core–shell–shell dots, 173
    CdSe quantum dots, 162, 170
    CdSe/ZnS QDs, 175
    CdTe quantum dots, 170, 172, 183
    first exciton absorption peak changes, 166
    homogeneous bandwidths, 170
    hybrid luminescence temperature nanosensor, 182
    Mn^{2+}-doped semiconductor quantum dots, 178, 181
    PbS/CdS/ZnS QDs emissions, 176, 182
    PbSe quantum dots, 167, 168
    PbS quantum dots, 166, 167
    QD luminescence, 166
    quenching and antiquenching, 172, 173
    single CdSe/ZnS core–shell quantum dot, 169, 172
    temperature-dependent dual emission, 180
    temperature-induced spectral shift, 170
    toxic heavy metals, 174
    two-photon luminescence microscopy, 174
    $ZnS\text{-}AgInS_2$ alloy quantum dots, 174, 175
Smart materials, 265
Sol–gel synthesis, 131
Solid-in-glass thermometers, 252
Solid-state method, 129
Solid state semiconductor detectors, 58
Solvothermal method, 131
Sonoluminescence, 39
Spatial resolution, 30, 216
Spectral-band radiometric thermometry, 18
Spectral-band thermometer, 27
Spirit-in-glass thermometer, 4
Spray and flame pyrolysis, 131
Stark effect, 270
Synchronous luminescence spectrum, 46

T

Tanabe–Sugano diagram, 114, 116
Temperature
  Carnot's theorem, 14
  CGPM, 16
  efficiency of ideal engine, 14
  first law of thermodynamics, 14
  kinetic approach, 14
  probability, 14
  qualitative definition, 13
  zeroth law of thermodynamics, 14
Temperature measurements
  biosystem temperature, 11
  determining factor, 1
  flows of liquid and gas, 11
  historical background
    air-thermoscope, 4
    Celsius scale, 8
    constant pressure, 7
    Fahrenheit scale, 7
    fixed points, 6
    Kelvin scale, 8
    Little Florentine thermometer, 4, 5
    luminescence thermometry, 10
    mercury, 5
    neutral temperature, 2
    Rankine scale, 8
    thermometer, 2
    thermoscope, 2
  infection, 1
  nanoelectronic devices, 11
  sensor market, 1
  surface temperature measurements, 11
  water temperature, 1
Temperature resolution, 30
Temperature-sensitive paints, 227
Temporal resolution, 31
Thermal barrier coating (TBC), 224
Thermistors, 25
Thermocouple, 13
Thermoluminescence, 38
Thermometer
  assorted temperature-dependent phenomena, 23
  classification
    bimetallic thermometers, 24
    infrared thermometers, 27
    liquid-in-glass thermometer, 24
    manometric thermometers, 26
    resistance thermometers, 25
    temperature range, 23, 24
    thermistors, 25
    thermocouple, 23, 25
  measurement errors
    precision index, 27
    random errors, 27
    result of measurement, 29
    systematic errors (bias), 28
  measurement method, 29
  measurement model, 29
  measurement principle, 29
  primary
    absolute, 16, 18
    acoustic gas thermometry, 13, 17
    constant-volume gas thermometry, 18
    dielectric constant gas thermometry, 19
    Doppler-broadening thermometry, 19
    Johnson noise thermometry, 19
    magnetic thermometry, 20
    relative, 16, 18
    spectral-band radiometric thermometry, 18
  sensitivity, 30
Thermoscope, 2
Three-dimensional total synchronous luminescence spectroscopy (3D-TSLS), 46, 48
Time domain measurements of emission decay times, 73
Time-resolved luminescence, 104
Tissue fluorescence, 238
Transient intensity decay law, 108
Triangular smoothing, 94
Triboluminescence, 38
Tryptophan, 238
Tunable lasers, 53
Two-photon luminescence microscopy, 174

U

Unlocalized extrinsic luminescence, 39

V

Vapor pressure thermometers, 26
Vibrational relaxation (VR), 36
Vitamins, 238

W

Water-in-oil microemulsions, 131
Water temperature, 1
Weighted smoothing function, 92

X

Xanthene dyes, 194

Z

Zero-crossing wavelengths, 100

CPI Antony Rowe
Chippenham, UK
2018-09-27 01:33